Studies in Computational Intelligence 408

Editor-in-Chief

Prof. Janusz Kacprzyk
Systems Research Institute
Polish Academy of Sciences
ul. Newelska 6
01-447 Warsaw
Poland
E-mail: kacprzyk@ibspan.waw.pl

For further volumes:
http://www.springer.com/series/7092

Thanasis Daradoumis, Stavros N. Demetriadis,
and Fatos Xhafa (Eds.)

Intelligent Adaptation and Personalization Techniques in Computer-Supported Collaborative Learning

 Springer

Editors
Prof. Dr. Thanasis Daradoumis
Depto. d'Informatica
Multimedia i Telecomunicacio
Universitat Oberta de Catalunya
Barcelona
Spain

Prof. Dr. Fatos Xhafa
Department of Languages and
Informatics Systems
Polytechnic University of Catalonia
Barcelona
Spain

Prof. Dr. Stavros N. Demetriadis
Department of Informatics
Aristotle University of Thessaloniki
Thessaloniki
Greece

ISSN 1860-949X
e-ISSN 1860-9503
ISBN 978-3-642-44768-6
ISBN 978-3-642-28586-8 (eBook)
DOI 10.1007/978-3-642-28586-8
Springer Heidelberg New York Dordrecht London

Preface

Contemporary research efforts in the Computer-Supported Collaborative Learning (CSCL) domain have clearly emphasized the need for building flexible, adaptable and intelligibly operating technology systems that could provide a personalized, more productive and satisfactory learning experience to all group learners. The key premise of this endeavor is that intelligent adaptation and personalization techniques can offer the basis for dramatically extending the affordances of the CSCL technology infrastructure and relevant pedagogical design so that learning gains (cognitive and metacognitive) are maximized through elegantly orchestrated peer interaction and support. This perspective lies at the crossroads of Adaptive Educational Hypermedia Systems, Intelligent Tutoring Systems and Computer-Supported Collaborative Learning, expanding the perspective of the fields and setting innovative research agendas which put forward challenging opportunities for constructively exploring the interdisciplinary landscape.

Indeed these research agendas integrate issues which can be approached from different but complementary perspectives and are of interest to researchers of diverse backgrounds such as learning scientists, educators, engineers/computer scientists, instructional designers, the Learning Design community and still others. It is only natural, therefore, that during the last years several research groups internationally have made significant advances in the field, exploring various research questions relevant both to the technological and pedagogical dimension of integrating and promoting intelligent, adaptive and personalization techniques in the CSCL context.

This volume makes a distinctive contribution and further extends our collective experience in the field by bringing together the scientific work and outcomes of fourteen such research groups. Common underlying aspect of all these efforts is the researchers' consistent attempt to empower technology systems with essential flexible functioning that strengthens users in carrying out complex teacher-student and student-student learning interactions. In general, these systems aim to help making pre-task interventions (such as facilitating group formation tasks), support in-task peer interactions and domain specific activities while also offering possibilities for implementing students' post-task assessment that helps modifying the activity flow. This is not, however, a trivial task and the inherent complexity in building and evaluating the efficiency of such systems is highlighted also by current research outcomes indicating that either providing no support at all (i.e. free collaboration) or unwittingly imposing unnecessary restrictions to group learners ("overscripting") may have detrimental effects on learning. Thus, to attain beneficial system operation, designers need to carefully consider and investigate a multitude of factors relevant to collaborative learning settings, the complex interactions among them and their possible impact on learning.

Within this framework of considerations the fourteen contributions in this book draw the reader's attention to different research directions and possibilities in the field; nevertheless, these diverse perspectives could be classified under three major themes: (1) *Design of Adaptive Learning Systems*, (2) *Interactive and Intelligent Learning Systems*, and (3) *Collaborative Learning Systems*.

More specifically:

(1) *Design of Adaptive Learning Systems:* extension of IMS-LD to overcome the limitations of IMS-LD modeling language to express adaptive interventions; use of adaptive techniques for adaptive CSCL scripts; use of collaborative scripting and adaptive patterns; presentation of case studies that use adaptive CSCL scripts.

IMS-LD is primarily a modeling tool which uses the metaphor of a theatrical play for describing a teaching-learning activity. Using IMS-LD developers can formally express a unit of learning, that is, a complete, self-contained unit of learning material and activities, such as a course, a module, a lesson etc. IMS-LD has become a de-facto standard in the CSCL field using concrete syntax and semantics which, however, can not adequately express the complex activity and data flow evident in the collaborative learning activities. Thus, extending the IMS-LD modeling capabilities and supporting interoperability is a major issue in current research agendas. In this volume, six contributions focus on the IMS-LD as well as the adaptive CSCL scripts and patterns to propose (a) the reuse of data flow designs in case of complex and adaptive collaboration scripts (Bordiés, Dimitriadis, Alario-Hoyos, Ruiz-Calleja, and Subert); (b) the combination of a Generic Service Integration system with an IMS Learning Design to provide a Unit of Learning (de-la-Fuente-Valentín, Pérez-Sanagustín, Santos, Hernández-Leo, Pardo, Kloos, and Blat); (c) concrete extensions to the IMS-LD specification, addressing a wide range of problems and omissions (König and Paramythis); (d) a framework for the integration of external and independent software components into IMS-LD through the use of a specific mediator component (Magnisalis and Demetriadis); (e) the implementation of the "adaptation pattern" approach in practice through the design and flexible operation of two prototype tools (Karakostas, Papamitsiou, and Demetriadis); and (f) the implementation of the "Students Team Achievement Divisions" (STAD) collaboration method as an online, adaptive collaborative design-pattern (Kordaki, Daradoumis, Fragidakis, and Grigoriadou).

(2) *Interactive and Intelligent Learning Systems:* exploration of the efficiency of interaction analysis methods that empower CSCL systems with adaptive capabilities; implementation of interactive and intelligent systems using agent technologies and formal languages.

Adaptation, when explored in the CSCL context, brings forth many significant and intriguing research questions related both to "behind the scenes" computational techniques (such as educational data mining and learning analytics methods) and also frontend (i.e. interface design) issues for making apparent to the user the results of adaptive operation. Moreover, empowering a system with both interactivity and intelligence and make it meaningful and useful to the user is not a trivial issue. Five chapters in this book cover relevant topics, such as: (a) investigating the relationship between adaptation and interaction analysis, with emphasis on asynchronous discussion platforms (Bratitsis); (b) introducing a specific interaction analysis tool (CoSyLMSAnalytics) to help teacher modify a typical Think-Pair-Share script (Petropoulou, Lazakidou, Georgiakakis and Retalis); (c) exploring the intelligence and interactivity as well as their alignment with the system's design and feedback so that to model users' expectations when interacting with the system (Benton, Altemeyer and Manning); (d) presenting the design of an intelligent monitoring agent that collects and aggregates information from a LAMS database (Chronopoulos and Hatzilygeroudis); and (e) presenting a system development approach that supports participants of a distance education forum by getting as input the discussion threads and outputs specific strings modelling the thread messages based on a formal language (Patriarcheas, Papaloukas and Xenos).

(3) *Collaborative Learning Systems:* analysis of collaborative learning interactions; assessment of collaboration quality; effectiveness of communities of practice.

When learners communicate and interact through various technological systems then a multitude of specific interactions emerge that need to be conceptualized and typified before developing computerized models that could enrich system operation and interventions. This theme emphasizes exactly the need for adequately modeling various aspects that concern the complex learner/user context and user-system interactions in a CSCL environment. Three contributions focus on the following topics: (a) Voulgari and Komis explore the massively multiplayer online games through a theoretical framework that helps analyze collaborative learning interactions; (b) Kahrimanis, Chounta, and Avouris employ an alternative analysis methodological approach to propose a rating scheme for the assessment of collaboration quality; (c) Kostas ad Sofos, research the literature and focus on defining a typology of critical elements for successful and sustainable Internet-mediated communities of practice.

Introduction

Adaptation and personalization have been extensively studied in CSCL research community aiming to design intelligent systems that adaptively support eLearning processes and collaboration. Yet, with the fast development in Internet technologies, especially with the emergence of new data technologies and the mobile technologies, new opportunities and perspectives are opened for advanced adaptive and personalized systems. The adaptation and personalization are posing new research and development challenges to nowadays CSCL systems. On the one hand, the adaptation should be focused in a multi-dimensional way (cognitive, technological, context-aware and personal). On the other hand, it should address the particularities of both individual learners and group collaboration. Therefore, the analysis and design of adaptive systems should deal with these new views in order to better support learners and teachers.

The ultimate aim of this book is to discuss the latest advances and findings in the scope of intelligent adaptive and personalized learning systems as well as the design and their implementation. The book also analyzes the new implementation perspectives for intelligent adaptive learning and collaborative systems that are brought by the advances in scripting languages, IMS LD, educational modeling languages and learning activity management systems. Given the variety of learning needs as well as the existence of different technological solutions, the book exemplifies the methodologies and best practices through several case studies and adaptive real-world collaborative learning scenarios, which show the advancement in the field of analysis, design and implementation of intelligent adaptive and personalized systems.

Main Contributions of This Book

Overall, the book covers the following research and development topics:

- *Design of Adaptive Learning Systems:* extension of IMS-LD to overcome the limitations of IMS-LD modeling language to express adaptive interventions; use of adaptive techniques for adaptive CSCL scripts; use of collaborative scripting and adaptive patterns; presentation of case studies that use adaptive CSCL scripts.
- *Interactive and Intelligent Learning Systems:* exploration of the efficiency of interaction analysis methods that empower CSCL systems with adaptive

capabilities; implementation of interactive and intelligent systems using agent technologies and formal languages.
- *Collaborative Learning Systems:* analysis of collaborative learning interactions; assessment of collaboration quality; effectiveness of communities of practice.

Organization of the Book

Consequently, the 14 chapters of this book are organized in three major areas as follows:

Part I: Design of Adaptive Learning Systems

Chapter 1: Osmel Bordiés, Yannis Dimitriadis, Carlos Alario-Hoyos, Adolfo Ruiz-Calleja, Andrés Subert. *Reuse of data flow designs in complex and adaptive CSCL scripts: A case study*
In this chapter the authors present a case study to overcome limitations of current approaches for data flow among CSCL activities. The authors have proposed an IMS LD solution to achieve reusability of data flow designs. The resulting solution is in addition interoperable. In the study a real-world complex CSCL script is considered in order to show the adaptive characteristics of the proposed approach.

Chapter 2: Luis de-la-Fuente-Valentín, Mar Pérez-Sanagustín, Patrícia Santos, Davinia Hernández-Leo, Abelardo Pardo, Carlos Delgado Kloos, Josep Blat. *System orchestration support for a collaborative blended learning flow*
The authors of this chapter have analyzed the new types of activities arising in CSCL due to the use of portable and interactive technologies. Then, the authors address the question of how the existing and new scenarios can be integrated to support collaborative processes without adding significant burden to the learners. The objective is therefore to efficiently organize and give structure to new types of complex collaborative blended learning scenarios. To achieve this goal the authors build a solution based on Unit of Learning suitable for instantiation with IMS Learning Design and complemented by a Generic Service Integration system.

Chapter 3: Florian König and Alexandros Paramythis. *Adaptive Collaboration Scripting with IMS LD*
In this chapter the authors propose an extension of IMS LD Language to overcome some limitations concerning the lack of support for comprehensive adaptation features. The proposed extension includes explicit representation of groups and corresponding collaboration contexts, flexible integration of communication and collaboration services, among others. The extension also provides a run time model and features to support event- and exception- handling. Examples are provided to show the advantages of the proposed extension for advanced collaboration scripts.

Chapter 4: Ioannis Magnisalis and Stavros Demetriadis. *Extending IMS-LD capabilities: A review, a proposed framework and implementation cases*

In this chapter the authors present a framework for the integration of external and independent software components into IMS-LD (Learning Design) based courses that cater for adaptivity. The proposed architecture introduces a mediator component as the key element to facilitate communication between Learning Design compliant e-courses and external tools that support collaborative learning (e.g. a forum, an agent, a service or a software component that provides a specific functionality). The authors provide example scenarios and also discuss some important issues toward integrating the adaptation pattern capabilities in IMS-LD compliant tools for collaborative learning design.

Chapter 5: Anastasios Karakostas, Zaharoula Papamitsiou and Stavros Demetriadis. *Prototype Tools for the Flexible Design of CSCL Activities based on the Adaptation Pattern Perspective*

The chapter presents the design and some preliminary evaluation data regarding two prototype tools (namely, FlexCoLab and PPR), which have been designed according to the prescriptions of the adaptation pattern perspective for promoting a flexible design of CSCL activities. Both tools aim to support teachers in the process of developing flexible designs of online collaborative activities by reusing and customizing adaptation patterns, according to the requirements of a particular learning situation. The authors present the theoretical background of the adaptation patterns approach, the design specifications of the two systems and student evaluation data from implementing an in-school collaborative learning activity supported by PPR.

Chapter 6: Maria Kordaki, Thanasis Daradoumis, Dimitrios Fragidakis, Maria Grigoriadou *Adapting the Collaborative Strategy "Students Team Achievement Divisions② in an Information Technology Work Place*

This chapter presents an innovative online adaptive collaborative design-pattern that implements the "Students Team Achievement Divisions (STAD)" collaboration method in a real world training-based scenario that takes place at an Information Technology work place, using the LAMS system. The approach used enabled to build a rich learning profile of the user that is subsequently employed to provide him/her personalized training, monitoring, scaffolding and evaluation.

Part II: Interactive and Intelligent Learning Systems

Chapter 7: Tharrenos Bratitsis. *Examining the Interrelation between the Interaction Analysis and Adaptation Research Fields within Communication-based Collaborative Learning Activities: Convergence, Divergence or Complementarity?*

In this chapter the relation among two important aspects in CSCL, namely Adaptation and Interaction Analysis, is analyzed. The research question posed in this work is either these two aspects can be seen as complementary or if they would rather converge/diverge at thru long run in CSCL. The author uses AI methods to examine and correlate the main constituents of adaptation and interaction analysis for the case of asynchronous discussion platforms. The objective is to highlight the similarities and links among adaptation and interaction analysis.

Chapter 8: Ourania Petropoulou, Georgia Lazakidou, Petros Georgiakakis, Symeon Retalis. *Making Adaptations of CSCL Scripts by Analyzing Learners' online Behavior*

The authors present a study on how to support teachers to create customized learning scripts in order to match needs of different learning strategies. These scripts are more appropriate to the learners' preferences and the learning context. To that aim, the authors suggest the use as a source for the scripts the learners' interaction data that is collected during an online learning process and analyzed using interaction analysis. A tool, called CoSyLMSAnalytics, is provided to support teachers at creating and modifying the learning scripts.

Chapter 9: Stephen Benton, Boris Altemeyer and Bryan Manning. *Behavioural Prototyping©: making interactive and intelligent systems meaningful for the user*

The authors in this chapter explore the intelligence and interactivity and their alignment with system's design and feedback. The concept of Behavioural Prototype© is introduced to characterize the interactive expectations and behaviour of users with the system.

Chapter 10: Themistoklis Chronopoulos and Ioannis Hatzilygeroudis. *The design of a teacher-driven intelligent agent system for supervising lessons in LAMS*

The authors of this chapter have presented an agent-based approach to support teachers in supervising and evaluating learners and activities in the Learning Activity Management System (LAMS). The activity monitoring is done through intelligent agents from LAMS data-base. On the other hand agents are used to notify users to support awareness. Finally the Systems feeds-back users with reports on ongoing activity.

Chapter 11: Kiriakos Patriarcheas, Spyridon Papaloukas and Michalis Xenos. *The text-based computer-mediated communication in distance education fora: A modelling approach based on formal languages*

The authors in this chapter are concerned with automating the interpretation of threads in asynchronous discussions. The proposed system is based on using content categories as a unit of analysis. The aim is to support participants of the discussion forum with updated information on the discussions carried out at the asynchronous forum.

Part III: Collaborative Learning Systems

Chapter 12: Iro Voulgari, Vassilis Komis. *Antecedents of Collaborative Learning in Massively Multiplayer Online Games*

Massively Multiplayer Online Games is one important type of implementing collaboration processes for goal-oriented activities and collaborative and social interactions. The authors of this chapter have presented a theoretical framework for linking

learning and collaborative learning principles with MMOGs. The authors also investigate through an exploratory and qualitative approach, features of the tasks, groups,
and player interactions that may support the emergence of collaborative interactions
and learning. Based on that analysis, the critical factors of effective collaboration are
identified.

Chapter 13: Georgios Kahrimanis, Irene-Angelica Chounta, Nikolaos Avouris.
Validating empirically a rating approach for quantifying the quality of collaboration
    The authors in this paper study the issue of how to effectively assess the suc-
cess of the collaboration in CSCL by quantifying some indicators of collaboration
processes. The authors have proposed a rating scheme for the assessment of collaboration quality by identifying first about 228 collaborating dyads in synchronous collaboration taking place in a problem-solving task.

Chapter 14: Apostolos Kostas, Alivisos Sofos. *Internet-mediated Communities
of Practice: Identifying a Typology of Critical Elements*
    Communities of practice have been identified in CSCL as means to achieve
specific goals on informal learning and professional development. However there
is a lack of a systematic theory or a blueprint for design of online communities.
The authors define a basic typology of various critical elements for successful and
sustainable Internet-mediated communities of practice, via a meta-analysis and
critical synthesis of related literature.

Targeted Audience and Last Words

The chapters of this book cover an interesting set of research and development issues in CSCL aiming to better support intelligent and effective collaboration processes. The book is suitable to researchers, developers and practitioners in CSCL
community interested in the analysis, design and use of intelligent techniques for
an effective collaboration among learners and groups of learners. The book also
covers the implications of the latest developments in networking and communication technologies such as mobile computing as well as advanced AI techniques to
design and build blended learning scenarios. One salient feature of the book is
dealing with the complex nature of collaboration from different angles and achiev
ing thus a comprehensive view of the different intelligent techniques van be used
altogether to analyze, design, develop, use and assess the collaboration. Therefore
the book is useful for a wide range of researchers and developers in CSCL and especially those interested in the multi-facet nature of the collaboration.
    Finally, academic researchers, professionals and practitioners in the field can
also be inspired and put in practice the ideas and experiences proposed in the book
for their specific goals.
    We hope that the readers find this book useful and help accomplish their goals.
Enjoy the reading!

Acknowledgements

This book follows the International Conference on Intelligent Networking and Collaborative Systems (INCoS-2010), held on November 24-26, 2010 in Thessaloniki, Greece. The editors of this book wish to thank the authors for their contributions to the book and the reviewers who have carefully reviewed the chapters and gave useful suggestions and constructive feedback to the authors.

We gratefully acknowledge the support and encouragement received from both Prof. Janusz Kacprzyk, the editor in chief of Springer series *"Studies in Computational Intelligence"*, and the Springer editors Thomas Ditzinger and Heather King during the edition of this book.

January 2012   The Editors

Prof. Dr. Thanasis Daradoumis
University of the Aegean, Greece /
Open University of Catalonia, Spain

Prof. Dr. Stavros Demetriadis
Aristotle University of Thessaloniki, Greece

Prof. Dr. Fatos Xhafa
Technical University of Catalonia, Spain

Contents

Part II: Interactive and Intelligent Learning Systems

Part III: Collaborative Learning Systems

Part I
Design of Adaptive Learning Systems

Part I

Design of Adaptive Learning
Systems

Reuse of Data Flow Designs in Complex and Adaptive CSCL Scripts: A Case Study

Osmel Bordiés[1], Yannis Dimitriadis[2], Carlos Alario-Hoyos[2],
Adolfo Ruiz-Calleja[2], and Andrés Subert[1]

[1] Department of Telecommunications, Electrical Engineering Faculty, Oriente University,
UO Santiago de Cuba, Cuba
obordies@fie.uo.edu.cu
[2] Department of Signal Theory, Communications and Telematic Engineering, School
of Telecommunications Engineering University of Valladolid, UVa Valladolid, Spain
{yannis,calahoy,adolfo}@tel.uva.es

Abstract. Existing Educational Modeling Languages (EML), and especially the
IMS LD specification, does not appropriately address data flow among Computer
Supported Collaborative Learning (CSCL) activities. Several solutions proposed
in the literature, have tackled the important dimensions of data flow automation or
the consistency of design, but they have not adequately covered the perspective of
reusing scripts. Current data and tools binding specifications do not establish the
dependencies between this setting and the structural design of the collaborative
data flow situations. This preclude a complete design particularization and raise
the issue of reusing data flow designs, which is especially important in the case of
complex and adaptive real-world collaborative learning scenarios. This paper
presents a case study in which the *LeadFlow4LD* approach, an IMS LD interoperable solution, is analyzed with respect to the reusability of data flow designs. Besides, the case study is performed using a real-world complex CSCL script in
order to illustrate the adaptive characteristics that could be taken into account.
Findings show limitations of the current approaches concerning reuse and structural design particularization of the data flow. Additionally an alternative solution
based on abstract workflow templates that is briefly outlined in this paper.

1 Introduction

"Computer-Supported Collaborative Learning (CSCL) is a research paradigm that
studies the role played by the Information and communication Technologies (ICT)
as a mediator in social interaction within a learning process" [28]. Effective social
interactions can be promoted or enhanced using CSCL scripts, which are computer
interpreted versions of collaborative pedagogical models [11]. A description of
effective CSCL scripts can be achieved through the definition of the **learning flow**
as a sequence of learning activities, complemented with the **data flow** (also referred as artifacts flow). The latter being a learning flow coordination mechanism
that satisfies the dependencies established among learning activities [32]. CSCL
scripts could be embedded in learning platforms or LMSs (Learning Management

T. Daradoumis et al. (eds.): Intelligent Adaptation & Personalization Techniques, SCI 408, pp. 3–27.
springerlink.com   © Springer-Verlag Berlin Heidelberg 2012

Systems), such as Moodle[1] or LAMS[2], in order to support real complex scenarios. Besides, LMSs should show some grade of flexibility, allowing the users or systems involved the collaborative learning situation to modify script features [12].

Typical CSCL scenarios incorporate in a single workflow activities, roles, tools and models of resources [27] that eventually, can be re-configured in order to increase the performance of pre-defined criteria [40]. Therefore, current approaches integrate adaptation mechanisms in collaborative scripts in which flexibility and a personalized learning are promoted [9]. However, they do not describe activities which are not directly related to knowledge creation, but those associated to coordination mechanisms like information exchange among activities, as it is required in more generally CSCL scenarios [32]. These mechanisms satisfy the dependencies that exist among learning activities concerning the artifacts produced or consumed by users, groups or tools [27]. The consistent design and automatic execution of these coordinating activities characterizing the data flow are expected to enhance the reliability of CSCL processes [35]. For instance, in a CSCL scenario several groups may co-exist and create different conceptual maps with the goal that in the next activity, these artifacts may be exchanged according to the *Peer review* assessment process [13]. If explicitly stated in the data flow, the aforementioned exchange of documents would be automatically performed. Thus, the assessment process would jeep in the pedagogical core of the learning design enactment and errors due to human intervention would be avoided.

Similarly to what is done with CSCL designs in a context of unskilled designers [20], collaborative data flow situations can be created from pre-existing designs, thus reducing implementation time and effort. Furthermore, the use of CSCL scripting patterns such as the Collaborative Learning Flow Patterns (CLFP) [19] and Adaptation Patterns (AP) [10] aims at reusing not only scripts code, but also experience and good pedagogical practices. However, a consistent and reusable definition of data flow situations is missing in Educational Modeling Languages (EMLs) [7, 46, 51], which are used for the creation of CSCL scripts. Specifically, IMS LD [23] that is considered as the *de facto* standard by part of the community can be used to model CSCL scenarios [20] or even predefined adaptive behaviors [5], but it has a limited expressiveness to represent coordination mechanisms, which are indispensable for data flow definition [27]. For instance, the data flow automation and the data flow design consistency and reuse are not covered [35].

Several approaches [33, 43] have been proposed to tackle the, so called, *Data flow problem in Learning Design* [7, 46, 51]. The current *Composition*-based approach [35], materialized, for instance in the **LeadFlow4LD** (Learning and data Flow for Learning Design) solution, is one among these proposals in which both IMS LD and workflow capabilities are jointly used in order to overcome this problem. Such proposal separate agreeably learning and data flows concerns and reuse dimension is promoted.

Despite the proposal of all these approaches, the reusability of data flow designs is still considered to be an important aspect that has not been fully covered [7, 46]. Existing instance-level designs do not establish the relationship between

---

[1] http://moodle.org/
[2] http:// lamsfoundation.org/

particular social contexts, data flow and the corresponding structural design when the activity sequence could be affected. Following the aforementioned *Peer review* example about *conceptual maps*, the assessment process could be implemented choosing one out of several exchange logics. There are many specific data flow collaborative situations (e.g. *teams to individuals cross review, inter-groups rotated cross review*), but only one generic model to describe it: the *Peer review* pattern description.

This paper deals with the *Data flow problem* in complex and adaptive CSCL scripts, and most specifically with the **reuse** of data flow designs by educational practitioners. This analysis is based on a real case study, whose design is guided by CLFPs and enriched through the use of an AP. Reuse or the definition of data flow *adaptation* mechanisms on the *Data flow problem* are studied using the complex computational scripts derived from the ***LeadFlow4LD*** *Composition*-based solution. The use of this approach over the other alternatives addresses its interoperability with IMS LD, the explicit separation between learning flow and data flow concerns and therefore the improved reusability. Initial evidence suggests paths to provide a new solution for the collaborative data flow reuse in adaptive complex CSCL settings based on **abstract workflow templates** [17]. We should point out that the paper is mainly focused on IMS-LD solutions, due to its importance as a *de facto* standard specification, as compared to other proprietary systems that support adaptive CSCL scripts, such as LAMS. However, we hope that the solution proposed in this paper could be partially extended in such systems, providing them with advantages regarding the reuse of data flow designs.

The rest of the document is structured as follows: section 2 presents some background concepts on CSCL scripts, adaptability and pattern-based design within the frame of the *Data flow problem*. Section 3 depicts the so-called ***MOSAIC*** case study, according to the ***LeadFlow4LD*** design methodology. Discussion about findings is provided in section 4, while section 5 outlines a potential solution for overcoming the limitations found in the case study. Finally, conclusions and future work are summarized in the last section.

2 Background

CSCL scripts are considered as "specific types of instructional support that allow setting and management of collaborative learning situations in order to achieve productive interactions among learners" [11]. This support can be applied in two granularity levels, as micro-scripts or macro-scripts. Micro-scripts support collaboration at a fine-grained level, providing specific and detailed instructions to learners involved in collaborative situations [50]. In contrast, CSCL macro-scripts are created to model coarse-grained pedagogical aspects such as the structure of individual and collaborative learning phases. In this paper we focus on macro-script taking into account their higher granularity and possibilities to be reused in different social contexts.

CSCL macro-scripts (from now on just CSCL scripts) can be formalized, and therefore they can be interpreted and executed by software systems [3]. CSCL scripts emerged as an alternative to the manual/person-based management of collaborative learning in order to reduce the likelihood of error prone situations, since in the latter, instructions may be forgotten, misunderstood, ignored or badly applied by the human users [48]. Besides the advantages of an automated coordination of the collaboration processes, computationally formalized collaborative scripts mediate in the social and cognitive processes effectively and efficiently as shown in the literature [50]. Finally, computationally interpretable CSCL scripts can be embedded within Learning Management Systems (LMS) and provide access to ICT tools [22].

From a computational perspective, CSCL scripts are considered as sequences of operators through which abstract data structures relating social structures, resources and products, are transformed during the script execution. These operators (e.g. [Products Matrix $\rightarrow$ Social Matrix] [11]) deal with important issues of CSCL, such as group formation, assignment of roles and resources, products creation or flexibility, and therefore they should be included in a specific CSCL script design. At a certain phase a group formation operator may transform a predefined group setting (social structure) based on the evaluation of products or artifacts created at an earlier stage. For instance, following the *Peer review* example, at the second stage of the assessment process, groups could be reconfigured to work individually or in pair and consequently the social structure and the particular exchange logic may change as well.

Nevertheless, the formalization of CSCL scripts is not a straightforward process [33, 26]. The difficulty is directly related to the high number of aspects and parameters, such as task or resources distribution, that should be configured across different social planes within a single workflow [11]. For unskilled practitioners, dealing with the technical difficulties related to *group settings, complex collaborative learning flow specifications* [33] and/or the use of *unfamiliar formal notations*, such as IMS LD [27], may represent a great challenge. Nonetheless, designers may also find difficulties when dealing with aspects related to adaptation mechanisms in order to prevent unexpected events that may occur during script execution [38]. But also, they have to face the definition of coordination mechanism as a mean of group interaction support that have an influence on the effectiveness and efficiency of collaborative learning [32]. One of such coordination mechanisms is the data flow that not only satisfies the dependencies among learning activities but also promoting or being affected by an adaptation mechanisms implementation [38]. Globally, educational practitioners face a complex task, when they have to design and provide support for adaptive and complex CSCL scripts.

The aforementioned technical difficulties related to the setting of collaboration parameters or operators can be addressed through the use of CLFPs [19]. These abstract representations of *good practices* structure the collaborative learning activities flows and have been formalized as IMS LD based templates [20]. Nevertheless, CLFPs do not focus on describing collaborative adaptive mechanisms [39] and IMS LD is limited in terms of flexibility and constrained by predefined adaptation behaviors [38]. As an answer, APs have been proposed as a framework aimed at supporting adaptive CSCL designs by means of alternative abstract solutions [25]. Such patterns also aim to formulate *good practices*, and help to develop adaptation strategies in collaborative contexts [30]. However, few patterns concerning CLFPs, APs or *assessment* patterns, actually describe collaborative data flow situations. For example, the *Peer review assessment* pattern describes the creation of certain products susceptible to be assessed, and suggests a set of *reviewing* methods [13]. At the *Pyramid* CLFP an artifact has to be created at the end of each level and therefore such an artifact creation is specified as an intrinsic script feature [24]. Nonetheless, the artifact flow among tools and users is not included in this pattern description. Another example refers the *Large group deliverable* AP [25], that suggests split a group and it deliverable into smaller and simpler pieces to make the learning objective affordable for learners, but do not model such a data flow situation.

The aforecited problem of adequately describing data flow in Learning Design (LD) has been reported extensively in the literature related Educational Modeling Languages (EML) and especially of IMS LD. This issue known as the *Data flow problem in LD* will be depicted in the next section along with the *Composition*-based approach, which currently is the most accepted solution to this problem.

2.1 The Data Flow Problem in Learning Design and the Composition-Based Approach

These subsections provide an overview of the data flow problem in LD, its dimensions and main approaches. The **Leadflow4LD** *Composition*-based solution is especially outlined, including the corresponding design methodology, since it will provide the basis for the case study described in section 3.

It was mentioned above, CSCL scripts should formally define **coordination** mechanisms in order to satisfy the dependencies among learning activities. For instance, the relationship between artifact flow and tools may be established by such dependencies [32]. But the *Data flow problem in LD* reveal some limitations on IMS-LD [7, 44, 46, 51], that may be extensive to some other EMLs. Several implementation attempts based on IMS LD [27, 36, 47] have been characterized by *hardcoded* interactions, as well as error-prone design and execution processes. Evaluation studies have revealed an increased cognitive load for both designers and performers i.e. teacher and learners. As an example, some cases demanded an ad-hoc *wired* design, in order to guide specific users to carry on the uploading and downloading tasks by hand [36], or the time-consuming and hazardous configuration of user-oriented HTML properties [47]. These experiences reveal as *desirable* features of further solutions the data flow **automation** as a method to limit the user responsibility achieving real learning activities. But, in a similar manner the design methodology should be **consistent** enough to reduce the designer cognitive load during data flow specification. Finally a separation of learning flow from dataflow is needed and the design **reuse** is fostered [35]. Specifically, the reuse dimension aims in this case at merging reusable learning flow definitions with specific data flow designs, with the minimal cost. As it is mentioned before, the artifact visibility associated to the use of monitoring services (IMS LD properties) and them assigned to specific roles preclude the overall reusability of designs. In this case the instantiation does not belong to the IMS LD specification but neither to the design process.

2.2 Data Flow Approaches and the Reuse Dimension

The solutions that have addressed these Data flow problem dimensions have followed three main approaches. Vantroys et al. [43] propose COW (*Cooperative Open Workflow*), driven by certain requirements like the support of collaborative activities and the reuse of existing courses or activity models. The proposal uses the *Model Driven Architecture* approach to create learning flow implementations out of IMS LD models. The *Mapping* approach proposes the execution of LD models through mapping IMS LD to the XML-based Process Description Language (XPDL [41]). Therefore, XPDL would be responsible of data flow and learning flow execution. However, the proposal requires extensions that affect the interoperability between XPDL and IMS LD standards. The model reuse is achieved through the separation between a *process/activity* model and an *instance* model. The workflow engine is responsible of creating specific threads for each student, mapping roles on users, and resources on tools. Nevertheless, the main drawback of the proposal is that does not cover the issue of the automatic construction of learning paths.

Meanwhile, Miao et al. considered that IMS LD cannot provide support to collaborative aspects and proposed a new CSCL script language in a *Substitution*-based approach [33]. They identified a number of limitations associated to the modeling of artifacts and data flow in collaborative contexts, as well as a complex control flow specification. It is proposed a non-standard language for CSCL scripts formalization, through which they provide a response to identified shortcomings. Hence, it is included an explicit definition of artifacts as an element that may be created and shared across activities, used by tools as an input/output parameters and also is possible to model complex scenarios. However, this proposal does not allow an interoperable solution regarding the IMS LD standard, the construction of data flow situations in a flexible manner neither the reutilization model are concisely tackled. Also in this case learning flow and data flow are defined together.

Alternately, the *Composition*-based approach [36] focuses on maintaining interoperability with IMS LD and states that the **data flow automation** or **consistency** dimensions can be tackled by combining IMS LD with workflow technology in an interoperable manner. Therefore, defining, executing or managing data flow and tools invocation can be supported by workflow technologies, while IMS LD is used exclusively, to support the learning flow definition. Once both parts of the design the learning and the data flow have been formalized, they are submitted to the respective compliant engines in order to be executed synchronously, following a *master-slave* coordination mechanism. In this case, the learning flow is the leader playing the *master* role and leaving the flow control to the data flow process. This way, dimensions such as design **consistency** and **data flow automation** are covered. This approach fosters the reusability separating the declarative and reusable design from the instantiation design. This decision promotes the interoperability with IMS LD, but also alleviates the stress between structural and instantiation designs. Similar to IMS LD, the workflow as goal-oriented language [8, 45] also exhibit a declarative nature like IMS LD [49], hence the instantiation setting must be remitted to an external XML-based specification. Through this specification data and tool instances relationship is set, but its dependencies with the structural workflow design and contextual setting is missing. For instance, according to the *Peer review* example developed at the introductory section, the *Peer review* model may be implemented according to many different data flow situations and hence different structural design. In this sense, workflow activities sequencing may be affected according to the artifact set order.

Attending to the coordination mechanisms reuse in collaborative scenarios several convergent ideas have been proposed. Miao et al. in [32] have proposed the use of high level templates in the context of reusing IMS LD based coordination mechanisms, taking in to account the time-consuming and error prone task that this represent. The idea separates the definition of these mechanisms into an abstract specification that should be understandable and simple, and the executable code. The system is then responsible to achieve the corresponding mapping process from one representation to another. Other efforts of the scientific community related to computational experiments (e.g. **WINGS** [17], **Kepler**, [34]), provide similar hints to complete the data flow description in a complex CSCL script life cycle. In general, the procedure of generating computational workflows begins with the creation of *declarative descriptions of the processes with a high level of abstraction*. Then, these abstract models are progressively particularized by incorporating available services or software components, data sets, and eventually the corresponding computing resources. This approach is promising with respect to the aforementioned problems of flexibility and reuse. However, it has not been employed in the educational context supporting CSCL scripts design but using such computational workflows for assessing student learning [29].

The table 1 sums up the features of the data flow specification approaches for collaborative situation that have been previously described. It can be observed that the common problem is that none of depicted approaches give a complete solution to a flexible and reusable specification of CSCL data flow situations. Computational workflow solution are not focused in learning designs *per se* and Miao et al. suggestion only point out to the use of abstract representation in single IMS LD designs.

Table 1 Characteristics analyzed of the main data flow approaches

Data flow design approaches	Abstract representation	Data flow situations	Educational application	Contextual mapping	*Design reusability*
Substitution [33]			✓	✓	
Mapping [43]			✓	✓	
Wired (IMS LD) [27]		✓	✓		
Composition [35]	✓	✓	✓		✓
Computational workflows [17]	✓	✓		✓	✓

Following the decisions taken according the ideas expressed in the Composition-based approach, the LeadFlow4LD has been conceptually proposed. The architectural solution and the design methodology associated will be depicted in the next section.

2.3 The LeadFlow4LD Design Methodology

As we mentioned before, the **LeadFlow4LD** (Learning and data Flow for LD) is a conceptual proposal that follows the *Composition* of the *Data flow problem in LD* and the instructional design [4] approach. This subsection will explain the reasons in which the selection of this solution to develop the case study depicted in section 3 and also will describe the steps that must be followed as document creation in order to achieve a formal description of the solution.

In this case the selection of this approach among the other alternatives is based on several desired characteristic that it have been discussed in the previous subsection, such as the interoperability with IMS LD and the separated concerns of learning flow and data flow, the relative higher reusability level of the composite designs respect to other current approaches and the possibility that CSCL patterns such as CLFPs, APs and *assessment* pattern have been supported. The general overview of **LeadFlow4LD** design methodology is explained as follows and depicted at the Fig. 1.

1. At the structural design phase both the **UoLF** (Unit of Learning Flow) and **UoDF** (Unit of Data Flow) documents should be formalized.

    1a. The Unit of Learning Flow (**UoLF**), uses IMS LD to specify the learning activity sequence, the roles to be played by the participants in each activity and the available resources for each role in each activity.

    1b. The Unit of Data Flow (**UoDF**), specifies the sequence of workflow activities as tool invocation activities, as well as the roles and associated

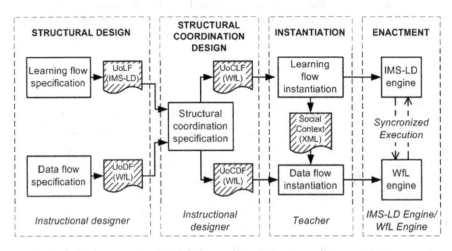

Fig. 1 General overview of LeadFlow4LD design methodology. Based on [35]

resources. The workflow roles and the learning flow roles must be the same, while resources refer to input/output artifacts linked to learning tools. LeadFlow4LD methodology proposes the use of a standard language for coding the data-flow structure document (e.g. BPEL, XPDL).

2. Once learning and data flows are defined, structural coordination aspects should be specified. Both streams are executed in a coordinated manner following a *master-slave* mechanism. Learning flow plays the master role and the data flow plays the slave one. In consequence, *coordination resources and activities* must be added in both flows in order to specify when the flow control switches from the learning flow to the data flow and backwards. The resulting documents are the Unit of Coordinated Learning Flow (**UoCLF**) and Unit of Coordinated Data Flow (**UoCDF**). The instructional designer is responsible to achieve these actions.

3. The third phase aims to specify the instantiation model of the compound design. The associated document reflects the particularities of the actual learning scenario in which learning and data flows should be enacted: users, groups and users to role assignment. The learning flow instantiation is described using a XML-based specification, which is supplied to complete the data flow instantiation phase. In this regard, the resulting specification depict that several copies of data flow sub-processes according to the number of users or groups that take place in the collaborative data flow situation and share the same tool. These documents do not follow a standard specification, and therefore, the interoperability of this solution is affected.

4. Finally, both particularized specifications, are submitted to the corresponding compliant engines (complaints to IMS LD and the workflow language used) to be executed automatically in a synchronized manner.

Next section describes the evaluation of a real-case study through the *Lead-Flow4LD* perspective.

3 MOSAIC Case Study

Throughout this section the *LeadFlow4LD* design methodology is conceptually evaluated in a complex realistic scenario that includes adaptive CSCL scripts. The use of a real case study [42] may provide important information about the issue of reusability of *Composition*-based designs.

The case study presented in this section is an enriched version of one performed in March 2007, where twelve PhD students from three Spanish universities were involved [36]. The main instructional goal of this experience was the interactive creation of a conceptual map on the topics of Grid services [14] and Service oriented computing [37] subjects. The MOSAIC scenario will be depicted as follow and illustrated at the Fig. 2.

At the beginning, the students read three basic documents about the subjects in order to build the/a first conceptual map. At second phase, the students were rearranged in *Experts* groups who read the same document to review the prior map in order to reach an intermediate agreement, and to create a new version of the conceptual map. The third stage, the student were rearranged to form three groups of four experts each to perform a similar activity. Later, according to a different collaboration pattern, the groups were combined to form two groups, which follow the same aforementioned protocol in order to achieve two more consistent conceptual maps (roles `Pyramid Level 2-1`, `Pyramid Level 2-2` and `Pyramid Level 2-3`). Finally, students reviewed the two previous maps, reaching a global consensus and delivering the final conceptual map. This learning design combines several well-know patterns, the *Pyramid, Jigsaw* [19] and the *Peer review* [13]. The three primer phases correspond to the *Jigsaw* representation in which the topic is subdivided in three sections (lectures). At the same time, the *Jigsaw* code is integrated in the first of three levels corresponding to the *Pyramid* pattern. This abstraction imply the achievement of a gradual consensus (level by level) among all participants involved in the corresponding collaborative scenario, in this case, mediated by using *Peer review* assessment processes in which products are exchanged and reviewed among participants in order to improve the products. Agreeably, in that case, some contextual features were known prior to the setup of the experience, for instance the number of participants and the specific learning tools [1].

The enriched version of this scenario includes, in a predefined manner, an adaptation mechanism for collaboration based on the *Group heterogeneity* adaptation pattern [24]. This design criterion responds to the need to incorporate in an IMS LD based CSCL script design, CLFP, *assessment* and AP patterns and reach what we have called *complex*, adaptive and realistic scenarios. The Group heterogeneity adaptation pattern is related to one of the most important mechanisms in collaborative learning, i.e. group formation [6]. Since the objective of this paper is to illustrate the inclusion of adaptation mechanisms in CSCL scripts in order to study the *Data flow problem in LD*, we are not going to refer to the wide literature underlying group formation. However, we should mention that there are several proposals in the literature that employ the aforementioned adaptation pattern. For example, in [31] a reference architecture called **MAPIS** (Mediating Adaptation Patterns & Intelligent Services) is proposed, in which predefined grouping parameters are set using IMS LD and group formation is handled by components implemented as web services. Also, among other approaches, the standard-based work developed in [39] is remarkable, in which intrinsic constraints identified from CLFP patterns are used to adapt predefined groups settings according to contextual particularities.

In our case, a new activity is added, thus promoting the group formation for the second *Pyramid* level. The teacher plays the role associated to this new activity and its objective is to evaluate the contributions of the *Jigsaw* groups, so that may select the most heterogeneous among different predefined grouping settings and therefore promote better interactions in the following phase. Each grouping configuration corresponds to a specific alternative route in the workflow that leads to specify a conditional structure. The overall structure of the enriched case is depicted in Fig. 2. The following subsection provides a description of the **Lead-Flow4LD**-based implementation of the current MOSAIC case.

3.1   A LeadFlow4LD Implementation of MOSAIC Scenario

The original MOSAIC case study was employed to analyze the eventual IMS LD limitations on the specification and execution of data flow situations [36]. Thus, the MOSAIC case study served as a basis to lay the foundations of the **Lead-Flow4LD** methodology as it was discussed in section 2. This subsection describes the formal process through which the enriched MOSAIC case is explained following the **LeadFlow4LD** methodology guidelines from the perspective of the documents generated. Using this procedure, several limitations related to the flexibility of **LeadFlow4LD** when specifying *complex* CSCL scenarios are identified.

1. At structural design phase both the **UoLF** (Unit of Learning Flow) and **UoDF** (Unit of Data Flow) documents are produced:

    1a. The first document defines the **learning flow specification** containing objectives, activities and tasks descriptions (according to the selected CLFPs). It is noteworthy that, as compared with the original case study, a new control activity (*Activity control Pyramid 2 grouping decision*) is now incorporated, in which the Teacher choose one from three grouping configuration at the *Pyramid* second level. In this case, the conditional structure must be represented as alternative acts associated to each specific grouping configuration. This new structure represents an implementation of the *Group heterogeneity* adaptation pattern. Unfortunately using IMS LD, a dynamic assignment of roles is only possible before the enactment. Therefore, in order to enable a single user or group to play one of three predefined roles corresponding to different *grouping configurations* (`Pyramid_Level_2x` roles in Fig. 3), they should be specified at design time. Each role is also assigned to a different affordable alternative act in a branching structure. The groups reach these acts through conditional structures defined by the *B-level* implementation type of IMS LD [21]. Furthermore, this design criteria is consistent with the relationship (one role per activity) which is prescribed by the LeadFlow4LD design methodology.

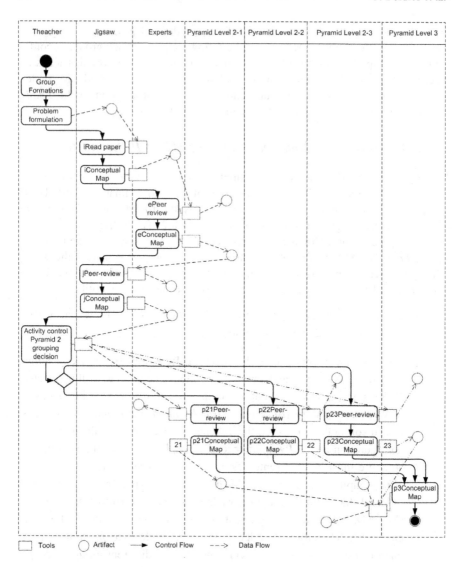

Fig. 2 A simplified version of the activity diagram of the case study comprising the learning flow at a declarative-level. Legend: i, e, j prefixes (individual, expert, jigsaw). p21, p22, p23 prefixes (pyramid second levels), p3 (pyramid third level). Dashed rectangles and dashed circles (tools and artifacts respectively)

1b. The **data flow specification** formally describes the workflow activities sequence according to the already defined learning flow. In this case, each learning activity is supported by a learning tool and thus, learning flow and data flow have a similar declarative-level flow structure. Consequently, tools or educational services are associated to workflow activities and resources are defined as input/output artifacts. Moreover, a conditional structure is incorporated to handle each grouping setting at the workflow level (represented by the tools numbered as *21, 22* and *23*).

2. The second step incorporates the **coordination resources specification** concerning both, the learning flow and data flow structural designs in order to make possible their synchronized execution at the enactment. Coordinating resources in the learning flow side are incorporated as learning objects. These objects encapsulate a client application through which communication is established with the workflow process interface. Similar to the learning flow specification, this step adds specific activities to the workflow structure to support the queries made by the learning flow clients. Each sub-structure corresponding to each different grouping setting incorporates their respective coordination resources (the coordination resources are not represented at Fig. 2 for simplicity).

3. Finally, the teacher should include at the **instantiation specification**, contextualized parameters. In this case, he includes information about users that take place in the course, and assigns the roles to be played in correspondence with these already specified at the structural specifications. The instance-level representation of data flow is partially displayed in Fig. 3 where only the data exchanges relative to a single user are considered. The other participant streams and respective interactions are not depicted in this diagram in order to keep the diagram simple as possible and readable.

The original MOSAIC case had specific requirements that impinged on its particularization. As we mentioned before, the designer team knew beforehand certain contextual information such as the number of users involved, and consequently the number of study material supplied to them. In addition, they deduced the group composition at each stage and assumed themselves the facilitator role. Based on the fact that there are 12 students who participated in the course several grouping decisions were already made. At the *Expert* phase belonging to the *Jigsaw* pattern, participants were organized in 3 groups of 4 users each. At third stage, students were reorganized into 4 *Jigsaw* groups of 3 learners each, encouraging effective interactions according to a social matrix rotation [11]. Finally, in the second level of the *Pyramid* the 4 previous *Jigsaw* groups were rearranged in two groups of 6 users each.

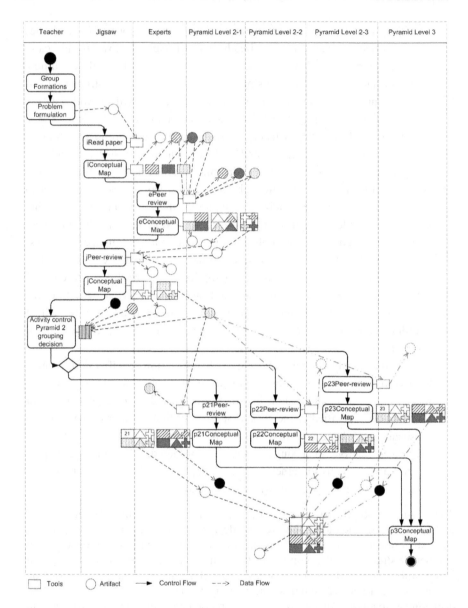

Fig. 3 A simplified version of the workflow activity diagram of the MOSAIC case study comprising the data flow at instance-level. The schema is focused on the interactions related to the learning tools represented by blank rectangles while the data it generates are represented as blank circles. Dashed, black and dotted rectangles and circles are tools and data corresponding to the peers.

The aforementioned decisions were conditioned mostly by the use of the IMS LD specification according to implement a collaborative data flow. The current enriched MOSAIC was performed according to the **LeadFlow4LD** design methodology. The instantiation specification document just reflects the relationship between data and tools.

4 Discussion

In this section, the most important issues that were identified from the application of the **LeadFlow4LD** methodology in the process of formalizing the MOSAIC scenario are highlighted and discussed. The *design over-scripting*, the *different component operationalizations* and the *missing structural instantiation* issues affects the abstract representation of complex CSCL scripts, its flexible instantiation and globally a high reusability level of designs.

Design over-scripting

This issue refers the contextual information the design team had in advance. The original MOSAIC case implementation was preceded by knowledge on how many people would be involved in the course. There are initial *structural and heuristic* constraints [1] that compromise the **reuse** and **creation** of IMS LD based scripts, but also the Composition-based CSCL designs. Similarly, the *hardcoded* implementation of the data flow interactions based on IMS LD is compromised as well.

However, through the implementation of the MOSAIC data flow situation according to **LeadFlow4LD** approach, the *generic* workflow structure is set in correspondence with the learning flow specification. For instance, the *ePeer review* workflow activities should be executed three times according to the number of artifacts to be checked (see Fig. 3). If these settings change the corresponding sub process structure must change as well. These changes require a repurposing effort that should be assumed by the *teacher* role. Note that a design instantiation entails a redesign of the supporting workflow-based data flow within the limits established by heuristic constraints.

Different component operationalizations

Reusing learning designs concerns the creation of complete designs from existing solutions [46]. These solutions can be CLFP templates, script components with different degrees of completeness, etc [20]. The issue refer to the so-called hybrid scenarios [46] whose include already operationalized contents and services. Current Composition-based approach describe designs with such of hybrid features.

At the MOSAIC case scenario, the *Pyramid* and the *Jigsaw* are combined to form an LD template. Nevertheless, it is necessary to integrate a predefined dynamic behavior based on the *Group heterogeneity* AP whose objective is to create heterogeneous groups in order to promote a deepening and widening understanding in a specific domain. An IMS LD based implementation of this adaptation

mechanism aim the creation of several branching paths where each route is associated to a corresponding group setting. At this point, many questions come up.
How many branches should be created? How late in the script life cycle this *context-dependent* AP should be implemented? Similar questions emerge when attempting to integrate the *Group heterogeneity* AP at the workflow-based specification of the structural data flow.

The alternative procedure of fixing the number of grouping settings at design
time the reusability of the resulting learning flow design is restricted as well. But
at the same time, the workflow structure design of the corresponding data flow
situation should already be set, thus prescribing its instantiation setting to very
specific context features. In this case the global **reusability** level of the UoL is
constrained by the most rigid component. Then as the proposed AP implementation based on IMS LD is restricted to pre-defined methods [30], the better time to
generate the executable code is once the learning flow design or UoL is instantiated (see Fig. 4).

Missing structural instantiation

An important concern about the LeadFlow4LD conceptual proposal, lies in the
fact that the data flow instantiation is not fully covered. The **LeadFlow4LD** solution proposes a non-standard instantiation specification (**iLeadFlow4LD**) through
which, users are associated to specific roles, and the link between data and tools
instances is set [35].

However, the **iLeadFlow4LD** specification [35] does not establish the relationship between the number of artifacts to be processed at a given activity and their
corresponding structural matching; neither provides support to the flexible specification of predefined AP-based adaptation mechanisms whose particularization is
context-dependent. For instance, as shown in the MOSAIC case instantiation diagram depicted in Fig. 3, the *ePeer review* activity aims the reviewing of 3 papers
(*striped, gray and dotted circles*) according to the chosen sharing logic used.

On the other hand, the explicit elements which establish the data binding, and
the manner through which such a collaborative data flow situation should be implemented are not specified. For instance, as the *ePeer review* workflow activity
does, the *Pyramid 2 grouping control decision* workflow activity instantiation is
structurally-dependent on the amount of artifacts to be revised and produced, and
how many times this activity should be repeated to complete the *checking* of previous contributions. It is clear that the *ePeer review* activity should be executed as
often as the number of artifacts to be processed (in our case, 3 files) following
either a sequential or a parallel structure. Similar situation occur in case of specify
a context-dependent adaptation mechanism as such integrated in the MOSAIC
script. If the size of the roles or the size of the groups at the level 2 of the Pyramid
(`Pyramid_level_2-x`) is smaller keeping the same number of users, then the
number of the corresponding sub-structures should be higher as well as the number of grouping settings.

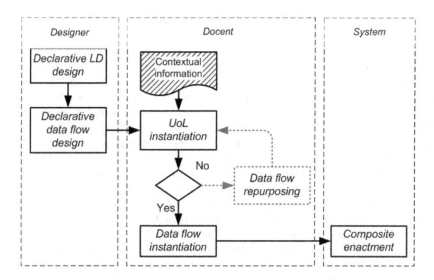

Fig. 4 A *LeadFlow4LD*-based design process incorporating the *Data flow repurposing* step, reachable in case of inconsistency between data flow setting and the structural data flow design.

Hence, if a teacher wants to reuse a *LeadFlow4LD*-based solution in a given context, he/she could be forced to go back in the step sequence defined by the design methodology in order to make the corresponding structural changes. The repurposing implies an increase of the cognitive loads for both designers and teachers. Additionally, this step increases the likelihood of error-prone situations at specification and instantiation times, especially in complex scenarios, which requires, from **teachers** high technical skills in workflow modeling. At the same time, this situation implies that the **designer** must know a priori contextual aspects, or restrict certain script parameters values, thus subtracting flexibility on design process. The Fig. 4 depicts such a situation in a *LeadFlow4LD*-based design process.

The above analysis reveals limitations of the *LeadFlow4LD* design methodology that affect the reuse of collaborative data flow designs. The declarative nature of workflow languages forces that all expected goals should be specified at design time, precluding a direct reuse from existing data flow solutions. Such characteristic increase the prescription, and restricts the reusability level of the overall composite (*hybrid*) design. So far, the main factors, affecting the reuse of adaptive and composite CSCL designs are the *design over-scripting, different component operationalizations* and the *missing structural instantiation*, and they point out to an alternative solution based on abstract representations of collaborative data flow.

5 An Abstract Workflow Template-Based Approach

The findings exposed in the previous sections point out to a solution whose requirements such as *minimal coupling between design at declarative level and instantiation level* should be satisfied, in order to increase the reusability of data flow designs for CSCL scenarios. In this approach the handling of abstractions related to data flow design is based on existing patterns and workflow-based data flow situations are automatically created. Similar requirements pointing out to fully-fledge technological support on learning process management depicted in [18]. In this case, the separation of concerns regarding abstract representation, runtime execution procedure and technological infrastructure are promoted. Technological support for learning processes is workflow-based (e.g. BPEL) and data or document exchange may be supported as well but the creation and reuse of collaborative data flow designs are not explicitly addressed.

In the context of scientific experimentation, the **WINGS** initiative uses semantic representations as **abstract workflow templates** through which the high-level structure of workflows is specified as a data-independent representation [16]. The structure is represented as a compact graph of nodes and links without including repetitions of components or dataset [15]. The *workflow instances* are created adding specification of the data to be computed, and finally these instances are mapped to a certain execution environment by binding tasks to available resources in order to create the so-called *executable workflows*. The implementation is based on ontology languages such as OWL (Ontology Web Language [2]) and an underlying reasoner, in order to reason about constraints on data and components, as well as to propagate those constraints throughout the workflow structure.

In principle, these features of the **WINGS** initiative respond to requirements identified from the findings of this paper: the automatic workflow creation based on abstract representations as templates and the reasoning about sequential or parallel execution of data and component collections. The first characteristic addresses the need that at a declarative level, data flow components of designs should be as less operationalized as possible. This way the *Designer* role player (instructional design or skilled teacher) should be also less aware about the specific contextual parameters, thus increasing the **design reusability**. The second feature provides an answer to the *missing structural instantiation* of the actual workflow-based *Composition* approach. In this case an unskilled practitioner (*Educator* role) do not have to pay attention on technical complexities associated to the *composite* CSCL designs instantiation, especially in workflow process modeling with which teachers may not be familiarized.

In our approach the *Designer* creates the workflow templates adding incremental information according to each designing intervention such as, the integration of CLFP-based templates or codes, AP implementations or anything else. This learning design information shapes the workflow template or **data flow template** defining variables (nodes or tool invocation components, links between nodes) and semantic constraints that define the variable and component properties. These constraints may serve as a basis to generate transformation rules aiming the creation or selection of data flow templates candidates. This process is fed with exchange logic and other contextual information such as the number of users or the group composition, provided by the *Educator*. But also, the data flow candidates creation may be supported by the use of existing designs or fragments available from the CSCL data flow situation repository. Based on these CSCL templates the data flow instances can be created including tool instance information and mapping the design to the final execution language (e.g. BPEL, XPDL). The learning design information and the first data flow template creation is responsibility of the generic *Designer* role. This role can be played by an instructional designer or a skilled teacher.

The learning design information supply and the creation of the abstract data flow template is responsibility of the *designer* role. Such a role can be played by an instructional designer or a skilled teacher without concern about specific contextual settings. In turn, the creation of data flow templates candidates and data flow instances is implemented by the reasoner playing the *System* role and the *Educator*. The proposal architecture for the novel solution is depicted in Fig. 5.

The design methodology described in section 3 should be also modified according to the proposed alternative solution schema

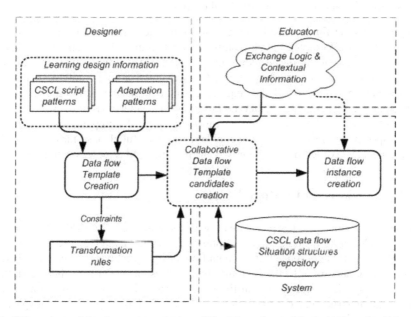

Fig. 5 Overview of the alternative solution of flexible and reusable designing of collaborative data flow situations, based in abstract workflow templates.

1. Structural design phase, both the **UoLF** (Unit of Learning Flow) and **UoDF** (Unit of Data Flow representation) documents are generated.
    1a. Similar to the original specification.
    1b. An abstract representation of the collaborative data flow situation should be created according to the structure depicted in Fig. 2, but adding

behavioral information or semantic constrains, such as the multiplicity of incoming data flow. The use of this information at instantiation time should be used in order to estimate how the involved activity sequencing is implemented or choose an already available collaborative data flow situation. Also supporting CSCL scripting patterns and AP could be enriched with an abstract and semantic representation of data flow situations (candidates).

2. Coordination document. The same methodology should be followed regarding the coordination representation. The latter representation should be enriched by adding coordination marks or tags that may be dynamically mapped into specific workflow activities during the instantiation phase.

3. Instantiation specification. The XML-based specification derived from instantiating the learning design aimed by the use of authoring tool, should be used in order to complete/refine the data flow representation. The social structure and the exchange logic choice determine it relation with tool and data instances, as well as with the structural design.

According to the findings and the current approaches tackling the reusability in collaborative and complex scenarios this incipient solution has been outlined. The approach points out the initial requirements and corresponding advantages but at the same time exhibit some limitations. Consequently, a new notation should be incorporated not only to fill the requirement of automatic reasoning about data flow semantic representations, but also to represent pattern-based data flow solution at an abstract level. The underlying technological complexities associated need a specific authoring tool not available until now. But also, so far the proposal do not includes explicit descriptions of tools incorporated as new constraints.

6 Conclusions and Future Work

The work presented in this paper deals with the so-called *Data flow problem in Learning Design* in the context of complex real-life, and adaptive CSCL scripts. It focuses on the issue of reusability of data flow designs that has not been adequately addressed, according to the literature review presented. On the other hand, the paper pays special attention to real-life CSCL scripts that may include non-trivial collaborative structures or adaptation mechanisms that respond to unexpected events. Therefore, this work aims to provide insights and point out to adequate solutions that may allow instructional designers and educational practitioners handle design, instantiation and enactment of complex CSCL scripts.

The approach adopted in this paper consists in the systematic analysis and critical review of the limitations regarding reuse of data flow designs within the context of a significant case study. The original MOSAIC case study, reported in the literature, was enriched in order to include adaptation mechanisms and served as the frame for the case study. Thus, we could analyze the effect the reuse dimension of the *Data flow problem* of adaptation features in CSCL scripts. On the other hand, we have employed the **LeadFlow4LD** design methodology in order to illustrate the existing approaches to the *Data flow problem*. Such a methodology is a representative of the *Composition*-based approach that aim maintain a high level of interoperability with the existing standards (IMS LD for the learning flow and BPEL or XPDL for the data flow) and therefore foster reuse and a wider acceptance by the community.

Findings of the case study show that there are several obstacles that have to be overcome with respect to reuse. The main problem that was detected is related to the coupling between design and instantiation, as well as difficulties in particularizing the structural and social information for specific contexts. These problems point out to the need to employ abstract workflow templates for the data flow designs, such as the ones employed by the **WINGS** initiative in the field of scientific computational experiments. In the paper, we have outlined the envisaged solution based on abstract workflow templates that contributes to a major reuse of the designs, while at the same time keeping maintains the advantages of the *Composition*-based approach with respect to automation and design consistency.

Another significant aspect of the paper is related to the pattern-oriented approach to CSCL scripts. The design of the enriched MOSAIC case study employs patterns regarding collaborative learning flow (such as the *Pyramid* CLFP), assessment (like the *Peer review* pattern) or adaptation (such as *Group heterogeneity*). These patterns allow the production of non-trivial scripts with features that are highly demanded by practitioners. Additionally, the patterns contribute to a wider adoption by the practitioners, hide technical complexities and promote reuse of good practices and therefore of the appropriate data flow designs.

Current work is related to the implementation of the proposed solution based on abstract workflow templates. The resulting prototype will allow us to show the expected advantages through the same enriched MOSAIC case study.

On the other hand, it is expected to advance in the direction of including a wider set of patterns and therefore creating *complex* real-world CSCL scripts. We will pay special attention to the adaptation patterns and their effect to the reuse dimension of the *Data flow problem in Learning Design*.

Finally, we expect to analyze the validity of the proposed approach through the use of the prototype and the design methodology in real educational contexts employing the complex and adaptive CSCL scripts produced through the aforementioned pattern-based approach.

Acknowledgements. This work was financially supported by a MAEC-AECID fellowship to the first author, as well as by the Spanish Ministry of Research and Innovation (TIN2008-03023/TSI) and the Autonomous Government of Castilla and León, Spain, project VA293A11-2. The authors would also like to acknowledge the contributions from Eloy Villasclaras and other members of the GSIC/EMIC Group.

References

1. Alvino, S., Asensio-Pérez, J., Dimitriadis, Y., Hernández-Leo, D.: Supporting the Reuse of Effective CSCL Learning Designs through Social Structure Representations. Distance Education 30(2), 239–258 (2009)
2. Bechhofer, S., van Harmelen, F., Hendler, J., Horrocks, I., McGuinness, D.L., Patel-Schneider, P., Stein, L.: OWL Web Ontology Language Reference. W3C Recommendation (2004)
3. Bote-Lorenzo, M.L., Hernández-Leo, D., Dimitriadis, Y.A., Asensio-Pérez, J.I., Vega-Gorgojo, G., Vaquero-González, L.M.: Toward Reusability and Tailorability in Collaborative Learing Systems using IMS-LD and Grid Services. Advanced Learning Technologies 1(3), 129–138 (2004)
4. Botturi, L., Stubbs, T.: Handbook of Visual Languages for Instructional Design: Theories and Practices. Idea Group Publishing, Hershey (2008)
5. Burgos, D., Tattersall, C., Koper, E.J.R.: Representing Adaptive e-Learning Strategies in IMS Learning Design. In: Proceedings of the International Workshop in Learning Networks for Lifelong Competence Development, TENCompetence Conference Sofia, Bulgaria, pp. 54–60 (2006)
6. Chen, W.: Supporting Teachers Intervention in Collaborative Knowledge Building. J. Network and Computer Applications 29, 200–215 (2006)
7. Dalziel, J.: Lessons from LAMS for IMS Learning Design. In: Proceedings of the Sixth IEEE International Conference on Advanced Learning Technologies, ICALT 2006, pp. 1101–1102. IEEE Computer Society, Washington, DC (2006)
8. Decker, G., Mendling, J.: Process Instantiation. Data & Knowledge Engineering 68(9), 777–792 (2009)
9. Demetriadis, S., Karakostas, A.: Adaptive Collaboration Scripting: A Conceptual Framework and a Design Case Study. In: Proceedings of the 2008 International Conference on Complex, Intelligent and Software Intensive Systems, CISIS 2008, pp. 487–492. IEEE Computer Society, Washington, DC (2008)
10. Demetriadis, S., Karakostas, A.: Introduction to Adaptive Collaboration Scripting. In: Daradoumis, T., Caballé, S., Marquès, J.M., Xhafa, F. (eds.) Intelligent Collaborative e-Learning Systems and Applications. Studies in Computational Intelligence, vol. 246, pp. 1–18. Springer, Heidelberg (2009)
11. Dillenbourg, P., Hong, F.: The Mechanics of CSCL Macro Scripts. International Journal of Computer-Supported Collaborative Learning 3(1), 5–23 (2008)
12. Dillenbourg, P., Tchounikine, P.: Flexibility in Macro-Scripts for Computer-Supported Collaborative Learning. Journal of Computer Assisted Learning 23(1), 1–13 (2007)
13. Dochy, F.J.R.C., McDowell, L.: Assessment as a Tool for Learning. Studies In Educational Evaluation 23(4), 279–298 (1997)
14. Foster, I., Kesselman, C.: Computational Grids. In: Foster, I., Kesselman, C. (eds.) The Grid: Blueprint for a New Computing Infrastructure, pp. 15–51. Morgan Kaufmann Publishers, Inc., CA (1999)
15. Gil, Y., Groth, P., Ratnakar, V., Fritz, C.: Expressive Reusable Workflow Templates. In: Proceedings of the 2009 Fifth IEEE International Conference on e-Science, E-SCIENCE 2009, pp. 344–351. IEEE Computer Society, Washington, DC (2009)

16. Gil, Y., Ratnakar, V., Deelman, E., Mehta, G., Kim, J.: Wings for Pegasus: Creating Large-Scale Scientific Applications Using Semantic Representations of Computational Workflows. In: Proceedings of the 19th National Conference on Innovative Applications of Artificial Intelligence, AAAI 2007, vol. 2, pp. 1767–1774. AAAI Press, British Columbia (2007)

17. Gil, Y., Ratnakar, V., Kim, J., Antonio González-Calero, P.A., Groth, P., Moody, J., Deelman, E.: Wings: Intelligent Workflow based Design of Computational Experiments. IEEE Intelligent Systems 26(1), 62–72 (2011)

18. Helic, D.: Technology-Supported Management of Collaborative Learning Processes. International Journal of Learning and Change 1(3), 285–298 (2006)

19. Hernández-Leo, D., Dimitriadis, Y., Asensio-Perez, J.I., Villasclaras-Fernández, E.D., Jorrín-Abellán, I.M., Ruiz-Requies, I., Rubia-Avi, B.: Collage, a Collaborative Learning Design Editor Based on Patterns Special Issue on Learning Design. Journal of Educational Technology and Society 9(1), 58–71 (2006)

20. Hernández-Leo, D., Harrer, A., Dodero, J., Asensio Pérez, J., Burgos, D.: A Framework for the Conceptualization of Approaches to "Create-by-Reuse" of Learning Design Solution. Journal of Universal Computer Science 13(7), 991–1001 (2007)

21. Hernández-Leo, D., Asensio Pérez, J., Dimitriadis, Y., Bote-Lorenzo, M., Jorrín-Abellán, I., Villasclaras-Fernández, E.: Reusing IMS LD Formalized Best Practices in Collaborative Learning Structuring. Advanced Technology for Learning 2(4), 223–232 (2005)

22. Hernández-Leo, D., Villasclaras-Fernandez, E., Asensio-Perez, J., Dimitriadis, Y., Retalis, S.: CSCL Scripting Patterns: Hierarchical Relationships and Applicability. In: Proceedings of the Sixth IEEE International Conference on Advanced Learning Technologies, ICALT 2006, pp. 388–392. IEEE Computer Society, Washington, DC (2006)

23. IMS Global Learning Consortium: IMS Learning Design Information Model. Specification, IMS (2003)

24. Karakostas, A., Demetriadis, S.: Adaptation Patterns in Systems for Scripted Collaboration. In: Proceedings of the 9th International Conference on Computer Supported Collaborative Learning, CSCL 2009, vol. 1, pp. 477–481. International Society of the Learning Sciences, Rhodes (2009)

25. Karakostas, A., Demetriadis, S.: Adaptation Patterns as a Conceptual Tool for Designing the Adaptive Operation of CSCL Systems. Educational Technology Research and Development 59(3), 327–349 (2011)

26. Kobbe, L., Weinberger, A., Dillenbourg, P., Harrer, A., Hamalainen, R., Hakkinen, P., Fischer, F.: Specifying Computer-Supported Collaboration Scripts. International Journal of Computer-Supported Collaborative Learning 2, 211–224 (2007)

27. Koper, R., Miao, Y.: Using the IMS LD Standard to Describe Learning Designs. In: Handbook of Research on Learning Design and Learning Objects: Issues, Applications and Technologies, pp. 16–48. IDEA Group (2007)

28. Koschman, T. (ed.): CSCL, Theory and Practice of an Emerging Paradigm. Lawrence Erlbaum Associates, Inc., Mahwah (1996)

29. Ma, J., Shaw, E., Kim, J.: Computational Workflows for Assessing Student Learning. In: Aleven, V., Kay, J., Mostow, J. (eds.) ITS 2010. LNCS, vol. 6095, pp. 188–197. Springer, Heidelberg (2010)

30. Magnisalis, I., Demetriadis, S.: Modeling Adaptation Patterns in the Context of Collaborative Learning: Case Studies of IMS-LD Based Implementation. In: Daradoumis, T., Caballé, S., Juan, A.A., Xhafa, F. (eds.) Technology-Enhanced Systems and Tools for Collaborative Learning Scaffolding. SCI, vol. 350, pp. 279–310. Springer, Heidelberg (2011)

31. Magnisalis, I., Demetriadis, S., Pomportsis, A.: Implementing Adaptive Techniques in Systems for Collaborative Learning by Extending IMS-LD Capabilities. In: Proceedings of the 2010 International Conference on Intelligent Networking and Collaborative Systems, pp. 70–77 (2010)

32. Miao, Y., Burgos, D., Griffiths, D., Koper, R.: Representation of Coordination Mechanisms in IMS Learning Design to Support Group-based Learning. In: Lockyer, L., Bennet, S., Agostinho, S., Harper, B. (eds.) Handbook of Research on Learning Design and Learning Objects: Issues, Applications and Technologies, pp. 330–351. IDEA group (2008)

33. Miao, Y., Hoeksema, K., Hoppe, H., Harrer, A.: CSCL Scripts: Modelling Features and Potential Use. In: Proceedings of the 2005 Conference on Computer Support for Collaborative Learning, CSCL 2005. International Society of the Learning Sciences, Taipei (2005)

34. Ngu, A.H., Bowers, S., Haasch, N., Mcphillips, T., Critchlow, T.: Flexible Scientific Workflow Modeling Using Frames, Templates, and Dynamic Embedding. In: Ludäscher, B., Mamoulis, N. (eds.) SSDBM 2008. LNCS, vol. 5069, pp. 566–572. Springer, Heidelberg (2008)

35. Palomino-Ramírez, L., Bote-Lorenzo, M., Asensio-Pérez, J., Dimitriadis, Y.: Lead-Flow4LD: Learning and Data Flow Composition-Based Solution for Learning Design in CSCL. In: Briggs, R.O., Antunes, P., de Vreede, G.-J., Read, A.S. (eds.) CRIWG 2008. LNCS, vol. 5411, pp. 266–280. Springer, Heidelberg (2008)

36. Palomino-Ramírez, L., Bote-Lorenzo, M., Asensio-Pérez, J., de la Fuente-Valentín, L., Dimitriadis, Y.: The Data Flow Problem in Learning Design: A Case Study. In: Proceedings of the 2008 8th International Conference on Networked Learning, NLC 2008, Halkidiki, Greece, pp. 285–292 (2008)

37. Papazoglou, M.P., Georgakopoulos, D.: Service-Oriented Computing: Concepts, Characteristics and Directions. Communications of the ACM 46(10), 25–28 (2003)

38. Paramythis, A.: Adaptive Support for Collaborative Learning with IMS Learning Design: Are we there yet? In: Proceedings of the Adaptive Collaboration Support Workshop, Adaptive Hypermedia 2008, AH 2008, pp. 17–29. Hannover, Germany (2008)

39. Pérez-Sanagustín, M., Burgos, J., Hernández-Leo, D., Blat, J.: Considering the Intrinsic Constraints for Groups Management of TAPPS and Jigsaw CLFPs. In: Proceedings of the 2009 International Conference on Intelligent Networking and Collaborative Systems, INCOS 2009, pp. 317–322. IEEE Computer Society, Washington, DC (2009)

40. van Rosmalen, P., Vogten, H., van Es, R., Passier, H., Poelmans, P., Koper, R.: Authoring a full life cycle model in standards-based, adaptive e-learning. Educational Technology & Society 9(1), 72–83 (2006)

41. Shapiro, R.: Process Definition Interface. XML Process Definition Language. Tech. rep., Workflow Management Coalition (2008)

42. Stake, R.: Multiple Case Study Analysis. The Guilford Press (2005)

43. Vantroys, T., Peter, Y.: COW, a Flexible Platform for the Enactment of Learning Scenarios. In: Favela, J., Decouchant, D. (eds.) CRIWG 2003. LNCS, vol. 2806, pp. 168–182. Springer, Heidelberg (2003)

44. Vignollet, L., Bote-Lorenzo, M., Asensio-Pérez, J.I., Dimitriadis, Y.: A Generic Specification of the Data-Flow Issue in the Learning Design Field. In: Proceedings of the 2009 Ninth IEEE International Conference on Advanced Learning Technologies, ICALT 2009, pp. 56–58. IEEE Computer Society, Riga (2009)

45. Vignollet, L., Charoy, F., Bote Lorenzo, M., Asensio Pérez, J.: Workflow Management and Learning Flow Management: Commonalities and Differences. IEEE Transactions on Learning Technologies 13(3), 35–37 (2010)
46. Vignollet, L., Ferraris, C., Martel, C., Burgos, D.: A Transversal Analysis of Different Learning Design Approaches. Journal of Interactive Media in Education (JIME), Special Issue: Comparing Educational Modelling Languages on the "Planet Game" Case Study Case Study (2), 1–10 (2008),
http://jime.open.ac.uk/article/2008-26/373
47. Villasclaras-Fernández, E.D., Hernández-Leo, D., Asensio-Pérez, J., Dimitriadis, Y., de la Fuente-Valentín, L.: Interrelating Assessment and Flexibility in IMS-LD CSCL Scripts. In: Proceedings of the Scripted vs. Free CS Collaboration: Alternatives and Paths for Adaptable and Flexible CS Scripted Collaboration Workshop, Organized in conjunction with the 8th International Conference on Computer Supported Collaborative Learning, CSCL 2009, pp. 39–43. University of Aegean, Rhodes (2009)
48. Villasclaras-Fernández, E.D., Hernández-Leo, D., Asensio-Pérez, J.I., Dimitriadis, Y., Martínez-Monés, A.: Towards Embedding Assessment in CSCL Scripts through Selection and Assembly of Learning and Assessment Patterns. In: Proceedings of the 9th International Conference on Computer Supported Collaborative Learning, CSCL 2009, vol. 1, pp. 507–511. International Society of the Learning Sciences, Rhodes (2009)
49. Vogten, H., Tattersall, C.: Implementing a Learning Design Engine as a Collection of Finite State Machines. In: Learning Design, A Handbook on Modelling and Delivering Networked Education and Training, Springer. Heidelberg (2005)
50. Weinberger, A., Kollar, I., Dimitriadis, Y., Mäkitalo Siel, K., Fischer, F.: Computer-Supported Collaboration Scripts. Perspectives from educational psychology and computer science. In: Technology-Enhanced Learning. Principles and Products, pp. 155–174. Springer, Heidelberg (2009)
51. Wilson, S.: Workflow and Web Services. White paper, CETIS (2005),
http://www.eframework.org/Portals/9/Resources/
SOAandWorkflow2.pdf

System Orchestration Support for a Collaborative Blended Learning Flow

Luis de-la-Fuente-Valentín[1], Mar Pérez-Sanagustín[2],
Patrícia Santos[2], Davinia Hernández-Leo[2], Abelardo Pardo[1],
Carlos Delgado Kloos[1], and Josep Blat[1]

[1] Telematics Engineering Department, Carlos III University of Madrid, Spain
[2] ICT Department, Universitat Pompeu Fabra, C/ Roc Boronat 138, 08018 Barcelona, Spain
{lfuente,abel,cdk}@it.uc3m.es, {mar.perez,patrícia.santos,
davinia.hernandez,josep.blat}@upf.edu

Abstract. Portable and interactive technologies are changing the nature of colla-borative learning practices and open up new possibilities for Computer Supported Collaborative Learning (CSCL). Now, activities occurring in and beyond the class-room can be combined and integrated leading to a new type of complex collabora-tive blended learning scenarios. However, to organize and structure these scenarios is challenging and represent a workload for practitioners, which hinder the adoption of these technology-enhanced practices. As an approach to alleviate this workload, this paper proposes a proof of concept of a technological solution to overcome the limitations detected in an analysis of an actual collaborative blended learning expe-riment carried out in a previous study. The solution consists on a Unit of Learning suitable to be instantiated with IMS Learning Design and complemented by a Ge-neric Service Integration system. This chapter also discusses to which extent the proposed solution covers the limitations detected in the previous study and how useful could be for reducing the orchestration effort in future experiences.

1 Introduction

Portable devices have impacted multiple aspects of our everyday life. In educa-tion, researchers and practitioners see the potential of this technology as a chance for expanding current educational scenarios and exploring innovative learning methodologies [15]. Particularly in the area of Computer Supported Collaborative Learning (CSCL), the introduction of portable devices opens up a new debate about how this discipline is going to evolve [22].

Significant research effort has been devoted to introduce portable devices in learning activities and to understand how they might enhance current educational settings. Theoretical studies such as the one by Spikol et al. [25] or by Sharples et al. [24] evidence the interest on these new types of learning practices. The first one provides designers with a framework to tackle the challenges of designing for innovative mobile learning activities, while the second one proposes a model to analyze these innovative practices.

T. Daradoumis et al. (Eds.): Intell. Adapt. & Personalization Tech. in CSCL, SCI 408, pp. 29–46.
springerlink.com

From a more experimental perspective, some works have started to explore the benefits of mobile and content delivery capabilities of this technology to generate learning settings enabling learners to work and collaborate at different spatial locations beyond the class. For example, Facer et al. propose a mobile gaming experience in which children are invited to understand the animal behavior in a savanna in direct physical interaction with this space [16]. The findings of this study show that this innovative experiment increased the self-motivation of children. Another work by Ruchter et al. describes an experiment using mobile computers as a guide for supporting environmental learning [23]. The results show that using these computers as mobile guides can lead to an increase in knowledge about the natural environment and an increase in students' motivation to engage in the educational environmental activities.

All these studies introduce a new concept of learning in which activities are no longer limited to the classroom space. A study by Park et al. states that "mobile learning activities could provide a better learning experience by establishing the conditions for optimal flow" [20]. This idea relates with the CSCL concept of orchestration. Orchestration is defined as the process of structuring and coordinating the actions of the course participants (the learning flows) for achieving potential effective learning outcomes [9]. According to Roschelle and Pea, "learning content's performance is optimized when it is orchestrated with a pedagogical sense" [22].

One of the proposals to organize and computationally support these learning flows is the so-called collaboration "scripts" [9, 12, 17, 18]. Scripts are based on the idea that free collaboration does not always produce learning. The rationale of these scripts is to structure collaborative learning processes in order to trigger group interactions that may be rare in free collaboration. When these interactions are technologically mediated they are called Computer Supported Collaborative Learning Scripts (or CSCL scripts). CSCL scripts manage resources and deliverables, define roles and phases and enable specific interaction in order to guide collaborative processes for producing situations of effective learning [9] by facilitating and reducing the coordination efforts of teacher and students [10, 11, 28, 29].

However, when these scripts combine activities supported by portable devices with activities taking place at different spatial locations, the orchestration process becomes more complex. In such as type of scenarios it becomes particularly challenging tracking students' progress [21]. These challenges hinder the establishment of the relations within activities and make the management of the collaborative learning flow more difficult. As a consequence, the orchestration of collaborative learning flows translates into an increase in the teaching staff workload.

The results of a previous work carried out by the authors of this chapter in an actual educational context evidence this workload [22]. The work presents a case study of a collaborative blended learning experience that combines mobile-based activities with in-class sessions. Despite the encouraging results, the enactment of these types of learning settings imposes a significant workload on the teaching staff. As a consequence, one of the conclusions of the study proposes automating some aspects of the enactment for future editions of the course.

The work presented in this chapter is based on the above-mentioned previous experiment. The goal is to present the proof of concept of a technological setting to automate some of the orchestration tasks of this learning flow. As a consequence, the teaching staff effort is expected to be reduced facilitating the replication of the course flow with a reasonable cost in future editions. With this aim, we created a scripted learning flow implemented in a Unit of Learning (henceforth simply UoL) for orchestrating the activities and automating management duties. The UoL is compliant with IMS Learning Design (IMS LD) [14] and extended with Generic Service Integration (GSI) [8]. As a conclusion, we discuss to which extend these technologies can overcome with the limitations detected and how useful might be in similar situations.

The rest of the chapter is organized as follows: Section 2 describes the experimental study in which this work is based on, gives an overview of the results obtained and exposes the limitations detected in the orchestration process. Section 3 describes the technological solution conceived to automate the orchestration process of this scenario. Section 4 presents the results of a simulation of the scripted flow proposed and discusses how this solution is envisaged to solve the limitations detected in the previous study and help reducing teaching staff workload on similar experiences. Finally, Section 5 presents the conclusions of this work and the future work lines.

2 Description of a Previous Experimental Study

This section is divided into three parts. First, the learning experiment carried out in a previous work by the authors of this chapter at the Universitat Pompeu Fabra (Barcelona, Spain) is presented [22]. Second, an overview of the main results obtained from the experiment is given. And third, the final subsection identifies the limitations regarding the orchestration process.

2.1 Description of the Experiment: Meeting the Campus Together

The CSCL experience was carried out with 74 first -year ICT engineering students enrolled in a mandatory course called Introduction to Information and Communication Technologies. The aim of the course is to give a global vision of the University and its resources, and an introduction to the professional world of ICT industry. The CSCL activity started the first day of the 2009-2010 academic years and continued during the next two weeks. The scenario was structured into three different phases following the learning flow defined by the Jigsaw Collaborative Learning Flow Pattern (CLFP) [1, 13].

The first phase consisted in an individual exploration of the campus. We named this phase "Discovering the Campus". To support this activity 46 Near Field Communication (henceforth NFC) tags were distributed around the 5 campus's buildings. These tags contained information about the place in which they were located. Students were equipped with NOKIA (N6131, N6212) mobile phones, which included an embedded Radio Frequency IDentification (RFID) reader for accessing the information stored in the tags. Students had 30 minutes to freely explore the campus. All the information regarding the sequence of tags accessed by each student was stored into a log file. After the visit, students had to fill in a Google Forms questionnaire indicating which buildings had visited and which seemed to them the most interesting.

The second phase was called "Explain the campus". In this phase, students were grouped in "Building's Expert groups". Each expert group was associated to one of the 5 campus buildings and was composed by 4 or 5 members randomly chosen from the students with similar building expertise level. To define the students' building expertise the teachers considered two sources of information: (1) the log files obtained during the exploration and (2) the answers to the Google Form questionnaire. Depending on the places of the campus visited (registered in the log files) and the preferences about the different buildings (indicated in the questionnaire) there were defined as experts in one or another building. The activity for these teams was to create a presentation explaining the main characteristics of the building assigned and upload it to the Moodle Platform of the University (henceforth Moodle).

Finally, the third phase was called "Reflect about the campus". For this activity, the teachers uploaded all the presentations from the previous phase to Moodle. Students had to access and review all the presentations and answer an individual test including questions about the whole campus. This last activity was carried out in a 25 minutes session in a classroom with PCs.

2.2 Overview of the Results of the Experiment

The results of the experiment show that the activity enacted was meaningful in terms of educational and motivational benefits. First, the results indicate that using mobile technologies in combination of other computational tools is a good mechanism to integrate all the activities into a unique learning setting that facilitates the students discovering the campus. Second, introducing an exploratory activity with mobile phones is shown a good way to foster the motivation of the students with regard to their studies, engineering research and teaching activities. And third, results show that the actual technology used during the activity was easily adopted and highly accepted by students and teachers.

Comments of teacher A at the end of the activity summarizes the aforementioned learning benefits and motivates its repetition in future courses: "*1. The activities are more significant to them (they experience the services of the University vs. they just hear about the services). 2. Students are active in the whole activity. Also, thanks to working physically with what they are learning, they have the opportunity to discuss with other students the buildings/services of their interest, to discover other buildings/services by explanations of their own classmates, etc. 3. Students make use of ICT technologies that they will be learning in their studies they are just starting (again enhances the significance of the activity)*".

Here we have summarized the main outcomes of the activity, since the main focus of this study is to propose alternative solutions to the limitations identified during the orchestration of the activity, which are presented in the next subsection. More details about the data of the experiment underlying these results are given in [22].

2.3 Limitations on the Orchestration Tasks

Two teachers and one researcher carried out all the orchestration processes of the case study. The activity was technologically supported (NFC tags, mobile phones, Moodle) but there was no system that automatically integrated the whole process. This translated into some of the orchestrations tasks being done by hand. In what follows we present a detailed explanation of teacher tasks in each phase.

The task for the teachers in the first phase was to store the log files once the students finish the visit of the campus. Due to the number of students and the number of available devices, some of the students had to share a device for the visit. To identify which data log belonged to which student, teachers annotated the time when a device was given to a student or pair of students. This information was used later to make the correspondence between the log files and each of the students participating in the experiment. The files were uploaded to a computer via Bluetooth connection.

For the second phase, teachers had to form the building's expert groups. As explained before, the expertise was measured taking into account the number of tags per building visited by each student and the preferences indicated in the questionnaire. This was the most complex and time-consuming task. As the teacher B commented after the experiment: "The most time demanding and difficult part of the activity was to organize the groups depending on the students' activity registered in the log files and the preferences answered in the questionnaires".

First, the teachers manually analyzed all the log files created during the visit. Due to the number of students (74) this part was very time consuming and the process had to be reviewed three times by the two different teachers and a researcher to avoid errors. The teacher calculated an amount of 3 hours invested in this task.

Second, in the analysis of the questionnaire answered after the visit, the building recommended by the student was taken as the preferred one. This was carried out approximately in 4 hours.

Besides, the students were divided into two groups corresponding to the regular lecturing sessions. For the experiment, students from both groups were randomly mixed. Combining people from these groups also posed some problems. On one hand, students could not contact easily their classmates because they did not meet face to face in the classroom. On the other hand, because the activity took place during the first two weeks of the course, there were students dropping out the course before the final presentation so some groups had to be rearranged. The teaching staff using e-mail for communication carried on all these group adjustments.

The comments of the teachers after the experiment evidence the complexity and limitation of this orchestration tasks. For instance, one of the teachers of the course highlights the group formation and communication as some of the most demanding issues: *"Once the whole activity was set-up, I think it was more a matter of complexity than of difficulty. The logistics was the more demanding is-sue: creating groups, informing students about the groups, orchestrating their tasks depending on the groups, managing and analyzing their outcomes in order to propose them the following tasks, managing their outcomes in order to facili-tate the assessment of their learning, etc" [Teacher A]. The other teacher stresses the need of an automatic tool facilitating the group formation: "We did not use any tool for creating the groups. It would have been very useful to have an auto-matic system to analyze the logs and the response to the questionnaires to create the groups"* [Teacher B].

In the third phase, the task of the teachers consisted in uploading the students' presentations to a public repository in Moodle and making students to complete the final test. The teacher organized the presentations per building and created one folder for each group in the public repository. The test was uploaded to the plat-form and the teachers had to control that all students had answered the test.

Finally, the teachers organized the workflow using Moodle. They used the plat-form to inform students of the steps for the next activities, and e-mail to inform when the description of a new activity was available. However, other activities in the course were also carried out in parallel during this period (and published in Moodle) and students had problems to have a unified view of the scenario.

Summarizing, the evaluation of the case study showed the following limitations of the activity flow:

1) Students' data analysis: Manually analyzing the log files was hard to carry out without errors. Also combining the preferences and the log file results for as-signing the students expertise was complex and very time demanding.
2) Expert group management: Creating and managing the expert groups was very time demanding because of the groups instability due to drop outs that characterize the first weeks of the course and mixing students from the two lecturing sessions.
3) Activity workflow: Moodle does not facilitate the integration of the activities to create an orchestrated view of the learning flow for both, the teachers and the students.
4) Scalability: Without technological support, these activities are very costly to carry out for a large number of students. The data analysis becomes very complex.

3 System for the Scripted Orchestration

A technological solution has been developed for dealing with the limitations hig-hlighted in Section 2.3. The proposal is to use a computational script as the or-chestration mechanism for automating the most demanding tasks. The IMS LD framework supports the authoring and deployment of the activity flow, resulting in a UoL that structures the learning flow of the scenario. Additionally, the proposed UoL (with minor changes) could be used for supporting analogous learning flows. This solution is a proof of concept to show that teaching staff workload can be significantly reduced in any learning situation which combines collaborative activities at different spatial locations supported by portable devices. This section overviews the technological framework that supports UoL as well as the translation process from the original course to its scripted version.

3.1 Course Flow Management Technologies

One of the best-established modeling languages used to computationally represent learning flows is the IMS Learning Design framework [27]. The vocabulary provided by the specification supports the use of a wide range of pedagogical models in the creation of learning courses, including collaborative and blended learning. IMS LD is constructed upon the metaphor of the theatrical play: different actors play different roles. Each role is assigned to a set of learning activities that may occur in sequence or in parallel, depending on whether they are organized in acts or structures. Each activity takes place in a given environment, which consists of a set of learning objects and/or services.

Collaborative learning is supported by means of the use of roles. That is, each course participant can be related to a different role, and the result is that different students may perform different activities at the same time. Furthermore, one course participant could be related to several roles. The combination of the emerging possibilities allows modeling complex collaborative learning models such as *jigsaw*, or *pyramid* [13, 27].

The IMS LD framework also supports the creation of adaptive content material [4]. The offered mechanism is based on the use of the so-called *properties* and *conditions*. That is, the author can define a set of properties, whose value will change during the course activities. The *conditions* evaluate such values to decide whether or not an action needs to be triggered. With such functionality, the course author can create several (maybe interlaced) learning flows and adapt the sequence of activities depending on the specific user's needs.

One limitation of IMS LD is the lack of integration with third party tools. The framework defines the use of a limited catalogue of services, which are *e-mail*, *conference*, *monitor* and *index*. In practice, available services are not able to support complex blended learning flows, where different tools are used in different scenarios.

Generic Service Integration (GSI) proposes a framework to include any kind of web-based tool in the context of IMS LD courses [6,8]. The integration covers authoring, deployment and enactment of courses. First, GSI provides a vocabulary for course authors to describe generic tools that will be used to support certain activities within the course. Then, when the UoL is uploaded to a compliant platform, the tool description is used to select a case-specific tool that matches with the expected functionality. Finally, the tool is instantiated and integrated in the enactment of the course, allowing the interaction of course participants with the third party tool. GSI also offers mechanisms for the information exchange of IMS LD and the external tool, so the course *properties* can be feed by information dynamically retrieved from the third party tool. Other initiatives that attempt to solve the integration problem in IMS LD have been described in the literature (e.g. [2,3,18]). However, a comprehensive review of the state of the art or the discussion bout the appropriateness of these solutions in the presented scenario is beyond the scope of this chapter. A detailed review of these solutions can be found in [6].

In the GSI model, the integration of third party tools is based on case-specific service adapters. In other words, each integrated tool requires a service adapter that translates IMS LD requests into case-specific requests constructed as specified by the tool's API. One of the existing service adapters called *GSpread*, supports the use and management of assessments through the use of Google Forms and Google Spreadsheets [7].

The integration of Google Spreadsheets[1] in a UoL can be summarized as follows: students access a questionnaire (an HTML form) through a hyperlink located in the environment of an activity; on the other hand, teachers own a spreadsheet populated by student's responses, where each row contains data from a single student. Teachers can manipulate the spreadsheet arbitrarily so that they produce a value suitable to be mapped to an IMS LD property. Then, IMS LD retrieves the data contained in the spreadsheet and the appropriate properties are updated.

The GSpread adapter uses the Google API[2] for documents and spreadsheets in order to execute the following actions:

- Before the activity flow starts, the adapter establishes a relationship between the teacher role and her corresponding Google identity. Such action requires manual intervention of the involved participants (the teacher) and uses the support of the SubAuth protocol [5].
- At the beginning of the activity flow, the adapter creates the external service instance (the spreadsheet) and relates it to the questionnaire contained in the UoL so that the answers of such questionnaire are stored in the spreadsheet, which is owned by the teacher.
- During the enactment of the activities, GSpread retrieves the information contained in the spreadsheet and parses it so that the relevant information is used to feed the IMS LD *properties*.

The inclusion of spreadsheets in IMS LD courses serves a double purpose. First, it provides support for assessment, the absence of which is one of the weaknesses of the specification. Assessment is made possible by including HTML questionnaires and using the responses to adapt the course flow. Second, it offers a well-known method to manipulate data, substituting the complex calculate element in IMS LD, which hinders the creation of mathematical formulas based on questionnaire responses.

For the purpose of the work presented in this paper, IMS LD has been used to define and deploy the activity sequence of the course. GSI has been used to integrate specialized data management tools as part of the learning flow. In particular, we have used Google Spreadsheets to administer students' data and to automate the group formation process. The resulting framework, that combines IMS LD and Google Spreadsheets with the integration provided by GSI, is referred as the *orchestration system*.

---

[1] http://spreadsheets.google.com
[2] http://code.google.com/intl/en-US/apis/spreadsheets/

3.2 The Scripted Course Flow

The orchestration system proposed in this document requires the course flow to be expressed with the IMS LD vocabulary. Such translation process imposes some minor changes to the course flow, which are required to adjust the sequence of activities to the particular needs of the technological framework. We present in what follows a description of the learning flow that resulted after the translation process, highlighting the particular elements that differ from the original flow explained in Section 2.1.

One restriction imposed by IMS LD is the need of computers to deliver the activity descriptions. In practice, it means that the students must go to the laboratory to perform the activities instead of being in any other place such as the library, or home. This fact limits the number of students that can participate in the activity. On the positive side, the orchestration method allows the course flow to be quickly instantiated and enacted, so several course instances can be held in the same day. The script was designed to support five working groups, whose number of members was set to five. As a result, 25 is the number of learners considered in the design of the learning script, whose enactment is expected to last 2 hours. A higher number of students can be supported by simply creating new course instances of the same course. The number of teachers is not restricted. We will refer to all teaching staff members as simply the teacher.

The course follows a blended learning approach: students receive the information through the computer; some of the activities are done on-line and the remaining ones are offline. An overview of the course flow is show in Fig. 1.

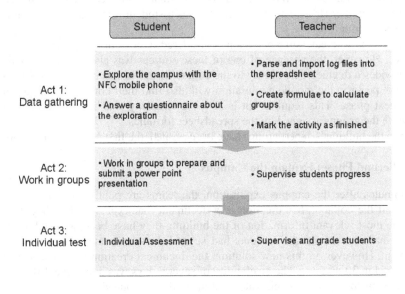

Fig. 1 IMS LD Mapping of the original flow.

3.2.1 First Phase: Discovering the Campus

During the first act, learners visit the campus and acquire the knowledge they will use in later activities. They perform the visit with a NFC mobile phone as described in Section 2.1. When a student returns back to the classroom, s/he has to obtain from the mobile phone the log file generated by his/her activity. Then, s/he uses the course interface to upload such log file into the server. Once finished, they fill in a questionnaire to show their acquired knowledge of the campus. After that, they have to wait until the teacher enables the next phase.

Meanwhile, the teacher has to supervise the learners' activities and track their completion. When all the students have completed both the exploration and the questionnaire, the teacher starts the group formation process. Such activity is divided in two main parts: (i) storing the data in the spreadsheet and (ii) creating groups.

The teacher will use two data sources to create the groups: the answers of the questionnaire and the information of the log files. The former is automatically stored in the spreadsheet, while the latter requires some preprocessing. The processing of the log files results in a comma separated value (.csv) file that contains the relevant information and can be easily imported by the spreadsheet.

Once the information has been loaded in the spreadsheet, each row contains numeric values that summarize the activity and the answers to the questionnaire of a single student. This summary contains, for each student: (1) the number of tags accessed per building and (2) the building expertise, which is the building with the maximum number of tags accessed. The teacher creates the formulas in the spreadsheet so that the output of the activity is finally produced. The calculated output is a number (from 1 to 5) assigned to each student representing the building's expert group.

The criteria to create students' teams considered data from questionnaires and log files. However, the absence of one of these sources was also supported. This fact provides a degree of flexibility to the course flow. For instance, students who could not perform the mobile exploration will also find their corresponding group in the next phase. This requirement is also supported by enabling the teacher to overwrite the groups assigned by the spreadsheet formulas.

When the building's expert groups have been created, the first phase is completed.

3.2.2 Second Phase: Explain the Campus

Few minutes after the campus exploration, the teams are published and the students start the collaborative creation of a document that explains what they consider the most relevant information of the building they have been assigned to. In the original experiment, the students had several days to create and submit the document. However, in this new solution, the document creation was expected to be finished in 30 minutes. The availability of several days was due to the difficulty for the teams to adjust their schedules and meet together outside the classroom. In the scripted version of the course, such scheduling problem does not appear, so the students have to finish the activity in the abovementioned 30 minutes.

The teams then upload the created documents to the course manager while the teacher tracks the process. When all the teams have finished their assignment (i.e. they have uploaded their documents), the teacher enables the third phase.

3.2.3 Third Phase: Reflect about the Campus

In this phase, the delivery of the previously submitted presentations requires no intervention from the teaching staff: the documents are directly accessible from the activity statement. Thus, students may review all the presentations and access to the final assessment task.

3.3 Technical Details of the Scripted Flow

Two types of participants take part in the course: learners and teachers. These are the roles defined in the UoL. Although the learners are divided into groups, there still is a single role for all of them. This is because roles are populated at the beginning of the course, and therefore at design time there is not enough information about the number of required groups. This division is performed in a later step using *properties*.

Both the answers to the questionnaire and the mobile activity logs are stored in a Google Spreadsheet. The use of Google Forms as the questionnaire provider guarantees that the answers will be stored in the spreadsheet without the need of human intervention, being each student's answer stored in a single row of the spreadsheet. The inclusion of the logs information follows a different path, shown in Fig. 2. When a student finishes the activity, s/he is requested to use the resulting log file as the value of a *file property*. All students' log files (that is, all *properties*) are stored in the same folder and are easy to manipulate. Furthermore, because files are related to their owners, it is also possible to easily identify which log belongs to which student. Moreover, the regular structure of the log files allows automatic parsing. A script specially developed for the case performs the log analysis and produces a file with comma separated values containing a summary of the events generated by each student. This summary can then be manually uploaded into the Google Spreadsheet.

The spreadsheet then contains all the data from the log files and questionnaires, where each row represents a single student. At this point, the teacher manipulates the data so that the output of the activity is finally produced. All values are calculated by the spreadsheet, which has been previously modified with the proper criteria. The formulas in the spreadsheet require numeric values, and as a consequence the original questionnaire was modified to include closed response questions to process the answers automatically. The questionnaire includes three types of questions: (1) a multiple choice option in which the students select the building they have visited, (2) a true-false question related to each building and (3) a Likert-scale question to evaluate each building. The use of closed response questions solves two problems: first, offers the possibility of automatically computing the students' preferences. Second, provides the teacher with an easy mechanism to evaluate the students' knowledge about the campus.

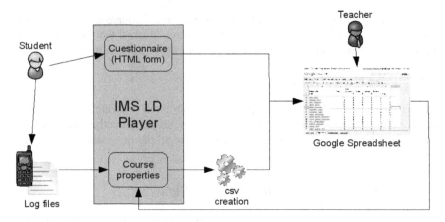

Fig. 2 Data flow for group assignment automation

Once the grouping phase has finished and no group changes are expected in the groups, the teacher marks the *activity* as finished. This action triggers data synchronization between IMS LD and the spreadsheet. That is, the IMS LD player, through the GSpread adapter, requests the spreadsheet's rows using the API offered by Google. The response is the XML-formatted data, from where the adapter parses and selects the relevant information, i.e. the group membership recommendation done by the spreadsheet. Such information, represented as a number and a character string, is the used to feed the *properties* of the UoL.

When IMS LD *properties* obtain their value, the corresponding *conditions* are evaluated and the course flow is properly adapted. There are two types of *properties* whose value is assigned:

- Each student has a *property* called group. The value is a number (from 1 to 5) that says in which team the student has been placed.
- Each group has a *property* called members, which contains the names of the team members.

The second phase of the course flow has been modeled as an IMS LD act: all course participants start at the same time. The act adapts its contents depending on which team the student has been related to.

There are three issues to be solved by the course flow:

1) Which task corresponds to each student?
2) How do students know who their partners are?
3) How do students submit their presentation?

To solve the first question, the course flow has been modeled with five different *activities*, one per building's expert team. The visibility of these activities is controlled by *property values*, so that only one of the *activities* will be shown to each student (see Figure 3). In practice, students receive the activity description that corresponds to their group, and they see no information about the other groups. Each activity description shows the *members* property of the group. Therefore, students are aware of who are their teammates.

```
<imsld:if>
  <imsld:is>
    <imsld:property-ref ref="prop-locpers-group-number" />
    <imsld:property-value>3</imsld:property-value>
  </imsld:is>
</imsld:if>
<imsld:then>
  <imsld:show>
    <imsld:learning-activity-ref ref="la-get-expert-tanger" />
  </imsld:show>
  <imsld:hide>
    <imsld:learning-activity-ref ref="la-get-expert-la-nau" />
    <imsld:learning-activity-ref ref="la-get-expert-rb" />
    <imsld:learning-activity-ref ref="la-get-expert-fabrica" />
    <imsld:learning-activity-ref ref="la-get-expert-tallers" />
  </imsld:hide>
</imsld:then>
<imsld:if>
```

Fig. 3 Conditions used in the IMS LD manifest to show and hide content

The submission of the collaborative document has been modeled as a *local property* whose value is set when students upload a file through a form included in the activity description.

Figure 4 shows part of the resource with a link to such *properties* published in the third phase of the course (that is, in the third act). In this phase, the delivery of the previously submitted presentations requires no intervention from the teaching staff and the documents of all the teams are available to be downloaded. Since file *properties* are directly accessible from the statement, students may review all the presentations and access to the final assessment task.

```
<h2>Grup Tanger</h2>
<div class="tanger-not-submitted">
    <p>No disponible encara</p>
</div>
<div class="tanger-submitted">
    <ld:view-property ref="prop-loc-tanger-file" property-of="self" view="value" />
</div>
```

Fig. 4 Use of the view-property element

The final assessment is an IMS Question & Test Interoperability (QTI) test[3]. Students access this test through a link in the UoL and login to the QTI server. The QTI test is composed of 5 questions: 3 common QTI questions (Multiple Choice, Yes/No and Multiple response) and 2 Google Maps-based QTI questions [20]. For these questions, students locate their answer in a Google Maps map. An example of QTI-GoogleMaps question is in Figure 5.

---

[3] IMS (2006). IMS QTI Question & Test Interoperability Specification v2.0/v2.1. Retrieved March 23, 2010, from HTTP://www.imsglobal.org/question/index.html

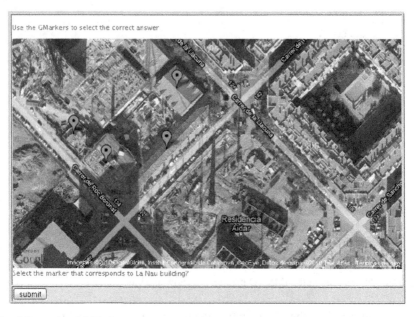

Fig. 5 Example of QTI-Google Maps question where students have to select which marker corresponds with La Nau building

4 Results Analysis of the Solution: A Simulation

This section analyzes whether the orchestration system proposed for the script enactment solves the limitations detected in the activity "Discovering the Campus": (i) the groups formation process, (ii) the expert groups management, (iii) the activity workflow, and (iv) scalability. With this aim, we propose an analysis that consists in a simulation of the script enactment with some of the data extracted from the empirical study to understand whether the solution proposed deals with the limitations detected.

Part of the students' data extracted from the empirical study presented in Section 2 has been used in the simulation. Specifically, the simulation has been performed with the 74 log files generated from the exploratory activity in the empirical study. The outcomes of the students' questionnaire performed after the campus exploration have not been considered since, in the proposed solution, the questionnaire has been modified and transformed to numeric questions.

The first step is the simulation of the expert groups formation process according to the information registered into the 74 log files. For this, three manual interventions are required: (1) to copy the log files to the folder where they will be parsed and processed, (2) to import the resulting *comma separated value* file to the spreadsheet and (3) to insert a set of spreadsheet formulas that capture the group formation criteria. Figure 6 shows the results of the analysis of the 74 log files.

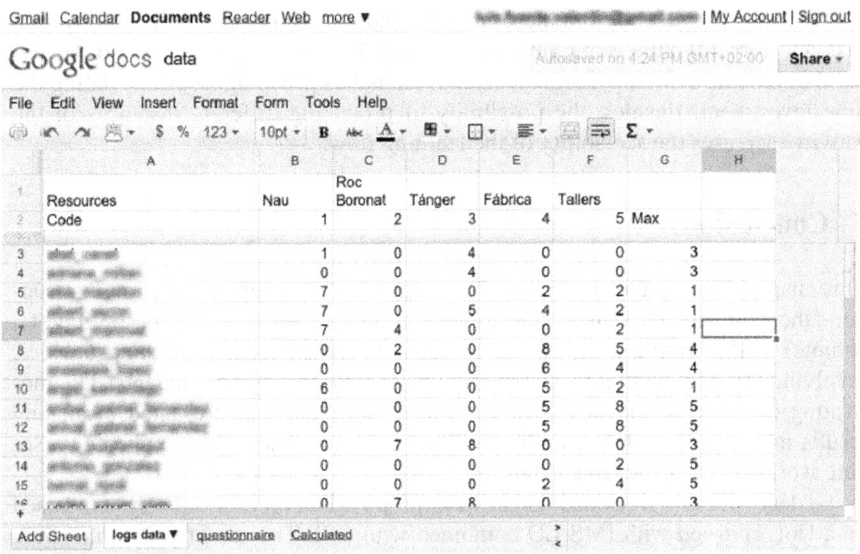

Fig. 6 Student activity data imported from the analysis of the 74 log files.

The second step is to simulate the workflow distribution among the potential students participating in the experiment. The activity tree and activity content, is adapted for each student who receives, at the end of the course, a complete view of the learning flow. The end of each phase has to be indicated by manually marking the phase as finished. This mechanism provides a control of the workflow on runtime.

The simulation shows the effectiveness of the orchestration system proposed to deal with the main limitations detected. First, both the module for automating log files analysis and the numeric questionnaires solve the main limitations of the students' data analysis for the first phase. On one hand, this solution strongly decreases the time spent for analyzing all the log files. On the other hand, this automatic approach might support the assignment of students' expertise by diminishing the number of errors when doing this process manually. Moreover, this approach also offers the possibility of modifying the automatic building assignment and to easily adapt the groups to the actual context of the activity.

Second, the scripted course has been designed to support 25 students and can be instantiated several times. Preliminary results on the course deployment showed a quite reasonable cost of the replication process. That is, the orchestration system makes the course as reusable as regular IMS LD courses are. Besides, the NFC tags can be reused from one course instance to another as well as for other similar experiences. It can be said that the solution is scalable to the extent that it is reusable, being each new course instance able to support 25 more participants.

Therefore, the results from the simulation shows that the proposed orchestration system offers a flexible semi-automatic system for analyzing log files and managing the students' building assignments that facilitates grouping tasks alleviating the time investment. Besides, the possibility of producing different instances of the activity increases the scalability of the learning flow.

5 Conclusions

This chapter has presented a proof of concept of a technological solution that supports the automatic enactment of learning activities requiring the orchestration of a complex collaborative learning flow, supported by different computing devices, involving different spatial locations and with a large number of students. The motivating example has been drawn from an experiment that presented promising results in terms of students' motivation and achieved learning but imposing a severe workload on the teaching staff.

To deal with this workload, is has been proposed an orchestration system based on a UoL codified with IMS LD combined with GSI. The use of GSI to integrate services in the context of the UoL allowed the learning flow to coordinate the use of different technologies such as NFC, Google Spreadsheets and QTI. In the designed course, a semi-automatic process of data acquisition and group formation complements the group-dependent scripted delivery of the learning material. The enactment simulation of the proposed script showed that this solution would provide significant reduction of teaching staff workload. The major limitations of the previous experiment disappear with the semi-automatic orchestration of the learning flow. One added value of the proposal is the simplicity of the replication process, which allows reusing the course flow with a reasonable cost. As a conclusion, the presented solution sheds some light on how technology can facilitate the orchestration process of complex and innovative collaborative learning using portable technology such as smart phones.

As next steps, we are working mainly into two main lines. On the one hand, the solution presented in this work has been already enacted as part the introductory activity of the engineering courses of the University Pompeu Fabra. The activity was successfully enacted and the data analysis is still under development. The evaluation results of this new experiment will complement the results of the simulation presented in this work to analyze whether the orchestration system proposed deals with the limitations detected when applied to an actual educational context.

On the other hand, we plan to study how the orchestration system proposed could be applied to other similar courses in order to understand the suitability of the solution to be applied into other learning contexts, thus extending the scalability of the course flow beyond the presented scenario.

Acknowledgments. This work has been partially funded by the Project Learn3 (TIN2008-05163/TSI) from the Plan Nacional I+D+I and "Investigación y Desarrollo de Tecnologías para el e-Learning en la Comunidad de Madrid" funded by the Madrid Regional Government under grant No. S2009/TIC-1650.

References

[1]   Aronson, E., Blaney, N., Stephan, C., Sikes, J., Snapp, M.: The jigsaw classroom. Improving Academic Achievement: Impact of Psychological Factors on Education, 209 (2002)

[2]   Bote-Lorenzo, M.L., Vaquero-González, L.M., Vega-Gorgojo, G., Dimitriadis, Y.A., Asensio-Pérez, J.I., Gómez-Sánchez, E., Hernández-Leo, D.: A Tailorable Collaborative Learning System that Combines OGSA Grid Services and IMS-LD Scripting. In: de Vreede, G.-J., Guerrero, L.A., Marín Raventós, G. (eds.) CRIWG 2004. LNCS, vol. 3198, pp. 305–321. Springer, Heidelberg (2004)

[3]   Bote-Lorenzo, M.L., Gómez-Sánchez, E., Vega-Gogojoa, G., Dimitriadis, A.Y., Asensio-Pérez, J.I., Jorrín-Abellán, I.: Gridcole: A tailorable grid service based system that supports scripted collaborative learning. Computers & Education 51(1), 155–172 (2008)

[4]   Burgos, D., Tattersall, C., Koper, R.: How to represent adaptation in e-learning with IMS - learning design. Interactive Learning Environments 15(2), 161 (2011), http://www.informaworld.com/10.1080/10494820701343736 (last visited July 2011)

[5]   de la Fuente-Valentín, L., Leony, D., Pardo, A., Delgado Kloos, C.: User identity issues in mashups for learning experiences using IMS Learning Design. International Journal of Technology Enhanced Learning 03, 80–92 (2011)

[6]   de la Fuente-Valentín, L.: Orchestration of learning activities through the integration of third-party services in IMS Learning Design, (Doctoral dissertation) University Carlos III of Madrid (2011)

[7]   de la Fuente-Valentín, L., Pardo, A., Delgado Kloos, C.: Using third party services to adapt learning material: A case study with Google forms. In: ECTEL 2009: Learning in the Synergy of Multiple Disciplines, pp. 744–750. Springer, Niza (2009)

[8]   de la Fuente-Valentín, L., Pardo, A., Delgado Kloos, C.: Generic service integration in adaptive learning experiences using IMS learning design. Computers & Education 57, 1160–1170 (2011)

[9]   Dillenbourg, P., Fischer, F.: Basics of computer-supported collaborative learning. Zeitschrift für Berufsünd Wirtschaftspadagogik 21, 111–130 (2007)

[10] Dillenbourg, P., Tchounikine, P.: Flexibility in macro-scripts for computer supported collaborative learning. Journal of Computer Assisted Learning 23(1), 1–13 (2007)

[11] Haake, J., Pfister, H.: Flexible scripting in Net-Based learning groups. In: Scripting Computer-Supported Collaborative Learning, pp. 155–175. Springer, Heidelberg (2007)

[12] Harrer, A., Hoppe, H.: Visual Modelling of Collaborative Learning Processes: Uses, Desired Properties, and Approaches. In: Handbook of Visual Languages for Instructional Design: Theories and Practices (2008)

[13] Hernández-Leo, D., Asensio-Pérez, J., Dimitriadis, Y.: Computational representation of collaborative learning flow patterns using IMS learning design. Educational Technology & Society 8(4), 75–89 (2005)

[14] IMS Learning Design specification (February 2003), http://www.imsglobal.org/learningdesign/ (last visited July 2011)

[15] Johnson, L., Smit, R., Levine, A., Haywood, K.: Horizon Report 2010. The New Media Consortium, Austin (2010)

[16] Joiner, R., Stanton, D., Reid, J., Hull, R., Kirk, D., Facer: Savannah: mobile gaming and learning? Journal of Computer Assisted Learning 20(6), 399–409 (2004)

[17] Kobbe, L., Weinberger, A., Dillenbourg, P., Harrer, A., Hämäläinen, R., Häkkinen, P., Fischer, F.: Specifying computer-supported collaboration scripts. International Journal of Computer-Supported Collaborative Learning 2(2), 211–224 (2007)

[18] Magnisalis, I., Demetriadis, S.: Modeling Adaptation Patterns in the Context of Collaborative Learning: Case Studies of IMS-LD Based Implementation. In: Daradoumis, T., Caballé, S., Juan, A.A., Xhafa, F. (eds.) Technology-Enhanced Systems and Tools for Collaborative Learning Scaffolding. SCI, vol. 350, pp. 279–310. Springer, Heidelberg (2011)

[19] Miao, Y., Hoeksema, K., Hoppe, H.U., Harrer, A.: CSCL scripts: modeling features and potential use. In: Conference on Computer Support for Collaborative Learning: the Next 10 Years!, pp. 423–432. International Society of the Learning Sciences, Taipei (2005)

[20] Navarrete, T., Santos, P., Hernández-Leo, D., Blat, J.: QTIMaps: A model to enable web-maps in assessment. Educational Technology & Society (in press)

[21] Park, J., Parsons, D., Ryu, H.: To flow and not to freeze: Applying flow experience to mobile learning. IEEE Transactions on Learning Technologies 3(1), 56–67 (2010)

[22] Pérez-Sanagustín, M., Ramírez-González, G., Hernández-Leo, D., Muñoz-Organero, M., Santos, P., Blat, J., Delgado, C.: Discovering the campus together: a mobile and computer-based learning experience. Journal of Network and Computer Applications (in press) doi:10.1016/j.jnca.2011.02.011

[23] Roschelle, J., Pea, R.: A walk on the WILD side: How wireless handhelds may change CSCL. In: Conference on Computer Support for Collaborative Learning: Foundations for a CSCL Community, pp. 51–60. International Society of the Learning Sciences (2002)

[24] Ruchter, M., Klar, B., Geiger, W.: Comparing the effects of mobile computers and traditional approaches in environmental education. Computers & Education 54(4), 1054–1067 (2010)

[25] Sharples, M., Taylor, J., Vavoula, G.: A theory of Learning for the Mobile Age. Learning through Conversation and Exploration Across Contexts. In: Bachmair, B. (ed.) Meidenbilgung in neuen Kulturräumen, pp. 87–99. Springer, Heidelberg (2010)

[26] Spikol, D., Kurti, A., Milrad, M.: Collaboration in context as a framework for designing innovative mobile learning activities. In: Ryu, H., Parsons, D. (eds.) Innovative Mobile Learning: Techniques and Technologies, Information Science, pp. 170–194 (2008)

[27] Tattersall, C.: Using IMS learning design to model collaborative learning activities. In: ICALT 2006: Proceedings of the Sixth IEEE International Conference on Advanced Learning Technologies, pp. 1103–1104. IEEE Computer Society, Washington, DC (2006)

[28] Tchounikine, P.: Operationalizing macro-scripts in CSCL technological settings. International Journal of Computer-Supported Collaborative Learning 3(2), 193–233 (2008)

[29] Weinberger, A., Kollar, I., Dimitriadis, Y., Makitalo-Siegl, K., Fischer, F.: Computer-Supported collaboration scripts. In: Technology-Enhanced Learning, pp. 155–173. Springer, Heidelberg (2009)

Adaptive Collaboration Scripting with IMS LD

Florian König and Alexandros Paramythis

Institute for Information Processing and Microprocessor Technology (FIM),
Johannes Kepler University, Altenbergerstraße 69, A-4040 Linz, Austria
{koenig,alpar}@fim.uni-linz.ac.at

Abstract. The IMS Learning Design specification is a widely known language that allows modelling of, amongst other learning designs, collaboration scripts in e-learning. Yet, it has been criticized for a number of shortcomings and specifically its lack of support for comprehensive adaptation features. We propose concrete extensions to the specification, which address a wide range of problems and omissions. The most important areas of modifications and amendments include: explicit representation of groups and corresponding collaboration contexts, as well as of artefacts as results of joint work; flexible integration of communication and collaboration services; a revamped script organization and sequencing model; a previously missing run-time model, with support for event- and exception- handling. The above are complemented by a wide range of adaptive interventions that can affect the script's progress at run-time, tailor it to changing circumstances, and support learners. Last but not least, sophisticated scenarios are made possible through support for non-traditional collaboration script elements: the possibility to represent human involvement in adaptation decisions, 'transactional' action processing, loops and branches for controlling action execution, and the declaration of re-usable action sequences and complex expressions. Further to the proposed changes, examples are provided that highlight the novel possibilities afforded by these changes for advanced collaboration scripts.

1 Introduction

Learning is regarded by many theorists as a social activity that can benefit from a collaborative setting [1]. In collaborative learning situations, certain interactions (such as discussion, mutual questioning, joint problem solving, conflict, collective sense-making and shared agreement) can trigger learning mechanisms and result in improved learning outcomes [2]. There is, however, no guarantee that beneficial interactions occur, even less so in distance learning, where lack of or limited 'real-world' contact amongst learners can negatively influence social- and group learning- patterns [3]. Especially in collaborative e-learning, support is needed to increase the probability that the desired interactions occur, because teams may not have worked together before, are usually formed for a comparatively short time, and individual learning goals are predominant. One way to support learners is by *scaffolding* their interaction in order to get group work going, mitigate disorientation and reduce cognitive load [4]. *Collaboration scripts* [5] implement such a

T. Daradoumis et al. (Eds.): Intell. Adapt. & Personalization Tech. in CSCL, SCI 408, pp. 47–84.
springerlink.com   © Springer-Verlag Berlin Heidelberg 2012

support strategy: They provide a detailed specification of the desired collaboration processes in a scenario, and can be used, for instance, to set up systematic differences among learners in order to trigger contentious interactions, or to make rich interactions for exchanging complementary knowledge necessary [2].

In computer-supported collaborative learning (CSCL), so called CSCL scripts [6] exist as computational representations of collaboration scripts. They specify how members of a group should collaborate to attain a task [7]. The most widely known formalization that supports modelling CSCL scripts is the IMS Learning Design (LD) specification [8], a learning process modelling language for formally describing designs of teaching-learning processes for a wide range of pedagogical approaches.

Traditionally, CSCL scripts in general, and IMS LD models in particular, have been rather static and inflexible with little possibility to change their execution characteristics at run-time or tailor them to new situations. Yet, these static collaboration scripts merely represent idealized scenarios and there is a danger of impeding fruitful collaboration by an overly coercive script (*over-scripting*) [5]. One strategy of adjusting scripts to learners and groups is to reduce their scaffolds over time, once learners have a good understanding of how to collaborate effectively – a process called *fading* [9]. More generally, *adaptive collaboration scripting* has been proposed [10, 11] and found to be an effective method [12] for adjusting scripts at run-time and targeting the individual needs of learners and groups.

The strengths of IMS LD lie in specifying personalized and asynchronous cooperative learning scripts. Areas of the specification that have received criticism include its insufficient expressiveness with regard to the collaborative learning process [6], and the absence of constructs that are vital in supporting adaptivity [13]: the language provides limited support to model group-based, synchronous collaborative learning and collaboration contexts [6]; the specification of services for providing communication and collaboration facilities (e.g., chat, wiki, whiteboard) is rather inflexible [14]; the activity sequencing model has been found to be constraining [15] and difficult to understand [16]; a very limited event model and an entirely missing run-time model allow for only rudimentary support for monitoring state changes and triggering adaptations; and, there exist only very few and basic primitives for modifying a collaboration process at run-time [13].

The work presented in this chapter uses that criticism as a starting point, while at the same time building upon proposals for extensions of the specification that have appeared in the literature (discussed in the next section), to propose a comprehensive set of modifications and additions to IMS LD, aiming to address many of its general shortcomings and lay the foundations for improved support for adaptivity in CSCL scripts.

The rest of this chapter is structured as follows: In section 2 we give a short introduction to CSCL scripts and how they can be improved by adaptation mechanisms. Section 3 discusses shortcomings of IMS LD and ideas for extensions that have appeared in the literature. Section 4 outlines the proposed modifications and extensions to IMS LD. In section 5, two collaborative learning scenarios that serve as examples throughout the chapter are described. The extensions that deal with improving the ability to model group work, specifically explicit groups, run-time member management, shared artefacts, group processes and collaboration contexts, are discussed in section 6. In section 7, a proposal is put forth for improved integration of services with a flexible selection, permission specification and instrumentation mechanism. Section 8 deals with new ways to model the script execution and react to state changes while a script is running. Adaptive interventions in the form of low-level adaptation actions on individual parts of the script as well as higher-level adaptations are discussed in section 9. Finally, section 10 summarizes the goals attainable through the proposed extensions, and provides an outlook on future work.

This chapter gives a summative account of work fragments of which have appeared in the literature before. Due to space constraints, not all findings can be discussed in full detail here. Readers are referred to the initial publications [17–19] for more in-depth information.

2 CSCL Scripts – Theory and Adaptation

Collaborative learning is defined by Roschelle and Teasley [1] as "… a coordinated, synchronous activity that is the result of a continued attempt to construct and maintain a shared conception of a problem." It combines aspects of constructionism (referring to the "construction of a model or object as an aspect of learning") and a focus on the discussion processes of socio-cultural learning [20]. A multitude of views exist on the mechanisms that make collaborative learning effective: beneficial cognitive processes (knowledge elicitation, internalisation, …) triggered by the extra activities (explanation, disagreement, mutual regulation, …) resulting from interaction [2]; increased accumulation and sharing of knowledge, as well as improved motivation and performance caused by social comparison [21]; mental conflict that can stimulate a cognitive restructuring by making learners aware of deficiencies in their understanding of the world or the problem at hand [22]; and, disagreement which produces cognitive dissonance, stimulating the individual to reduce this dissonance by social communication, presenting alternatives, requesting and giving explanations or revising a position [23].

Computer-Supported Collaborative Learning (CSCL) focuses on how collaborative learning can be supported by technology to enhance peer interaction and work in groups, and how collaboration and technology facilitate sharing and distributing knowledge and expertise among community members [24]. Support is provided in two main ways [25]: (a) through computer-mediated communication (CMC) where technology provides the (synchronous) communication channels required for collaboration; (b) as guidance and constraint for the actions of participants in order to promote beneficial interactions. Such support is required primarily for two reasons: First, the restrictions of electronic communication media create a cognitive overhead and require explicit coordination steps between collaborating learners, which can impede learning processes [26]. Second, learners rarely interact effectively without explicit prompting or guidance [27].

There are two complementary approaches to supporting collaboration [5]: *anticipatively* structuring the collaborative process in order to favor the emergence of productive interactions, or *retroactively* regulating interactions, akin to what tutors do. Different support strategies have been identified along these two extremes [2]:

- *setting up initial conditions*: group size / membership, task assignment, etc.
- *over-specifying the 'collaboration contract'*: scenarios that create systematic differences among learners aiming to trigger conflictual interactions or that make necessary rich interactions for exchanging complementary knowledge; such a scenarios are commonly called 'collaboration scripts' [5]
- *interaction rules in the medium*: usage rules in the communication/ collaboration medium for shaping the interaction
- *monitoring and regulating the interactions*: monitoring of interactions and collaboration behavior with subsequent pedagogical interventions as guidance

The scenarios represented by collaboration scripts can be defined with certain components (groups, participants, roles, activities and resources) and mechanisms to represent the dynamics: distribution of participants over groups (group formation); distribution of roles, activities or resources over participants (component distribution); and, how these two processes are distributed over time (sequencing) [28]. Although not readily apparent in the above itemization, collaboration scripts can encompass strategies from the whole spectrum of collaboration support: So-called *macro scripts* influence collaboration more indirectly and from an educational perspective [29]. They set up conditions aimed at eliciting favorable activities and productive interaction, and push learners to engage in those activities, but don't give specific support. Macro scripts typically describe longer time segments, in which they orchestrate activities which spread over more social planes [30]. In contrast, *micro scripts* implement the psychological perspective of support by directly scaffolding the interaction process per se, for example by prompting participants to interact with their peers in specific ways [29]. Both approaches can be combined, for example by making micro scripts part of a macro script.

In CSCL, collaboration scripts can be implemented into an electronic learning environment to guide the learning process, either through presenting feedback on interaction episodes (in order to allow reflection on collaboration success), explicit advice to learners, provision of cooperation opportunities (e.g., tools) or genuine support in the form of predefined collaboration structures (e.g., roles, activities, sequences, tools, material). The goal of CSCL scripts is to automate the support and control of group interaction [26]. Educational modeling languages (EML) allow representing CSCL scripts in a computational format suitable for execution in an electronic learning environment. These languages provide a well-structured terminology for describing instructional sequences [29]. IMS Learning Design [8] is an implementation of such an EML.

In non-electronic form, collaboration scripts may just capture the core pedagogical intention and leave the instructor enough flexibility to adapt them once they are employed. However, in order to achieve automated support in electronic environments, CSCL scripts have to be more stringently defined, which can leave less flexibility for instructors and participants alike. In that respect, micro and macro scripts represent differing degrees of *coerciveness* and *flexibility* [31]: micro scripts are more coercive because they try to influence or even constrain participants, but they are also more dynamic through their continued support; macro scripts are less coercive, but their support remains largely static after the initial setup. An undesired consequence of high coerciveness or overly static scripts can be *over-scripting* [5] where natural collaboration is constrained or inhibited by a script that is applied in a non-flexible way. As a countermeasure, the degree of support by the script can be adapted based on the quality of collaboration exhibited by the participants. This concept, called *fading* [32], is the process of gradually reducing (or, if necessary, increasing) the level of scaffolding provided by the system. As participants learn how to better collaborate, their need for support decreases and the system can ensure that it does not constrain them by adaptively fading the scaffolding measures.

Adapting collaboration scripts at the time of their application, something that effective instructors do quite naturally, is a topic of ongoing research in the area of CSCL. Many CSCL scripts have been implemented in learning environments in a 'hard-wired' way that allows little re-use or adaptation to changing circumstances [26]. Therefore, *adaptive collaboration scripting* has been proposed [10, 11] in order to allow (automatic) adjustment of scripts at run-time so as to better target the individual needs of learners and groups. According to Walker et al. [33] the move towards adaptive collaboration support is a natural progression of research into collaborative learning: Starting in the *conditions stage*, early work focused on controlling the conditions in the collaborative setting and their effects on learning outcome. In the *interactions stage*, interest has shifted to the collaborative interactions as the decisive factor. How to support collaboration with fixed assistance (scaffolding, scripts) to overcome the lack of spontaneous beneficial collaborative behavior became the topic of the *fixed support stage*. To balance between coercive scaffolding (like scripts) and more loosely structured measures (like collaboration training), adaptive support has finally been proposed. In this *adaptive support stage* of collaboration support, interactions are analyzed and modeled as they occur, and the (content of the) assistance is based on these models.

The possible adaptations for supporting collaborative learning can be broadly categorized into ones related to the establishment of collaboration and those dealing with collaboration process support [13]. Adaptive support for the *establishment of collaboration* can come in many forms. One of the most prominent is (semi-) automatic group formation based on learners' personal- and learning- characteristics and preferences (such as prior collaborative interactions, propensity to collaborate and participate as well as current engagement, load and availability) [13]. The pedagogical approach chosen for a specific collaborative setting might pose additional constraints such as the number of members per group [34] or the composition (homogeneous, heterogeneous) with respect to learners' properties like domain knowledge [35]. A similar application of such grouping methods is peer matching, where one or more suitable partners are recommended in an ad-hoc manner, for example to provide immediate help or delegate a task [36]. Apart from grouping, there are other areas where adaptation can support collaboration establishment like task assignment [12], collaboration context setup and configuration of available communication / collaboration media [34].

Once a collaborative setting has been enacted, the system can intervene in the form of *collaboration process support*. Theoretically, all aspects of the setting, as modeled by the domain model (i.e., the script), can be affected by adaptation actions. However, there are (practical) constraints, and ideally the pedagogical validity is checked before applying adaptations. A number of adaptations have been identified as suitable for run-time interventions [13]: adjusting the group size, recommending or assigning roles, changing the availability of elements such as activities, services or artefacts, and modifying the activity structure. Other possible, low-level adaptations are turning features of services on/off, giving feedback, changing permissions of roles and switching between different modes of floor control in moderated communication/collaboration settings. High-level adaptation measures have been proposed in the areas of interaction support (e.g., encouraging learners to make contributions or listen to others), decision support for groups [36], and negotiation support [37]. In addition, adaptation patterns have been identified: supporting novice learners in larger groups to make them more confident to participate; and, adjusting the scenario challenge level for expert participants in order to offer an interesting learning experience to them too [35].

Support for adaptation can be implemented in multiple ways, which can also be combined for even greater flexibility:

- adaptation algorithms as part of the system's implementation of a script
- possibility for human instructors / participants to influence the script execution
- adaptation rules as part of the EML modeling the CSCL script

In this chapter, the latter two approaches are followed in order to model adaptive CSCL scripts. The presented work builds upon IMS LD, which provides limited support for expressing adaptation rules and enacting them at run-time.

3   Related Work

Since its creation, the IMS Learning Design specification has been scrutinized in a multitude of publications and has been used as basis for modelling collaborative (and adaptive) learning scenarios. A number of extensions have been proposed to address problems in the language and provide missing functionality. One area on which work has concentrated is collaboration scripting, where a number of shortcomings have been identified [6]. To start with, a fundamental hindrance in this area is that modelling groups independently from role(s) is only possible with limitations in IMS LD [29], and manipulating groups at run-time is not supported. Miao and Hoppe [38] propose an extension for modelling groups and operations for member management. However, they do not provide a high-level specification mechanism for run-time grouping, which would allow authors and adaptation engines to readily express the semantics of group (membership) management.

With regard to collaboration contexts, IMS LD does not provide the means for explicitly specifying how the members of a group interact within an activity, other than through a so-called 'conference service'. Hernández-Leo et al. [39] propose the 'groupservice' for specifying group collaboration spaces with communication facilities, floor control, different interaction paradigms and awareness functionality. Yet, this suffers from the same limitations as the original IMS LD specification: the actual collaboration happens outside the script's specification and control. Miao et al. [6] propose running multiple instances of activities, if required by the respective social plane (one per role/group/person) to allow, for example, groups to work in parallel on the same problem. Their approach, however, is still not geared towards high expressive flexibility and describes only how this changes the process model and not how it is reflected in the collaboration environments. Even with these multi-instance activities, IMS LD has insufficient support for artefacts (i.e., results of, or inputs to joint work) and no straightforward mechanism to model their flow between activities [40]. One suggested model for artefacts, which is also proposed to some extent by Caeiro et al. [14], is to implicitly derive the data flow from the specification of creation-/consume- activities.

IMS LD's modelling support for the use of services is limited in its expressiveness – a shortcoming addressed by Valentin et al. [41] in their *Generic Service Integration* approach: It allows for automatic service selection according to keywords specifying the expected service behaviour, author-defined alternative services, and explicit permission specification. Permissions are defined for different actions (read/write/administer) and object types (contribution, context), which, considering the wide range of possible services, is still rather limited. Collaboration contexts are created by specifying the appropriate multiplicity (one-per-group, one-per-user, one-for-all, one-per-role) of a service but are not reflected in the process model. The possibility to specify an action that should be performed after a service has been closed hints at rudimentary event handling.

IMS LD's activity sequencing capabilities have been described as "quite simplistic" [42], and fine-grained splitting or synchronization of the control flow is not readily expressible [6]. Caeiro et al. [14] address some shortcomings by introducing a different model for the *role-part* construct in IMS LD and propose having conditional transitions between acts to allow more complex sequencing. Miao et al. [6] reiterate the need for advanced transition and routing constructs, guided by approaches in the area of workflow management, and present a comprehensive collection of ideas on how to improve support for adaptive collaboration scripting through a more flexible sequencing model with workflow semantics.

The features discussed until now form the basis for collaboration scripting. With respect to support for adaptivity, means to monitor the run-time state of the script and intervene accordingly are required. IMS LD provides some support for adapting a scenario, of which Towle and Halm [43] give an overview. The description of how to implement adaptations, however, already shows limitations of IMS LD like the difficulty to express complex interventions, hard-coded and unstructured adaptation rules, no easy re-use of rules and lack of a run-time model.

In more detail, IMS LD is missing a model of the run-time state of the script, has a very constrained event model, and supports only a limited number of mechanisms to modify aspects of the collaboration process while the script is running [13]. Miao et al. [6] propose ways for improving support for adaptive collaboration scripting, through access to more run-time state information and with operations concerning activities, artefacts, roles, groups, persons, transitions, environments and relations between them. However, a complete event model has not been included in any extension proposal yet.

On the practical side, Berlanga and García [44] present a framework with tests on learning style and knowledge, student modelling, and strategies for tailoring a scenario to individual students expressed in adaptation rules. Zarraonandia et al. [45] describe an approach for introducing small variations into learning design scripts at design- and run- time. The proposed adaptation actions modify activities, environments, the structuring of these elements, resources, properties, and completion criteria. Advanced adaptations like the introduction of new roles or acts, complex conditions and major structural changes are, however, not discussed.

Other suggestions for adaptation actions in the literature include varying the group size, recommending or assigning (changes in) roles for participants, modifying the activity structure (e.g., adding / removing / reordering tasks) and determining the availability of elements (activities, services, artefacts) [13]. Paramythis and Cristea [46] present some more requirements for adaptation languages in the area of collaboration support, none of which are supported by IMS LD: workflow- or process- based reasoning, temporal operators, policies for grouping, invoking system facilities / manipulating system state (e.g., initiating communication sessions) and support for 'provisional' adaptation decisions.

Finally, ideas for easier development of adaptive collaboration scripts have been proposed as well: action and expression declarations for creating re-usable fragments of 'code' [6]; a mechanism for defining complicated expressions and actions; and, the need for a loop control structure for action declarations [38].

4 Proposed Extensions to IMS LD

The shortcomings identified in the IMS LD specification and the various approaches to extending this modelling language described in literature (see previous section) provide a basis for the range of extensions necessary to comprehensively support adaptive collaboration scripting. These requirements have been validated and adjusted through the analysis of widely known collaboration scripts, two of which – JigSaw and TAPPS – will serve as examples in this chapter (see section 5). In addition, the primitives used to implement generally acknowledged control-flow patterns [47] have served as a model for an alternative to the inflexible process sequencing constructs in the original IMS LD specification. The feature taxonomy for selecting services (see section 7) was created based on literature from the area of computer-supported collaborative work (CSCW) [48–52].

The rest of this section provides an overview of the proposed extensions to IMS LD as a result of the analysis and meta-modelling efforts mentioned before. A more detailed description can be found in later sections and in previous publications [17–19]. Our aim has been to cover both collaboration support requirements (groups, contexts, artefacts, etc.) and requirements with regard to adaptivity (run-time state modelling, event handling, interventions, etc.) We specifically set out to not only provide the syntactic means for describing adaptive collaboration scenarios, but also enhance the semantics of the modelling language. This should make it easier for developers to understand and use the language constructs and for run-time engines to detect patterns in the execution and effect meaningful adaptations.

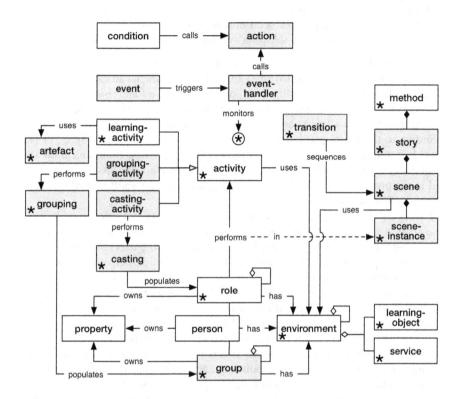

Fig. 1 Partial, high-level view of the information model with extensions in grey

A simplified model of the run-time objects and their inter-relations can be seen in Fig. 1. Elements added to or changed from IMS LD are shown in grey. Collaboration scripting capabilities have been extended as follows:

- Groups can be explicitly modelled (*group*), either statically or, more interestingly, via the specification of constraints for run-time creation of dynamic groups.
- A choice of different policies gives control over the apportioning of participants, for populating groups at run-time (*grouping*).
- Group/Role members can share an *environment* that acts as a common group/ role workspace.
- Assigning roles to participants (seen in these proposed extensions as the inverse activity of casting participants into roles) is possible at run-time via policies similar to those for grouping (*casting*).

- Grouping and casting operations can be sequenced by means of two new activities: *grouping-activity* and *casting-activity*.
- Artefacts are introduced to explicitly model results of (collaborative) work. Their flow between learning activities is defined by referencing them as input-, output- or transient- artefacts.
- The *service* specification has been enhanced to support a wide range of services, constraint-based auto-selection and fine-grained permissions.

In the modified semantics of the new model, the *method* can have multiple *story* containers (corresponds to the IMS LD element *play*), which can each have multiple *scene* elements (corresponding to a combination of *act* and *role-part*). Scenes are connected by (conditional) *transitions* allowing for flexible sequencing of activities; they tie actors (members of *role* objects) to activities, and are instantiated according to the social plane: one *scene-instance* for all actors, per group, or per actor. In combination with *environments*, which may provide communication and collaboration services, collaboration contexts for groups and classes can be created at the process level. Workflow semantics are used for splitting and synchronizing the flow of action.

To implement advanced adaptation features in a script, knowledge of its run-time state is required. The proposed extensions herein feature a run-time model with fine-grained access to all elements of the script during execution. This model also contains pure run-time data, such as information on participants. More flexible and expressive definitions of adaptation rules are made possible through an event handling mechanism with event-condition-action (ECA) semantics (*event-handler*). Each object shown with an asterisk in Fig. 1 can be monitored by an event handler. Events are also generated by services, which allows reacting to behaviour of participants within tools in execution environments that are external to the one in which the script is managed. This is important as a lot of interaction between participants happens in such tools (e.g., text chats, wikis, whiteboards, etc.)

Event handlers can trigger a multitude of new adaptation *actions* for modifying a scripted scenario at run-time: manage the life-cycle of objects, their attributes and relation to other objects; adapt the environment (resources, services) and the control flow; and, adjust adaptations themselves on a meta-level. Exception states can be detected and reacted upon by a hierarchy of handlers. Finally, advanced adaptation features allow human involvement in adaptation decisions, enable transactional action processing, as well as loops and branches for controlling action execution and make the declaration of re-usable action sequences and complex expressions possible.

All of the above extensions and modifications are described in detail in sections 6 to 9.

5 Examples

To explain our extensions in the context of examples, we will use two well-known collaborative learning scenarios, JigSaw [53] and Thinking Aloud Pair Problem Solving (TAPPS) [54], that are often implemented as CSCL scripts. Both examples require features that cannot be modelled readily with the original IMS LD specification. Table 1 summarizes these requirements, the limitations of IMS LD, and in which section of this chapter the extension supporting the requirement is described.

Table 1 Limitations of IMS LD support for unique requirements of scenarios.

Script	Requirement	IMS LD limitations	Extension
JigSaw	run-time grouping	no concept of groups	section 6.1
	run-time casting of leader	roles assigned before start	section 6.1.1
	explicit (flow of) artefacts	only properties	section 6.2
	(group) collaboration contexts	not in process model	section 6.3
		no group workspaces	and 6.4
	flexible event handling	unstructured set of conditions	section 8.2
	exceptions like 'empty role'	no remediation possible	section 8.3
	introduction of extra activity for faster learners at run-time	only simple adaptations	section 9
TAPPS	service selection by criteria	3 types of 'conference' service	section 7.1
	role rotation	only manual switching between fixed roles during run	section 9
	looping executions of activity	rigid sequencing model	section 6.3
	arbitrary time-based events	limited timeouts	section 8.2
	exceptions in service access	no failure mode specified	section 8.3
	fading of scaffolding in service	no adaptations in services	section 9.1.3

In *JigSaw*, learners are grouped into mixed (in terms of ability, knowledge, etc.) groups of 5-6 members and a group leader is appointed. The material is split into 5-6 topics and each group member gets assigned to one. Learners read their material individually and then come together in 'expert groups' (one per topic) to discuss their parts and how to present them to their JigSaw groups. They re-unite in the JigSaw groups, present their topic, and group peers can ask questions. In the end, each group collaboratively solves an assignment that involves all topics. The instructor collects the assignments and provides feedback.

In *TAPPS*, students are paired and given a list of problems. One gets assigned the role of the 'problem solver'; the other becomes the 'listener'. The problem solver tries to solve the problem while thinking aloud. The listener follows, identifies errors and asks questions. After solving a problem, they exchange roles and move on to the next problem.

As already mentioned, these two collaborative learning scenarios will be revisited throughout the chapter, to illustrate the necessity and effects of proposed extensions and amendments.

6 Modelling Group Work

The basis for adaptive collaboration scripting is solid support for modelling collaborative learning scenarios and collaboration processes. The IMS Learning

Design specification has certain shortcomings in that area (see section 3), which were dealt with by extending the language in order to introduce groups, run-time member management, explicit artefacts, and collaboration contexts.

6.1 Groups

In IMS LD, grouping participants can only be simulated by assigning them to predefined, static roles. The set of roles is defined before the script starts and membership does not change thereafter. In our approach, groups can be directly modelled, and assignment of participants to both groups and roles is possible at run-time. There are statically defined groups, which already exist when the script starts, and dynamic groups, which are created at run-time according to certain constraints (e.g., random assignment based on preferred group size). This makes it possible to, on the one hand, define groups that are a fixed part of the script's pedagogics and are required for the execution (intrinsic constraints [55]). On the other hand, a dynamic arrangements of participants in groups at run-time can deal with varying numbers of learners (extrinsic constraints [55]) while still having influence over the groups' size and composition.

Every (sub-)group has a name and can have a group environment to give a group its own 'group space' with communication facilities and shared material. In the JigSaw example, there would exist a group for each topic (the 'expert group') with the material provided in the respective group environment. The properties of group environments are described in more detail in section 6.4.

6.1.1 Run-Time Member Management for Groups and Roles

The creation of dynamic groups and the assignment of participants to (static or dynamic) groups can be specified in a *grouping* (see Fig. 1). A finished grouping consists of a set of groups and a mapping of participants to them. Groupings may be requested from an external provider (*provided grouping*) or created at run-time (*runtime grouping*). The first option allows re-using groupings that exist, for example, in the Learning Management System (LMS) running the script. With the second option, static groups can be used, or dynamic ones created and populated with members. In both cases, usually a whole set of groups (called a *partition*) is populated at once, for example in the first step of the JigSaw scenario (although it's possible to have a grouping for just one group). This set of groups is defined either through references to static groups, which have already been specified in the script (see Fig. 2a), or indirectly through constraints that control the automatic creation of dynamic groups at run-time (see Fig. 2b).

The number of dynamic groups and/or the size of each group can be constrained to an exact number, or to a range, or required to be divisible by a specific value. Additionally one can specify a set of proportions to create, for example, groups with the size ratios of 1:2:4. In the case of the TAPPS example, for instance, participants would need to be grouped at run-time into an arbitrary number of dynamic groups with an exact value of 2 for the size in order to get pairs.

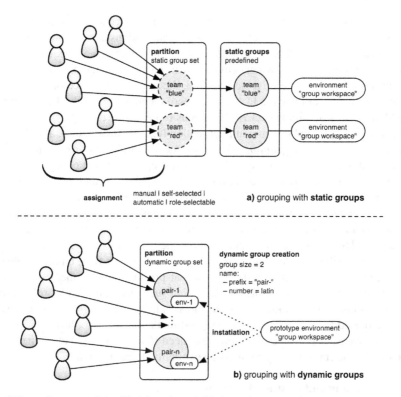

Fig. 2 Exemplary grouping with (a) static and (b) dynamic groups.

The parts of the specification described until now allow for defining the groups (static or dynamic) and their origin (provided or created at run-time). The actual assignment of participants to groups is specified in a *grouping method*:

- *manual*: members of a certain role are tasked to perform the assignment
- *self selected*: participants can choose in which group they would like to be
  - *direct*: participants are immediately assigned to chosen group
  - *prioritized*: assignment by balancing the priorities given by participants
- *automatic*: performed without human intervention
  - *random*
  - *by grouping service*: (external) grouping strategy[1] with set of parameters
  - *by existing grouping*: distribute/concentrate members of previously established groups based on a given mapping
- *role-selectable*: grouping methods from which, at run-time, a member of a certain role can choose one to perform the grouping

---

[1] The interface is implementation-specific and could, for example, call a web service.

Two methods warrant closer attention: First, using a grouping service makes it possible to employ arbitrarily complex grouping algorithms and provide them with the required data. In the JigSaw example, where mixed groups are required, an algorithm could create them when provided with parameters detailing the relevant attributes (knowledge, interests, etc.) Second, the re-grouping based on an existing grouping is useful for example in JigSaw, where the members of the JigSaw group should be equally distributed across the groups representing the topics.

Finally, in order to allow human intervention (e.g., by instructors), a *review* can be defined for a *runtime grouping*. The run-time system has to provide means for members of a specified role to review and accept/reject the result of a grouping.

Similar to run-time member assignment for groups, roles can be populated during execution in a process called *casting*. The casting defines for one or more roles the *casting method*, which offers choices similar to the *grouping method*. An addition here is a new *by vote* mode, where all eligible participants can vote candidates into a role. The automatic *by casting service* method works like the *by grouping service* as described above, also with one addition: for each role, multiple role requirements can be used to specify criteria that prospective role members should fulfil. In JigSaw, this could contain information such as that the leader role should have an authority level above a certain threshold. Like for a grouping, a *review* can be required for a casting as well. Depending on the scenario, however, not all roles can be cast at run-time. Any roles that need to be populated before the script is started by the run-time engine, must be explicitly defined as *initial roles*.

6.1.2 Grouping and Casting Activities

Like for learning activities, there needs to be an element for describing acts of grouping participants or casting them into roles, that can be referenced when defining the sequencing of a learning design. Grouping participants and casting them into roles, however, conceptually fits neither in a *learning-activity* nor in a *support-activity*. Therefore, two new activity types are introduced: a *grouping-activity* and a *casting-activity* (see Fig. 1). Both reference a *grouping* or a *casting* respectively and can use a certain *environment*. These activities trigger the creation of groups and the populating of groups and roles. Disbanding groups and removing roles is part of the adaptation actions, because it needs to be modelled specifically for a certain script and doesn't lend itself easily to a descriptive approach.

6.2 Artefacts

In the original IMS LD specification, the result of work performed by participants is stored in, and retrieved from, properties. For participants, access to properties is possible via any referenced XHTML document through the use of special *global-elements* tags. The script itself, however, does not contain any explicit information about which result had to be created in which activity, and where else this result was used.

In our approach, artefact objects can be specifically defined in a *learning-activity*. If an artefact should be visible in a certain activity, it can be referenced as a *used artefact*. There are three modes of how an artefact can be used:

- an activity can have an *output* artefact as a result;
- artefacts of type *input* are only shown and cannot be modified; and,
- *input-output* artefacts can be both accessed and changed.

Artefacts are owned by the entity that created them (participant, group or class). The default permissions (*read*, *write* and *append*) belong to the *owner*. However, it is also possible to give specific permissions to either *nobody*, *everybody* in the current activity, or members of a certain *role*. In the JigSaw example, students would create an output artefact as a solution to the assignment in one activity. In the next activity, this artefact could be used as an input artefact, and the instructor role could be given read access. The feedback artefact created by the instructor would be handled in a similar way.

6.3 Sequencing- and Execution Flow- Specification

In complex scenarios and in adaptive collaboration scripts, which require flexible execution flows, the sequencing semantics of the IMS LD specification have been found to be constraining [15] and difficult to understand [16]. Due to the lack of state-of-the art concurrency control features, one would often have to resort to custom-made mechanisms employing properties and conditions. Our approach aims at simplifying the constructs for sequencing, while at the same time supporting collaboration processes, allowing arbitrary transitions between activities and employing concurrency controls from the area of workflow management.

IMS LD's metaphor of a theatrical play is replaced by the following concept (see also Fig. 1): in a *story* there are *scenes*, sequenced by directional transitions connecting them. Each scene has *actors* defined by their role, who perform a task: an *activity* or alternatively a nested *story*. In addition to the *environment* within the activity, one can be specified for each scene as well, possibly as an override.

This model of sequencing has been influenced by common workflow control-flow patterns [47]. These represent generally acknowledged solutions to frequently encountered sequencing requirements in flow-oriented process models. Torres and Dodero [15] have found IMS LD to be lacking with regard to expressing certain workflow patterns. The extended scripting language supports the patterns' basic building blocks, which allows expressing advanced sequencing and synchronisation structures. Individual *scene* elements are connected by transitions to form a general directed graph, leading from the start scene to the end of the story. This provides maximum flexibility and makes it possible to model loops (required for example in the TAPPS scenario, see Fig. 3), which are not supported by IMS LD.

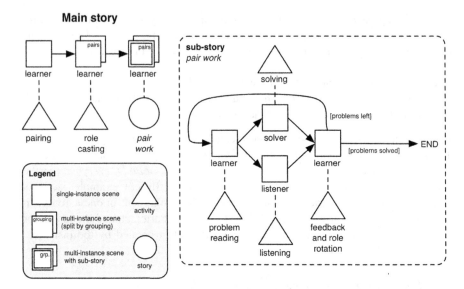

Fig. 3 Simple representation of the TAPPS script [54] in the new sequencing model.

Transitions between scenes can be either unconditional (progression is always possible) or conditional (progression depends on a certain condition). Multiple transitions can originate from a scene or lead to it. This allows the modelling of scenarios with optional or parallel activities where the flow of execution splits and re-joins to synchronize. A specific flow split mode and flow join mode can be set for each scene to model this behaviour:

- *all following* (AND-split), *single following* (XOR-split), and *multiple following* (OR-split);
- *all preceding* (AND-join), *single preceding* (XOR-join), and *multiple preceding* (OR-join).

Each split mode requires a specific configuration of (un-) conditional transitions (see [17]). There are other constraints as well to ensure that there exists an executable path through the script and to prevent deadlocks; these are intended to be checked by authoring tools.

To enable support for modelling group work at the level of learning processes, multi-instance scenes are introduced. When a scene is run as a single instance, all actors share the same activity and, through it, environment and artefacts (which would represent a class context). Alternatively, a scene can be defined to have multiple instances at run-time: one per person (individual scene) or one per group (collaborative scene). For JigSaw, the reading and expert group scenes would be split by topic grouping, and the topic presentations and assignment solving scenes would be split to give each JigSaw group a separate instance. In TAPPS, pairs act in their own, shared instances as well. The multiplicity also defines how artefacts are aggregated or distributed: artefacts created in the activity of a multi-instance scene are collected and made available when they are used in the activity of a subsequent single-instance scene. Artefacts originating from a single-instance scene and used in a multi-instance scene, are copied to each instance. Other transfers are only possible when the multiplicity of both scenes matches.

6.4 Environments and Collaboration Contexts

In IMS LD, activities are performed in the context of environments, containing learning objects (external resources), services (programmatic resources like a chat) and other environments. Each environment in the script source translates to one instance of that environment at run-time. Participants of an activity are all placed in the same environment specified for that activity. The only way to have different contexts is by assigning participants to different roles and have environments specific to each role-part, which is not possible if the number of such 'groups' is not known beforehand. Dynamically creating workspaces for different sets of participants is not part of the original specification but is vital for collaboration contexts.

In the extended learning design model, environments can be declared for learning activities and scenes to specify a work context for when activities are instantiated. In order to support collaboration contexts at various levels, the number of scene instances created from a scene definition at run-time can be configured (see section 6.3): (a) one for all actors, (b) one per participating group, or (c) one per person. Therefore, an environment defined for the learning activity and/or scene merely acts as specification for the individual environment instances created for each scene instance. If an environment is specified in both activity and scene, the components of the two are merged in the environment instance. This instantiation mechanism is also used in the context of groups and roles: Each group or role can have an environment as its own 'workspace'. These environments can be specified for each static group and role individually. Alternatively, environments declared in groupings or castings act as specifications for the environment instances to create for each assembled group or casted role.

In the real world, as an instructional scenario proceeds, items such as learning material, assignment results and feedback accumulate. This can happen at three main levels: for each single participant, in groups, and in the class. In the extended IMS LD model, this behaviour is mimicked: The contents of environments encountered during the run of a script are accumulated at each level on which they were introduced: (a) for a scene, of which only one instance is run (i.e., class context), the contents of the scene and activity environments are added to the workspace of each role that is specified as an actor in that scene; (b) with one instance per group, the environment is added to each group's workspace; and, (c) when running one instance per person, it is added to each personal workspace. In the JigSaw example, this means that when participants have convened in the JigSaw groups, they can still use the services (e.g., communication services) and consult the material of their expert group, of which they are members as well.

An illustration of the workspaces during and exemplary "brainstorming" scenario can be seen in Fig. 4. For each social plane (individual, group and class context) there exists a workspace. Depending on the number of distinct instances of such a context, more than one instance of a workspace exists: a single class workspace; one for each group; and, one for each participant. In each activity, material (e.g., "task description") and services (e.g., "text chat") are added to the work space of the context, in which the activity takes place. Material and services remain available to the participants throughout the scenario (symbolized by the arrows). Additionally, the group workspace is created with an initial set of services ("members" and "text chat"), independent from environment components encountered. The flow of an artefact and the necessary operations ("merge" and "distribute") when it crosses the boundaries of a social plane are also shown.

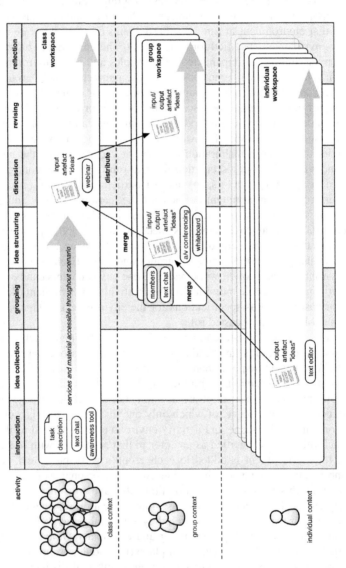

Fig. 4 Workspaces and collaboration contexts on different social planes.

Both the aforementioned accumulation mechanism and the merging of scene/activity environments require the definition of precedence rules. Like in IMS LD, parts of an environment can be hidden or made visible via configuration options in the script source or by adaptation rules. Precedence of such properties is defined by a partial ordering of the relevant objects, shown in Fig. 5: Settings for environments of objects further to the right have precedence over settings made in the context of objects to the left. Environment settings of a learning-activity, for example, are overridden by those in a scene, which in turn are overridden by those for specific instances of a scene.

Changes made to a more general environment specification propagate to specific environment instances (if not overridden there). Fig. 5 shows how changes are propagated from left to right: Changes to a scene's environment specification, for example, propagate to instances of that scene, and possibly to class, group or personal workspaces. For instance, changing the visibility of an item in a grouping's environment affects the workspaces of all groups originating from the grouping. This mechanism allows for managing many environment instances at once.

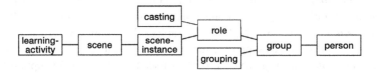

Fig. 5 Environment precedence and propagation relation.

7 Services

As mentioned in section 3, the definition of services in IMS LD has been criticized as being rather inflexible. To support the plethora of services existing in today's learning and collaboration platforms, a generalizable approach to specifying the requirements for the requested service is needed. Additionally, the wide range of possible applications coupled with a diverse set of functions provided by the underlying tools mandate a fine-grained permission model.

7.1 Service Selection

Our extensions build upon the *Generic Service Integration* approach of Valentin et al. [41] and provide three basic methods of selecting a service to be used in the script's scenario: (a) by service type descriptor, (b) by using constraints, or (c) directly via a URI (for services independent from the run-time engine). A fourth possibility is to prepare a set of choices using the previous specifications methods, and have a human participant select one to instantiate at run-time.

Using the first method, a service registered with the run-time engine can be referenced by a service descriptor like *service:chat*. Common services have been defined under the general *service* namespace (*service:wiki*, *service:chat*, …) The engine is expected to support these services directly and map them to appropriate implementations. In addition, platform-specific services must be registered with the run-time engine by an administrator and can be referred to by a compound descriptor comprising the name of the platform (e.g., *sakai, blackboard*) and a general tool name (e.g., *forum, wiki*) or the name of a specific product (e.g., *wimba, elluminate*). It is also possible to specify a list of requested descriptors from which the first available service is chosen, the rest serving as fallback services.

With the second method, a service is requested by setting constraints and letting the run-time engine find a suitable one automatically. In the TAPPS example, this could be used to request an explicit, synchronous, audio-oriented communication service for the thinking aloud sessions. Constraints are either hard (must not be violated) or soft (may be violated, but the constraint solver still tries to avoid that), and they can be weighted with a value between 0 and 1 (default 0.5). The constraints can be broadly classified into the following categories:

- *function*: communication, coordination, consensus, cooperation, collaboration
- *interaction*: synchronicity, interaction mode (implicit, explicit) [48] and multiplicity (1-to-1, 1-to-some, 1-to-many, many-to-many)
- *action coordination*: free, moderated, floor control policy (see section 9.1.3)
- *awareness*: social/group-structural/action/activity awareness indicators [49, 50]
- *interaction modality*: text, graphics-2d, graphics-3d, audio or video
- *artefacts*: (MIME) types supported for display or creation
- *shared environment*: what-you-see-is-what-i-see [48], tele-pointer support [51]
- *privacy*: private data/workspace, anonymous/hidden participation [48, 52]
- *technical*: number of supported users, session persistence

In IMS LD, all services are instantiated before the script starts. This early binding is still possible when the service has been specified via a registered descriptor or directly via a URI, and, under certain conditions, also for services defined by constraints. Late binding of services, one of the prerequisites for adaptation, is applied when constraints refer to run-time information (e.g., *supported users* constraint set to number of participating users), or when the service should be selected manually. In the later case, an adaptation rule might take over this task and use knowledge about the participants to tailor the service choice to their needs.

7.2 Functions

For specifying the permissions of roles in relation to a service (see section 7.3), modelling the events that a service can generate (see section 8.2), and defining how adaptation actions can control it (see section 9.1.3), a model of each service's functionality is required. The run-time engine, where each supported service has to be registered, needs to know which functions a service provides to its users. In particular, the engine needs: a name for referring to the function; a way to effect function-specific permissions; instrumentation to be notified upon function use; and, optionally, an interface to call the function with any required parameters.

To keep the integration and configuration effort at a reasonable level and prevent over-specifying the interface to services, a small set of general functions has been defined, to which the basic functionality of any service should be mapped. Functions exist for:

- managing content and contributions of participants (*new, read, revise, delete, moderate, archive*);
- managing and interacting with sub-divisions of the interaction space (*create/remove/enter/leave* section);
- carrying out (out-of band) communication (*communicate*); and,
- session management for synchronous services (*start/end/join/quit* session).

If the service supports floor control, the following functions can represent primitives for expressing various policies [48]: *request-floor, release-floor, assign-floor, revoke-floor*.

The functionality of tools commonly found in LMSs can be mapped to the above functions, as shown in Table 2. Other functions of a service that cannot be readily mapped (e.g., *vote* in a polls tool) must be separately registered with the run-time engine. Other than that, there is no difference in usage between a general and a service-specific function.

Table 2 Example mapping of specific service functionality to the model's predefined functions.

Function	Wiki	A/V conferencing	File store	Whiteboard
new	create page	make utterance	upload	draw
read	view page	listen/view	download	view
revise	edit page	–	edit	alter contribution
delete	delete page	–	delete	erase
moderate	publish/reject	–	publish/reject	publish/reject
archive	export	record meeting	–	export
communicate	comments	text chat	comments	comments, chat
session		meeting	–	session
section	category	private room	directory	–

7.3 Permissions

Instead of relying on fixed categories such as *observer, participant, moderator* and *conference-manager*, as defined in IMS LD, an approach for fine-grained control over which role may perform which operation is proposed. Permissions are assigned to roles in the script. They can be specified by either defining a set 'from scratch' or by using (and possibly extending/overriding) a service-specific profile that provides reasonable default permissions. Permissions are defined for specific operations in a service, expressed by an operation identifier (rendering a service and a function the context of the permission). A scope (*own* or *any*) can also be optionally defined that, for instance, distinguishes between the right to operate on one's own data or anyone's data. For each operation, a positive (*allow*) or negative (*deny*) permission can be set.

Due to the various ways in which a service can be specified (see section 7.1), it may not be known at design time which one will be instantiated. Since operation identifiers refer to services, however, this could create an implicit dependency between the two. This potential problem is mitigated either: (a) by defining permissions for the functions of all services that can potentially be instantiated (for example, when using a list of registered descriptors for requesting a service); or (b) through the use of wildcards for the service part of the operation identifier, when the set of possible services is not known a priori (e.g., so that the role "participant" can be given the permission "read" for all services in an environment).

Permissions are inherited throughout the hierarchy of roles. This allows for giving a limited set of permissions to general roles like *learner* or *staff* and providing extra permissions to more specific roles like *group-leader* or *administrator*. To resolve conflicts arising when permissions are both inherited and defined in a role, there are two resolution rules: *deny* permissions have precedence over *allow* permissions; and, permissions with more specific identifiers override those with more wildcards or a wildcard at a more general position in the operation identifier.

8 Monitoring Execution and Reacting to State Changes

The extensions discussed so far are not specifically related to adaptation but merely form the basis for comprehensively modelling collaboration scripts. In order to adapt a script at run-time, means are needed to monitor its execution, determine its current state (and the state of its individual components), react to state changes and effect adaptation actions. This section deals with all those aspects, except the adaptations themselves, which are discussed in section 9.

8.1 Run-Time Model

IMS LD allows only very limited access to run-time information via operators like *is-member-of-role*, *complete* or *time-unit-of-learning-started*. Therefore, run-time modelling is usually done by storing the script's state in properties. However, these custom-made models are not standardized or readily re-usable across scripts. Just like the run-time state of a computer system (users, files, network connections, etc.) can be queried by programs through well-defined interfaces to the operating system, the run-time state of a script (participants, current activities, artefacts, running services, etc.) needs to be accessible through a built-in mechanism.

The run-time model can be accessed through operands in script expressions (e.g., in a condition or a calculation) and in other places where a value is required. A query (called a *run-time model access expression*) specifies which specific value(s) to get from which part of the model. All script elements, including their values, attributes and relations can be queried. References are automatically resolved and the respective elements behave as if they contained the referenced ones. In addition, run-time-only elements and attributes are accessible via the model (e.g., *members* in a *learner* role). One (exclusively) run-time element, reachable for example via *members*, is *person*, which represents a participant.

The following exemplary run-time model access expressions illustrate the range of information that can be queried from a running learning design:

- all members of the top-level learner role
  `roles.learner.members`
- the environment used in the "discussion" learning activity
  `learning-activity[@id = "discussion"].environment`
- groups of the "pairing" grouping which are not pairs (e.g., caused by uneven number of participants)
  `grouping[@id = "pairing"].group[@members.count != 2]`
- output artefacts of completed activities that are owned by group "lawyers"
  `activity[@state = "completed"].artefact[@type = "output" &&`
  `@owner = group[@id = "lawyers"]]`

Directly altering (values in) the run-time model risks violating its internal integrity; it is, therefore, recommended that predefined adaptation actions be used for that purpose (see section 9). Furthermore, a new operator *exists* is introduced, which can be used to check whether an object referenced by a run-time model access expression is present. To ensure safe access to the model, in addition to the aforementioned operator, we have introduced a convention whereby accessing non-existing elements in the model can cause exceptions (see section 8.3). In turn, some operators of the IMS LD specification like *is-member-of-role* can be removed, as their results are now represented in the run-time model.

8.2 Event Handling

Fundamental to adaptation rules are the triggers that cause the rule's action(s) to be executed. In IMS LD, most rules are expressed as conditions, which are triggered by Boolean expressions referring to properties. Some problems exist with this approach:

- State changes that should trigger an action need to be recorded in properties, with one property required for each dimension of the state model, even though most information is contained in the run-time state of the script anyway.
- Conditions lack semantics because the properties contain no semantic information about their purpose, except in their name, which cannot be used for automatic inferences.
- Properties and conditions form an unstructured base of state and adaptation information, which makes it difficult to understand and use them.
- Values of properties can only be changed when certain script components (activities, act, play, method) complete, or users change them manually. This places big restrictions on the types of state changes that can be detected altogether.

To address these shortcomings, extensions to IMS LD are proposed, replacing the *on-completion* event hook (in *activity, act, play* and *method*) with a general *event-condition-action handler* concept. Each handler can process events specific to the context in which it is specified, can have a filter expression to specify the exact conditions of its triggering, and has a list of actions to perform once triggered. The main benefits of this approach are:

- State changes do not need to be modelled in properties but can be 'detected' by event handlers.
- Events have defined semantics, which allows reasoning about them by components other than the event handlers themselves.
- Event handlers are defined at the element which they monitor, which models their semantic relation and improves readability
- Event handlers can be defined for most elements of a learning design script, which allows for fine-grained reactions to state changes. It is, for example, possible to perform an action when a JigSaw group delivers an artefact as a result of their problem solving activity.

Two types of events are distinguished in this model: script and service events. *Script events* originate from the objects shown in Fig. 1 with an asterisk (*). The following script events cover the relevant changes in the state of these objects:

- *activity, scene-instance, scene, story*: started, paused, resumed, cancelled, completed, completed-repeatable, participant-joined/-left
- *transition*: fired
- *learning-activity*: artefact-delivered
- *artefact*: created, read, updated, locked, unlocked, deleted
- *role*: addition or removal of member/subrole/property
- *group*: addition or removal of member/subgroup/property
- *casting*: permitted-methods-shown, started, casted, casted-with-exceptions, review-started, rejected, finished, finished-with-exceptions
- *grouping*: permitted-methods-shown, started, grouped, grouped-with-exceptions, review-started, rejected, finished, finished-with-exceptions
- *learning-object*: reference-changed
- *service*: deployed, un-deployed, participant-joined/-left
- *environment*: learning-object-added/-removed, service-added/-removed

Service events originate from services and are closely tied to the service's functionality. Registered general and service-specific functions (see section 7.2) define which events exist for a service. There is a wide range of possible data that can be returned from a service by an event. To prevent over-specification, only a reference to the service, a timestamp, the participant causing the event, and (if applicable) identifiers for the session and/or section and the affected data in the service's workspace are mandated. Modelling mechanisms requiring more detailed information should be kept outside the script source, where they can have custom interfaces to services, receive a wide range of data and employ arbitrarily complex algorithms to create user and group models. Via a connection with the run-time engine, these mechanisms could populate the run-time model of a script with their models.

Similar to events, IMS LD provides a way to complete activities (and sequencing elements) after a time limit. In our approach, this mechanism is generalized with so-called *timers*. For each element that can produce events, timers can be specified to trigger a *timer event*, which can in turn trigger arbitrary actions via an event handler. Recurring timers can be defined to trigger with a certain frequency, either until a certain time or for a certain number of iterations. Timers can be used, for example, to perform a remedial action in the TAPPS scenario when one member of a pair doesn't react for some time. It would also be possible to regularly give feedback to participants or to rotate roles in a group every week.

8.3 Exception Handling

With the shift to more adaptive and dynamic collaboration scripts, a lot of data is processed at run-time and actions may constantly modify the scenario. This could lead to situations and states that were not anticipated while authoring the scenario and could result in a lock-up or termination of the script.

Similar to event handlers, we propose to extend IMS LD with a mechanism to handle those exception states. There is, however, an important difference between events and exceptions: Events can be handled when the script author deems it important or helpful for detecting state changes. Exceptions, on the other hand, cannot be left unhandled; there must be at least a default handler. If, for instance, the instantiation of a service (e.g., a wiki) fails, this could be handled by trying a different service, or altering the scenario so that the service is not needed any more. If no such custom exception handling is defined, a default handler must take appropriate actions such as stopping the script and notifying an administrator.

Exception handlers can be defined for individual objects, which allows them to intercept problems immediately where they appear and provides them with maximum context. Default handlers can also be specified for the whole script, and, as a last resort, have to be provided by the run-time engine.

Exceptions can happen in a lot of cases, but in general the following categories and exemplary instances of exceptions can be distinguished:

- *participant*: required roles not populated, required groups not populated or disbanded, participants did not act in the required way (e.g., voting for a leader)
- *action*: constraint violation, rejected by reviewer, can't be performed on object
- *sequencing*: sequencing model incomplete while story is running
- *data*: type mismatch, run-time model not accessible, expression invalid
- *constraint solving*: constraint not satisfiable (e.g., when grouping, casting)
- *external*: service not reachable, operation in service failed

In JigSaw, an exception might arise when grouping constraints are not satisfiable or a group leader drops out – both cases requiring remedial actions. The TAPPS script might be disrupted when the communication service is unavailable.

Defining exception handlers works in a similar manner to specifying event handlers. However, they cannot have filter expressions, because they need to be triggered every time one of the specified exceptions occurs. Each exception has a reference to the object in which it was caused, and a human-readable description of the cause, which can be used for example in related notifications. An exception handler also defines whether exceptions (after being handled) are propagated to the default handlers on the script or run-time engine level. This makes it convenient, for instance, to handle exceptions where they occur and use a "catchall" handler at script-level to send notifications to instructors or administrators (see Fig. 6 for an example of such a handling hierarchy).

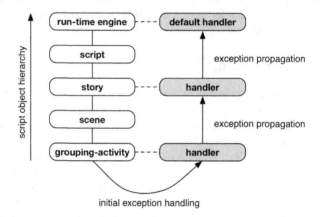

Fig. 6 Exemplary exception handling hierarchy with handlers on three levels.

9 Adaptive Interventions

With the information contained in the run-time model and the notifications of state changes provided by events, the only step missing is the actual intervention or adaptation of the script at run-time. Modelling these is made possible with additional basic adaptation actions, which effect changes on the script's objects, and advanced mechanisms that make developing higher-level operations possible.

9.1 Basic Adaptation Actions

The original IMS LD specification provides only a limited number of actions: showing/hiding objects, changing the value of properties and giving feedback. Comprehensive support for adaptive interventions, however, requires a wide range of additional actions in order to change the relevant parts of the run-time model and effect changes on the execution of the learning design. The necessary actions on the objects representing the main script constituents (activities, groups, participants) at run-time can be broadly categorized as follows:

- *object* actions: create, modify and destroy objects
- *relation* actions: modify attributes of objects or processes in relation to other objects or processes (membership, ownership, permissions, visibility, …)
- *control flow* actions: start, stop, and modify the script process

- *environment* actions: manage services, perform human-like actions in the environment
- *adaptation control* actions: manage the adaptations themselves

These actions cover the life cycle of objects, relations between them, control of active processes, adaptations of services and basic support for meta-adaptivity.

9.1.1 Object Adaptations

Adapting objects as such deals with managing their life cycle and adjusting those attributes and parameters that do not represent relations to other objects. All objects (except participants) can be created at run-time by adaptation actions. Mandatory attributes and relations must be provided as parameters to the creation action; optional attributes may be provided as well. In the JigSaw example, some participants might finish the individual reading of the material on their topic earlier than others. However, the topic group activity cannot start before every member of each group has finished. An additional activity where they can get additional background information might be introduced for those who finished early. The required objects (activity, scene, reference to learning material, ...) have to be created at run-time and put in place (see other adaptation actions further down).

Most attributes of objects can also be modified at run-time, yet some constraints exist: required multiplicity of attributes; range of valid values; and, changeability depending on the current state of the object. If an action violates such a constraint, an exception is generated. As an example, the size of the JigSaw groups could be adapted before the grouping activity: By making the groups smaller, fewer topics need to be considered and the demands on the collaboration should be less. In larger groups the effort for collaboration can be higher, but may at the same time be more fruitful.

Destroying objects is theoretically possible, yet in practice they are often tightly interconnected (see Fig. 1), requiring prior "detachment". Constraints may prevent objects from being destroyed, for example when they are part of an on-going process (e.g., running scene instance). Checking all constraints and severing all connections to other objects would require a lot more operations than a single action, and may not be feasible to specify at design-time without detailed knowledge of the specific situation. One approach of removing objects from the immediate context of a run, but leaving them in place to retain referential integrity, would be to prepare alternatives to the object at design-time or at run-time (e.g., an alternative role or a path in the sequencing of the story). The alternatives need to be connected to objects just as the original ones, so in our example the new role must be added as actor to all the places where the original role was used, and the alternative path may be connected to the main path via conditional transitions. Next, the "active" elements are transferred to the alternative object (this can also be done during the rest of the script run). For roles, this means moving the members to the new role. The new sequencing path can be activated by making sure that conditional transitions from preceding scenes direct the control flow along it, instead of along the original path. What remains is replacing any references to the original object (e.g., in conditions) with ones to the new one. Ideally, the object would be ready for destruction afterwards, but this step is not always necessary.

9.1.2 Relation Adaptations

Adaptations to the relations between objects deal with all the attributes of objects that link them to other objects. In the extended information model, the following categories (and exemplary relations) can be identified:

- *containment*: environment–learning-object/service
- *hierarchy*: role–sub-role
- *membership*: group–person, role-person
- *task*: scene–activity
- *actor*: scene-instance–role/group/person
- *environment*: activity–environment
- *sequencing*: scene–transition–scene
- *ownership*: person–property
- *permissions*: service–role
- *artefacts*: learning-activity–artefact
- *visibility*: environment–learning-object/service

For managing relations, element-specific *set* operations and, depending on the multiplicity, *add* and a *remove* operations are required. Role members, for example, could be defined by *set-members*, or changed, one by one, with *add-member* and *remove-member*.

Continuing the example of an additional individual activity in JigSaw for fast participants (see previous section), the extra learning material needs to be referenced as a learning-object and put into an environment (*containment* relation). This environment needs to be connected with the newly created activity (*environment* relation). The activity needs to be connected to a scene (*task* relation), which has to be made part of the flow of execution (*sequencing* relation).

In JigSaw, during the work in the JigSaw group, the communication channel(s) available to the topic groups in their workspace could be hidden if the participants spend more time there than with the actual group work (*visibility* relation). Learners in the TAPPS script could be supported through the introduction of a "contentious topics" artefact, if it has been detected that they spend unusually long in each "round" of the script. By letting them take notes of unsolved items, which they can resolve later, they might progress more smoothly.

In other scripts, the author might want to (at run-time) introduce roles (e.g., "question poser", "answerer") similar to the TAPPS script, in order to provide more structure to the collaboration (*membership* relation). Examples for adaptation of permissions in services can be found in the next section.

9.1.3 Environment Adaptations

In IMS LD, services are 'black boxes' that can be neither monitored nor manipulated. However, a lot of the interaction happens in the environment, specifically in services. Therefore, adaptively supporting users of services is a way to advance from collaboration establishment to active collaboration process support [13].

Adaptation can first happen when the service is automatically selected, taking into account needs and characteristics of participants and groups. This decision might be based on non-pedagogical criteria (e.g., whether participants' systems provide a camera and a microphone for audio/video-conferencing) or pedagogical criteria (e.g., past behaviour/experience with a certain kind of collaboration service). In addition, if an artefact needs to be created in an activity, a supporting service might be chosen based on whether it allows creating such an artefact.

While a service is running, the permissions of roles can be adapted, for example to give a group leader moderation rights (e.g., when the number of concurrent utterances in a text chat is disproportionately high). Permissions can also be used to selectively enable specific functionality. Similarly, coordination mechanisms like floor control can influence the interaction in communication and collaboration tools. The floor control policy of a service may be adapted, for instance, in order to make interaction more free-form and minimize the scaffolding (thus fading the support offered by the script), or, conversely, to switch to a more coercive policy. Simple policies can be specified with appropriate permissions for the *request-floor*, *release-floor*, *assign-floor*, *revoke-floor* functions (see section 7.2). Alternatively, invoking these functions directly allows for direct intervention. The TAPPS scenario could start with a policy that enforces the roles of 'problem solver' and 'listener', and, once participants progress, switch to a more free-form interaction.

In synchronous interaction services, means to start and end sessions and invite participants allow to set up ad-hoc communication/collaboration contexts. For example, if it has been detected that a group needs help, a communication session (e.g., chat) could be started and its members and an instructor automatically invited. Performing adaptive interventions in the environment in a manner akin to that of humans can be used to provide feedback and content scaffolding. This requires adaptation actions that directly put content into running services, such as by invoking the functions *new* (e.g., create a content item) and *create-section* (e.g., create a conference room). This makes it possible to express interventions like:

- give feedback or advice in a text chat, for example in the form of phrases like "Invite others to participate" or "Reflect with teammates about ..." [56];
- provide structures like forum topics or folders in a file space to help participants organize their collaboration; and,
- provide content scaffolding by creating new items in a glossary or (template) pages in a wiki.

9.1.4 Control Flow Adaptations

Adapting the control flow deals with modifying the sequencing of activities in scenes and stories, instantiation and completion of sequencing elements, and controlling their run-time state.

In our model, transitions sequence the scenes in a story by connecting them and modelling simple sequences, branches or joins of the control flow, and loops. Possible adaptations to transitions are: (a) setting a condition, to make the transition conditional; (b) modifying an existing condition; and, (c) removing the condition, to convert to an unconditional transition. Another adaptation action is to change how many preceding scenes (one, multiple, or all) need to transfer the control flow to start a scene. All of these adaptation actions are required when new scenes are introduced in a script, such as in the JigSaw example introduced in section 9.1.1.

The scene instantiation configuration used to model collaboration contexts (see section 6.4) may also be changed by an adaptation action, but only before a scene has started. If, for example, the number of participants is too big, they might need to be grouped, and the scenes of a script changed to run at the group level. Conversely, if there are too few participants for a sensible grouping, the scenes could be run in a class context. Completion criteria can be changed at run-time as well, which is important if, for instance, an artefact is adaptively introduced. In that case, delivering the artefact can be defined as criterion for completing an activity.

Finally, there are adaptation actions to directly control the state of the control flow: *start*, *pause*, *resume*, *complete*, and *cancel*. Depending on the target object they have different effects, with those on scene objects being as follows:

- Starting a scene creates the scene instances, instantiates the environment, transfers existing artefacts from preceding activities according to the specification and starts activities whose control mainly rests with the system and not the participants (grouping-activity, casting-activity).
- Pausing a scene (instance) disables the associated event handlers, suspends the timers, locks the artefacts currently used in it, and makes the state change visible to participants.
- Resuming a scene (instance) reverses the above process.
- Completing a scene stops all its instances, marks it as complete, notifies the participants of the completion and passes the control to outgoing transition(s).
- Cancelling a *scene* works in the same manner, but marks it as cancelled.

9.1.5 Adaptation Control Actions

These actions are different from the others, because they do not adapt those parts of a learning design script that directly influence its participants but rather modify the specification and run-time enactment of adaptive interventions themselves. They are necessary for implementing meta-adaptivity (i.e., adaptation of a system's adaptive behaviour), also referred to as a second-level adaptation cycle [57].

Through access to the run-time model of the script, which also contains the configuration of many adaptation mechanisms, the adaptation behaviour of the script can be changed at run-time. Conditions, for example, can be modified and enabled or disabled. The same is true for event handlers, which can also be added to (and removed from) objects that support them. For triggering them, *event* objects can be created and 'injected into' objects supporting event handlers or directly into a specific event handler. Depending on its filter expression, a handler will react by executing its actions.

For implementing adaptation control actions, a meta-model of the script, its rules, event handlers and actions is needed. It also has to be noted, that these actions only provide the infrastructure for realizing interventions. The complex problem of assessing the resulting effects and the reasoning processes required for deciding when and how to modify the adaptation strategy are not part of this work.

9.2 Advanced Adaptation Mechanisms

The basic adaptation actions described in the previous section form the foundation for run-time interventions in scripts. Yet, they act on a low level and only effect singular changes. In order to create adaptations on a higher conceptual level, make them more re-usable, and let human supervisors influence the interventions, extensions for modelling advanced adaptation mechanisms have been devised.

9.2.1 Provisional Adaptations

Human involvement in run-time decisions is already possible in grouping, casting (see section 6.1.1) and service selection processes (see section 7.1). However, script authors may also want to express that adaptation actions are not automatically enacted, but are left for a human (e.g., an instructor) to decide. In JigSaw, switching to an interaction model with rotating roles in order to scaffold the collaboration could then be ultimately decided by an instructor. Like other actions, a provisional action can be triggered by an event or a condition. Subsequently, one or more persons can decide whether to enact it, and the system takes over again to execute this choice. These decisions can be binary (enact specific action or not) or multi-valued (enact one of many choices).

Any list of actions can be made provisional by specifying details like who is asked and what happens if this person did not respond, and can then be used like a 'normal' action. Each action list needs to have a description that explains its intention and effect. In combination with context information by the system, these descriptions are used to generate the information presented to the decision makers, who are identified via their roles. The following modes of decision are supported: *one* member of the role(s) can decide alone; *all* members of the role(s) have to agree; or, a simple *majority* of role members suffices.

One problem with provisional adaptations is that they may never be reacted upon, because notifying the decision makers was not possible, or they did not respond, or they could not agree. To remedy this, a time-out can be defined, after which one of the following fall-back options are put into effect: the adaptation is cancelled; a default choice (specified in the choice of actions) or the default option of the single action (execute/do not execute) is enacted; or, the decision is escalated to a different role and a new timeout starts with different fall-back options.

9.2.2 Action Control Structures

Advanced control over the execution of sets of adaptation actions requires two structures known from traditional programming languages: branches and loops. For modelling branches, the structure of the original IMS LD conditions [8] with the required *if-then-else* scheme can be re-used. With loops, a list of actions is repeated as long as an expression evaluates to *true*. The *while-do* scheme is used.

Apart from these 'traditional' control structures, a container construct for actions is needed. Many higher-level adaptations require individual steps that would violate constraints (such as referential integrity) when applied one by one, but the end result (once all have been applied) is valid again.

In order to make this kind of adaptation possible, actions can be grouped in an action block. Each action block works like a transaction: it is regarded as an atomic operation, which either completes in its entirety or fails because its accumulated effects violate constraints, which are only checked after executing the block. Action blocks contain nested actions and can be used in all places where actions can be used. An example where this encapsulation might be necessary is the introduction of an additional activity in JigSaw for participants that finish reading their material early. Adding a new activity with a respective environment and giving certain participants the opportunity to perform it requires many single adaptation steps which need to be effected as a whole.

9.2.3 Composite Actions and Named Expressions

Most basic adaptation actions work at a rather low level, dealing with just one object and performing one operation (not counting side-effects these may have on other objects). These atomic operations must be combined to result in meaningful, higher-level effects. This requires composite actions and named expressions as combinations of (composite) actions and (named) expressions, which makes it possible to define often-required operations once and re-use them across a script. Authoring tools or run-time engines could provide libraries of such operations.

Composite actions require a unique name with which they are invoked, a list of actions to perform and (optionally) parameters. When invoked, the actions are performed in sequence. Parameters allow abstracting the declaration from concrete situations in which the action is used.

Named expressions work in a similar manner, the only difference being that instead of the list of actions, the respective expression needs to be declared. Parameters can be used as well. Upon invocation, the named expression is evaluated and its result is returned.

As an example, a composite *rotate-artefact* action could be created to effect a circular exchange of the referenced artefact among those (single participants or whole groups) currently performing an activity, something that is currently not possible in IMS LD. Similarly, a *rotate-roles* action could rotate roles among participants of a certain casting (which must have assigned more than one role) or the members of a set of groups. For example in TAPPS, there is a casting for the 'problem solver' and 'listener' roles in each pair. With the *rotate-roles* action, the roles can be switched between the members of each pair.

10 Summary and Outlook

We have presented extensions and alterations to the IMS Learning Design specification that aim to better express collaboration scripts and allow for more comprehensive adaptation to individual learners and groups. The new elements of the information model support explicit modelling of groups (either statically defined, or dynamically created) and group environments. Assignment of participants to groups can be controlled through a number of policies; the same is true for the assignment to roles. Activity types for these grouping- and role casting- operations have been introduced. For representing collaboration contexts, we have extended the environment model to support workspaces for whole classes, individual groups and single users. A flexible service model has been introduced, which allows fine-grained service selection and permission configuration. Adaptation actions have been introduced for manipulating objects and their relations at run-time, and for effecting interventions in collaboration environments, specifically in services.

The extensions also include a run-time model, providing access to all elements of the script and pure run-time data (such as information on participants), while it is being executed. Access to the run-time model is possible in expressions, making some operators obsolete. Artefacts can be modelled explicitly. In the context of a learning activity, artefacts may be used as input, output or transient elements, optionally with access permissions for different participants. An extensible event handling specification with event-condition-action semantics has been introduced, and new actions for adapting the scenario at run-time have been defined. The new event model allows for a wide range of state changes in run-time objects and services to be detected and used to trigger actions. Finally, the original sequencing model has been replaced with a more flexible one, which supports running multiple instances of activities according to the social plane (class, group, individual), offers (conditional) transitions for arbitrary sequencing of activities and uses workflow semantics for synchronizing and splitting the flow of action.

Current work is focusing, on the prototypical implementation of the extended model presented herein, and on further refining and enriching the model. Specifically, we have started to implement the specification in an executable model, with run-time components capable of running scripts and effecting adaptations at run-time. Implementation is targeting the Sakai e-learning platform [58], and aims to integrate and make use of the services, user interface and infrastructure provided by the platform. The developed prototype will then be employed in real-world student-based evaluations, where we will seek to establish the impact of the newly enabled types of adaptive support on the collaborative learning process.

In parallel to the aforementioned development efforts, we are working on fine-tuning the current features of the information model and on providing new ones to enhance its expressiveness towards adaptivity. To start with, the run-time model needs to be thoroughly specified, providing clear semantics for every element, its (data) type and multiplicity (a requirement also related to the model's implementation). Furthermore, the run-time model needs to be made accessible from the client side (assuming the script is 'executed' on the server side), in a way that conceptually extends (and ideally replaces) the global-elements of IMS LD. The main idea is to provide an implementation-independent interface specification similar to the SCORM run-time API [59], to allow for bi-directional data transfer between the learning design engine and the client (application). Another requirement is access from within the learning design script to external models [60], such as the personal learner models of participants in the learning management system. Making these models accessible via the respective element in the run-time model (e.g., person) would render them immediately usable by the mechanisms described so far.

Further to the above, in order to ascertain the expressiveness and suitability of the extensions in practice, we are currently assessing them against a representative set of existing scripts and collaboration patterns. In more detail, the proposed extensions are currently under evaluation using criteria such as the ones described in [61]. Additionally, a representative selection of existing, well-known collaboration scripts is being used to assess whether all required features are supported and can be expressed. The results of this evaluation will guide further extension efforts.

Beyond the theoretical aspects of the extended specification and the implementation-oriented ones of the supporting run-time engine, our interests in the area extend to the development of tools that allow authors to take full advantage of the potential of both the specification and the engine. This includes the provision of tool-based support for authoring CSCL scripts in the extended specification, with a focus on facilitating the utilization of common strategies, patterns, and templates even by educators with little prior experience in the area.

Acknowledgments. The work reported in this chapter has been supported by the "Adaptive Support for Collaborative E-Learning" (ASCOLLA) project, financed by the Austrian Science Fund (FWF; project number P20260-N15).

References

1. Roschelle, J., Teasley, S.D.: The construction of shared knowledge in collaborative problem solving. In: O'Malley, C. (ed.) Computer-Supported Collaborative Learning, pp. 69–97. Springer, Heidelberg (1995)
2. Dillenbourg, P.: What do you mean by collaborative learning? In: Dillenbourg, P. (ed.) Collaborative-Learning: Cognitive and Computational Approaches, pp. 1–19. Elsevier, Oxford (1999)
3. Paramythis, A., Mühlbacher, J.R.: Towards New Approaches in Adaptive Support for Collaborative e-Learning. In: Proceedings of the 11th IASTED International Conference on Computers and Advanced Technology in Education (CATE 2008), pp. 95–100. ACTA Press, Crete (2008)
4. Zumbach, J., Schönemann, J., Reimann, P.: Analyzing and supporting collaboration in cooperative computer-mediated communication. In: Proceedings of the 2005 Conference on Computer Support for Collaborative Learning – Learning 2005: The Next 10 Years!, pp. 758–767. International Society of the Learning Sciences, Taipei (2005)
5. Dillenbourg, P.: Over-scripting CSCL: The risks of blending collaborative learning with instructional design (2002), http://hal.archives-ouvertes.fr/hal-00190230/en/
6. Miao, Y., Hoeksema, K., Hoppe, H.U., Harrer, A.: CSCL Scripts: Modelling Features and Potential Use. In: Proceedings of the 2005 Conference on Computer Support for Collaborative Learning – Learning 2005: The Next 10 Years!, pp. 423–432. International Society of the Learning Sciences, Taipei (2005)

7. O'Donnell, A.M., Dansereau, D.F.: Scripted Cooperation in Student Dyada: A Method for Analyzing and Enhancing Academic Learning and Performance. In: Hertz-Lazarowitz, R., Miller, N. (eds.) Interaction in Cooperative Groups: The Theoretical Anatomy of Group Learning, pp. 120–141. Cambridge University Press, London (1992)

8. IMS Global Learning Consortium, Inc.: Learning Design Specification (Version 1.0 Final Specification) (2003),
   http://www.imsglobal.org/learningdesign/

9. Pea, R.D.: The social and technological dimensions of scaffolding and related theoretical concepts for learning, education, and human activity. The Journal of the Learning Sciences 13, 423–451 (2004)

10. Rummel, N., Spada, H., Hauser, S.: Learning to collaborate while being scripted or by observing a model. International Journal of Computer-Supported Collaborative Learning 4, 69–92 (2009)

11. Rummel, N., Weinberger, A., Wecker, C., Fischer, F., Meier, A., Voyiatzaki, E., Kahrimanis, G., Spada, H., Avouris, N., Walker, E., Koedinger, K.R., Rosé, C.P., Kumar, R., Gweon, G., Wang, Y.-C., Joshi, M.: New challenges in CSCL: Towards adaptive script support. In: Proceedings of the 8th International Conference of the Learning Sciences, vol. 3, pp. 338–345. International Society of the Learning Sciences, Utrecht (2008)

12. Demetriadis, S., Karakostas, A.: Adaptive Collaboration Scripting: A Conceptual Framework and a Design Case Study. In: International Conference on Complex, Intelligent and Software Intensive Systems (CISIS 2008), pp. 487–492. IEEE Computer Society, Los Alamitos (2008)

13. Paramythis, A.: Adaptive Support for Collaborative Learning with IMS Learning Design: Are We There Yet? In: Proceedings of the Adaptive Collaboration Support Workshop, held in Conjunction with the 5th International Conference on Adaptive Hypermedia and Adaptive Web-Based Systems (AH 2008), pp. 17–29. L3S Research Center, Hannover (2008)

14. Caeiro, M., Anido, L., Llamas, M.: A Critical Analysis of IMS Learning Design. In: Proceedings of CSCL 2003, pp. 363–367. Kluwer Academic Publishers, Bergen (2003)

15. Torres, J., Dodero, J.M.: Analysis of Educational Metadata Supporting Complex Learning Processes. In: Sartori, F., Sicilia, M.Á., Manouselis, N. (eds.) MTSR 2009. CCIS, vol. 46, pp. 71–82. Springer, Heidelberg (2009)

16. Hagen, K., Hibbert, D., Kinshuk, P.: Developing a Learning Management System Based on the IMS Learning Design Specification. In: IEEE International Conference on Advanced Learning Technologies (ICALT 2006), pp. 420–424. IEEE Computer Society, Los Alamitos (2006)

17. König, F., Paramythis, A.: Towards Improved Support for Adaptive Collaboration Scripting in IMS LD. In: Wolpers, M., Kirschner, P.A., Scheffel, M., Lindstaedt, S., Dimitrova, V. (eds.) EC-TEL 2010. LNCS, vol. 6383, pp. 197–212. Springer, Heidelberg (2010)

18. König, F., Paramythis, A.: Collaboration Contexts, Services, Events and Actions: Four Steps Closer to Adaptive Collaboration Support in IMS LD. In: International Conference on Intelligent Networking and Collaborative Systems (INCoS 2010), pp. 145–152. IEEE Computer Society, Los Alamitos (2010)

19. König, F., Paramythis, A.: Closing the Circle: IMS LD Extensions for Advanced Adaptive Collaboration Support. In: International Conference on Intelligent Networking and Collaborative Systems, pp. 421–426. IEEE Computer Society, Los Alamitos (2010)

20. Laurillard, D.: The pedagogical challenges to collaborative technologies. International Journal of Computer-Supported Collaborative Learning 4, 5–20 (2009)

21. Soller, A., Lesgold, A.: Modeling the Process of Collaborative Learning. In: Proceedings of the International Workshop on New Technologies in Collaborative Learning (NTCL 2000), Awaji-Yumebutai, Japan (2000)

22. Gweon, G., Rose, C., Carey, R., Zaiss, Z.: Providing support for adaptive scripting in an on-line collaborative learning environment. In: Proceedings of the SIGCHI Conference on Human Factors in Computing Systems, pp. 251–260. ACM, Montréal (2006)

23. Constantino-González, M.A., Suthers, D.D.: An approach for coaching collaboration based on difference recognition and participation tracking. In: The Role of Technology in CSCL: Studies in Technology Enhanced Collaborative Learning, pp. 87–113. Springer, Heidelberg (2007)

24. Lipponen, L.: Exploring foundations for computer-supported collaborative learning. In: Proceedings of the Conference on Computer Support for Collaborative Learning: Foundations for a CSCL Community, pp. 72–81. International Society of the Learning Sciences, Boulder (2002)

25. Suthers, D.D.: Technology affordances for intersubjective learning: a thematic agenda for CSCL. In: Proceedings of the 2005 Conference on Computer Support for Collaborative Learning – Learning 2005: The Next 10 Years!, pp. 662–671. International Society of the Learning Sciences (2005)

26. Haake, J.M., Pfister, H.-R.: Flexible Scripting in Net-Based Learning Groups. In: Fischer, F., Kollar, I., Mandl, H., Haake, J.M. (eds.) Scripting Computer-Supported Collaborative Learning, pp. 155–175. Springer US, Boston (2007)

27. King, A.: Scripting Collaborative Learning Processes: A Cognitive Perspective. In: Fischer, F., Kollar, I., Mandl, H., Haake, J.M. (eds.) Scripting Computer-Supported Collaborative Learning, pp. 13–37. Springer US, Boston (2007)

28. Kobbe, L., Weinberger, A., Dillenbourg, P., Harrer, A., Hämäläinen, R., Häkkinen, P., Fischer, F.: Specifying computer-supported collaboration scripts. International Journal of Computer-Supported Collaborative Learning 2, 211–224 (2007)

29. Dillenbourg, P., Jermann, P.: Designing Integrative Scripts. In: Fischer, F., Kollar, I., Mandl, H., Haake, J.M. (eds.) Scripting Computer-Supported Collaborative Learning, pp. 275–301. Springer US, Boston (2007)

30. Häkkinen, P., Mäkitalo-Siegl, K.: Educational Perspectives on Scripting CSCL. In: Fischer, F., Kollar, I., Mandl, H., Haake, J.M. (eds.) Scripting Computer-Supported Collaborative Learning, pp. 263–271. Springer US, Boston (2007)

31. Suthers, D.D.: Roles of Computational Scripts. In: Fischer, F., Kollar, I., Mandl, H., Haake, J.M. (eds.) Scripting Computer-Supported Collaborative Learning, pp. 177–187. Springer US, Boston (2007)

32. Wecker, C., Fischer, F.: Fading scripts in computer-supported collaborative learning: the role of distributed monitoring. In: Proceedings of the 8th International Conference on Computer Supported Collaborative Learning, pp. 764–772. International Society of the Learning Sciences, New Brunswick (2007)

33. Walker, E., Rummel, N., Koedinger, K.: CTRL: A research framework for providing adaptive collaborative learning support. User Modeling and User-Adapted Interaction 19, 387–431 (2009)

34. Dillenbourg, P., Baker, M., Blaye, A., O'Malley, C.: The evolution of research on collaborative learning. In: Learning in Humans and Machine: Towards an Interdisciplinary Learning Science, pp. 189–211. Elsevier, Oxford (1996)

35. Karakostas, A., Demetriadis, S.: Adaptation patterns in systems for scripted collaboration. In: Proceedings of the 9th International Conference on Computer Supported Collaborative Learning, vol. 1, pp. 477–481. International Society of the Learning Sciences, Rhodes (2009)
36. Masthoff, J.: An Exploration of Adaptive Collaboration Support. In: Paramythis, A., Weibelzahl, S. (eds.) Proceedings of the Adaptive Collaboration Support Workshop, held in Conjunction with the 5th International Conference on Adaptive Hypermedia and Adaptive Web-Based Systems (AH 2008), pp. 5–7. L3S Research Center, Hannover (2008)
37. Dillenbourg, P., Baker, M.: Negotiation Spaces in Human-Computer Collaborative Learning. In: Proceedings of the International Conference on Cooperative Systems (COOP 1996), Juan-Les-Pins, France, pp. 187–206 (1996)
38. Miao, Y., Hoppe, U.: Adapting Process-Oriented Learning Design to Group Characteristics. In: Proceeding of the 2005 Conference on Artificial Intelligence in Education: Supporting Learning through Intelligent and Socially Informed Technology, pp. 475–482. IOS Press, Amsterdam (2005)
39. Hernández-Leo, D., Asensio-Pérez, J.I., Dimitriadis, Y.A.: Computational Representation of Collaborative Learning Flow Patterns using IMS Learning Design. Educational Technology & Society 8, 75–89 (2005)
40. Miao, Y., Burgos, D., Griffiths, D., Koper, R.: Representation of Coordination Mechanisms in IMS Learning Design to Support Group-based Learning. In: Lockyer, L., Bennett, S., Agostinho, S., Harper, B. (eds.) Handbook of Research on Learning Design and Learning Objects: Issues, Applications and Technologies, pp. 330–351. IDEA Group, Hershey (2008)
41. de la Fuente Valentin, L., Miao, Y., Pardo, A., Delgado Kloos, C.: A Supporting Architecture for Generic Service Integration in IMS Learning Design. In: Times of Convergence. Technologies Across Learning Contexts. pp. 467–473 (2008)
42. Dalziel, J.: From Re-usable E-learning Content to Re-usable Learning Designs: Lessons from LAMS (2005), http://www.lamsinternational.com/CD/html/resources.html
43. Towle, B., Halm, M.: Designing Adaptive Learning Environments with Learning Design. In: Koper, R., Tattersall, C. (eds.) Learning Design. A Handbook on Modelling and Delivering Networked Education and Training, pp. 215–226. Springer, Heidelberg (2005)
44. Berlanga, A.J., Garcia, F.J.: A Proposal to Define Adaptive Learning Designs. In: Proceedings of the International Workshop on Applications of Semantic Web Technologies for Educational Adaptive Hypermedia (SW-EL 2004), pp. 354–358. Technische Universiteit Eindhoven, Eindhoven (2004)
45. Zarraonandia, T., Dodero, J.M., Fernández, C.: Crosscutting Runtime Adaptations of LD Execution. Educational Technology & Society 9, 123–137 (2006)
46. Paramythis, A., Cristea, A.: Towards Adaptation Languages for Adaptive Collaborative Learning Support. In: Proceedings of the First International Workshop on Individual and Group Adaptation in Collaborative Learning Environments (IGACLE 2008) held in Conjunction with the 3rd European Conference on Technology Enhanced Learning (EC-TEL 2008). CEUR Worhshop Proceedings, Maastricht, The Netherlands (2008), ISSN 1613–1673,
http://CEUR-WS.org/Vol-384/FULLPAPER-p6.pdf
47. Russell, N., Arthur, van der Aalst, W., Mulyar, N.: Workflow Control-Flow Patterns: A Revised View. BPMcenter.org (2006)

48. Reinhard, W., Schweitzer, J., Völksen, G., Weber, M.: CSCW Tools: Concepts and Architectures. Computer 27, 28–36 (1994)
49. Carroll, J.M., Neale, D.C., Isenhour, P.L., Rosson, M.B., McCrickard, D.S.: Notification and awareness: synchronizing task-oriented collaborative activity. International Journal of Human-Computer Studies 58, 605–632 (2003)
50. Gutwin, C., Greenberg, S., Roseman, M.: Workspace Awareness in Real-Time Distributed Groupware: Framework, Widgets, and Evaluation. In: Sasse, A., Cunningham, R.J., Winder, R. (eds.) People and Computers XI (Proceedings of HCI 1996), pp. 281–298. Springer, London (1996)
51. Ellis, C.A., Gibbs, S.J., Rein, G.: Groupware: some issues and experiences. Communications of the ACM 34, 39–58 (1991)
52. Ellis, C., Wainer, J.: A conceptual model of groupware. In: Proceedings of the 1994 ACM Conference on Computer Supported Cooperative Work, pp. 79–88. ACM, Chapel Hill (1994)
53. Aronson, E., Blaney, N., Stephin, C., Sikes, J., Snapp, M.: The Jigsaw Classroom. Sage Publishing Company, Beverly Hills (1978)
54. Lochhead, J., Whimbey, A.: Teaching analytical reasoning through thinking aloud pair problem solving. New Directions for Teaching and Learning, 73–92 (1987)
55. Dillenbourg, P., Tchounikine, P.: Flexibility in macro-scripts for computer-supported collaborative learning. Journal of Computer Assisted Learning 23, 1–13 (2007)
56. Constantino-Gonzalez, M.A., Suthers, D.D., de los Santos, J.G.E.: Coaching Web-based Collaborative Learning based on Problem Solution Differences and Participation. International Journal of Artificial Intelligence in Education 13, 263–299 (2003)
57. Totterdell, P., Rautenbach, P.: Adaptation as a problem of design. In: Browne, D., Totterdell, P., Norman, M. (eds.) Adaptive User Interfaces, pp. 59–84. Academic Press, London (1990)
58. Sakai: Sakai Project. Sakai Foundation, http://www.sakaiproject.org
59. Advanced Distributed Learning Initiative: Sharable Content Object Reference Model (SCORM 2004) 4th edn., Version 1.1 – Run-Time Environment (2009), http://www.adlnet.gov/Technologies/scorm/
60. Abel, F., Heckmann, D., Herder, E., Hidders, J., Krause, D., Leonardi, E., Van Der Slujis, K.: A Framework for Flexible User Profile Mashups. In: Proceedings of International Workshop on Adaptation and Personalization for Web 2.0 (AP-WEB 2.0 2009), CEUR Workshop Proceedings, Trento, Italy, pp. 1–10 (2009) ISSN 1613-0073. http://CEUR-WS.org/Vol-485/paper1.pdf
61. Caeiro-Rodríguez, M., Llamas-Nistal, M., Anido-Rifón, L.: Towards a Benchmark for the Evaluation of LD Expressiveness and Suitability. Journal of Interactive Media in Education, 1–14 (2005)

Extending IMS-LD Capabilities: A Review, a Proposed Framework and Implementation Cases

Ioannis Magnisalis and Stavros Demetriadis

Aristotle University of Thessaloniki, Greece
P.O. Box 114, 54124, Thessaloniki, Greece
{imagnisa,sdemetri}@csd.auth.gr

Abstract. In this article we present a framework for the integration of external and independent software components into IMS-LD (Learning Design) based courses that cater for adaptivity. Our framework comprises a design specification and an implementation of adaptations in CSCL (computer-supported collaborative learning) oriented and standards based architecture. The architecture allows combining existing research on explicit representations of collaborative learning processes (i.e. learning designs) with the availability of existing and tested collaborative learning tools (e.g. a forum in a virtual learning environment (VLE), an agent, a service or even a software component that provides a specific functionality). The architecture allows controlling the learning tools either by a human or a pedagogical agent and thus enables adaptive interventions to the flow of the learning activity. A mediator component is the key element in the proposed architecture. To prove the soundness of the architecture and the flexibility of its implementation example scenarios are illustrated. In these scenarios IMS-LD based modeling and Coppercore engine are used to implement adaptations by setting IMS-LD properties according to input from an external Moodle forum tool. The whole process is mediated by an integration component provided to the teacher as a Moodle resource. Finally, we highlight what would be important issues toward integrating the adaptation pattern capabilities in IMS-LD compliant tools for collaborative learning design.

1 Introduction

Lately, there has been an increasing interest for introducing adaptive capabilities in systems for collaborative learning 1. Brusilovsky and Paylo in 2 identify at least three distinct technologies that implement some type of adaptation regarding the collaborative learning activity: adaptive group formation and peer help, adaptive collaboration support, and virtual students. Efforts to implement adaptive techniques in computer-supported collaborative learning (CSCL) in order to improve the learning experience have been systematically reported in the literature providing encouraging evidence on the impact of adaptive methods to enhance student learning ([3], [4], [5], [6]). Despite promising evidence, however, it is a fact that the systems that adaptively support collaboration are at an early stage. One reason

T. Daradoumis et al. (eds.): Intelligent Adaptation and Personalization Techniques, SCI 408, pp. 85–108.
springerlink.com   © Springer-Verlag Berlin Heidelberg 2012

for this is the fact that the currently available interoperability standard IMS-LD is not capable of expressing the computationally complex and demanding constructs necessary for implementing adaptive techniques [7].

In this work, we address exactly this issue, that is, how IMS-LD, a de-facto standard in the CSCL area, could facilitate the required adaptive behavior of a CSCL system through communication with external software components (such as web services, communication tools or virtual learning environments (VLEs).

We present specific design case studies (as a proof of concept) exemplifying how the key issues of the adaptation pattern approach can be expressed using the IMS-LD modeling language. In the following, we firstly provide a background in the area of adaptive and IMS-LD based CSCL architectures. We discuss mainly the IMS Learning Design (IMS-LD) modeling language, relevant tools and specifications and literature identifying their advantages and limitations. Next we present our proposed framework consisting of: a) the IRMO (Input, Rules, Model, Output) design specification-methodology that aims to be a guide on how a teacher analyses and models a pedagogic design including points of possible variations (i.e. adaptation patterns), and b) a reference architecture -called MAPIS- using technological (e.g. web services) and pedagogical standards (i.e. IMS-LD). Then, we provide evidence on the value and usability of our approach presenting a prototypical implementation of a scenario that involves specific peer interaction (PI) support strategy. We conclude with a discussion of our experience and future plans emphasizing how a software component can be engineered to facilitate the application/design of such adaptation patterns in IMS-LD format.

2 Background

2.1 AICLS Systems and Adaptation Patterns

In the context of technology-enhanced learning, system designers have tried to systematically exploit the modeling potential of computers and develop systems that support learners through adaptive or intelligent operation. Adaptive systems are model-based systems. An adaptive educational system (AES) is mainly a system that aims to adapt some of its key functional characteristics (for example, content presentation and/or navigation support) to the learner needs and preferences 2). Thus an adaptive system operates differently for different learners, taking into account information accumulated in the individual or group learner models.

Introducing adaptive characteristics gave birth to the strand of Adaptive Hypermedia Systems (AHS), a significant subset of which is Adaptive Educational Systems (AES) with systems like AHA 57, InterBook and WebCOBALT 2. Respectively the strand of Intelligent Tutoring Systems (ITSs) appeared with systems like ELM-ART, KBS-Hyperbook and SQL-Tutor 2). According to Brusilovsky and Peylo 2, ITS traditionally focused on Curriculum Sequencing, Intelligent Solution Analysis & Problem Solving Support, while AES focused strongly on Adaptive Presentation & Navigation Support.

The above approaches aim principally on helping the individual learner. Recently research efforts have focused on introducing adaptivity and intelligence in the context of computer-supported collaborative learning (CSCL) bringing together AESs and ITSs on one hand and CSCL systems on the other. Additionally, there is initial evidence that adaptation advances the learning effects of CSCL (e.g.3).

Computationally supported adaptive and intelligent operations are increasingly integrated in the design of CSCL systems in an effort to maximize the user-tailored support provided to group learners, focusing both on improved domain learning and development of collaboration skills. In general, creating adaptive/intelligent systems for CSCL is considered to be more demanding than creating respective systems for individual study, since apart from the pedagogical aspects one must also take into account aspects related to social relations and group dynamics 58.

In our recent work (see 38) we have coined the acronym "AICLS" (Adaptive and Intelligent Collaborative Learning Support) system as a general term to denote the broader research area of adaptive and/or intelligent systems that aim specifically to support the collaborative learning activity. The intervention of an AICLS can target either of two layers:

1.   Layer 1: Preparation of the activity (pre-task intervention, such as group formation). This layer deals with pre-task issues.

2.   Layer 2: Support of the activity itself (in-task intervention) providing domain knowledge-type support or peer interaction type support 38. This layer deals with in-task issues.

From our point of view, we have emphasized the need for a generalized conceptual framework of adaptive scripting, relevant to all types of collaboration scripts, as a basis for formalizing the design of flexible adaptive interventions to support group learning (39, 40). Research has consistently emphasized that collaborating students might fail to engage in productive learning interactions when left without teachers' support (e.g. 41). Consequently, collaboration scripts have been proposed as a means to structure the collaborative activity by didactic scenarios and engage all students in fruitful learning interactions (e.g. 42, 43).

Nevertheless, adjusting the script level of granularity and flexibility emerges as an important issue that affects the outcome of scripted collaboration (43). We have argued elsewhere (7, 8, 39, 40) that a solution to the script flexibility issue could be the integration of adaptive characteristics to systems for scripted collaboration by means of integrating "Adaptation Patterns" (APs) to the design. An AP captures some core idea of pedagogical value on how to adapt the collaborative learning activity when specific conditions occur. Therefore, an adaptation pattern is essentially an abstraction based on teachers' key ideas regarding adaptivity (as synonym to flexibility) during collaborative learning. An adaptation pattern is a process which takes into account the user, a group and/or script model (or other modeled entity) and adjusts certain aspects of the collaborative activity in order to maximize student engagement, satisfaction and, consequently, the learning outcomes (39, 40). We have also proposed and exemplified elsewhere (Karakostas & Demetriadis, in 39 & 40) a design methodology (DeACS) for identifying adaptation patterns to be embedded in adaptive scripting systems.

Naturally, the important technical challenge is how to link the core non-adaptive pedagogical design of the script (currently supported by various non-standardized script editors) with adaptive design functionalities. Our position on this is that adaptation patterns can be built either as software add-ons or web services that are invoked by a script editor when available (i.e. the software extends its functionalities depending on the available add-ons library or list of web services). The teacher then could integrate the selected adaptation pattern at the appropriate point of the computerized script representation and parameterize the properties and methods of the pattern as desired. In this way the "adaptive logic" can reside at a separate software component taking advantage of modifying the adaptive strategy without touching the original pedagogy pattern expressed with IMS-LD.

In this work we are proposing a framework for building IMS-LD based adaptations. As a first step, we have previously proposed that the structure of any elementary adaptive function in a CSCL system can be modeled based on the IRMO methodology (8). IRMO can be thought of as a modeling tool for the designer who wishes to analyze an AP consisting of four major components, namely: Input, Rule(s), Model(s) and Output. Input refers to one or more parameter(s) which are monitored by the system during runtime and trigger the enactment of the adaptation (these could be, for example, a student assessment outcome, a group deliverable, the synthesis of a group, etc.). Rule(s) implies input processing: one (or more) rules (of the form: "IF Input satisfies condition THEN the Output is ADAPTED") are applied to input. The Model part defines which (one or more) entities of the collaborative activity are necessary to be modeled in order for the adaptation to function properly (these entities could be learner or group characteristics, collaboration script aspects, activity phases, material, etc.). Finally, Output refers to the result produced when applying Rule to Model according to some Input. The Output could be, for instance, a change in the synthesis of the group, the material provided to individual learners, the sequence of the activity phases, the roles of the learners, etc. In general, the Output results to an updated representation (internal and/or external) of the activity.

IRMO resembles in a sense with AHA 57 architecture as one can locate a one-to-one correspondence of basic elements: a) Input of IRMO is linked to User requests of AHA, b) Model of IRMO corresponds to the user and domain model of AHA, c) Rule of IRMO is the Adaptation Model-rules of AHA And d) Output of IRMO corresponds to served output of AHA. However, AHA -which is a simple Web-based adaptive engine, that was originally developed to support an on-line course- was designed and used in individual learning environments and up to now deals with no issues embedded in CSCL learning scenarios (e.g. no user-group modeling is dealt with in AHA).

However, if an adaptation intervention is to be reusable it has to be expressed using a common modeling language 'understandable' by CSCL systems. Our next step, therefore, is to explore how the IRMO modeled structure of an adaptation can be expressed using standards and implemented through architectures that support IRMO adaptive logic.

Having said the above, it is clear that a number of issues should be considered when different types of adaptation need to be supported with some formalization method, such as IMS-LD. In the following section we discuss what the Learning Design (IMS-LD) standardization can offer to formally express the adaptive design of collaborative learning activities and review the efforts towards providing guidelines (i.e. architectures or frameworks) for realizing AICLS systems.

2.2 A Review of IMS-LD Based Adaptation Attempts

Learning Design (IMS-LD) is primarily a modeling tool which uses the metaphor of a theatrical play for describing a teaching-learning process 9. Its main components are: metadata, roles, acts, environment, role-part (i.e. activities of actor, who does what, when and how), sequence of activities within a role-part, conditions and notifications (interactivity and control over a live learning design as a form of event driven messaging system within an IMS-LD player). Through IMS-LD tool one can formally express a unit of learning (UoL), that is, a complete, self-contained unit of education or training, such as a course, a module, a lesson etc. To be usable by computers, Learning Design has to be given a concrete syntax and semantics. Thus, we come to Learning Design specification, called IMS-LD (10, 11). IMS-LD specification consists of three levels of implementation and compliance and each level is mapped to separate XML Schemas in three levels.

It is stressed that a researcher must consider IMS-LD as a de-facto standard in the CSCL field. There is a debate in CSCL community as according to 12 IMS-LD lacks expressiveness, while according to 31 and 32 this is not fully true. IMS-LD is criticized for its capabilities to express adaptive behavior. Paramythis (12) concludes that IMS-LD offers: (1) No support for modeling groups, (2) no support for modeling artifacts (e.g. a vote, an argument, an answer etc), (3) poor support for dynamic features modeling, (4) poor support for modeling complicated control flow, (5) poor support for modeling social interaction, (5) no exchange of information across UoLs, (6) poor modeling of services and their characteristics (additional services maybe "name-spaced" into the IMS-LD specification), (7) acts within plays cannot be re-sequenced or structurally modified. More limitations are identified by Towle and Halm in 13 including: (1) difficulty of supporting multiple rule interactions (e.g. student profile with multiple characteristics); (2) lack of user/group driven activity ordering.

However, the approach taken in IMS-LD specification is not to define a single large schema with a core of mandatory elements and numerous optional elements, but rather to define a complete core that is yet as simple as possible, and then to define possible levels of extension that capture more sophisticated features and behaviors. Analyzing the IMS-LD structure Burgos et al. 14 identify three levels of support that the specification can offer to various types of adaptation: (a) well supported (for learning flow, content, evaluation and interactive problem solving support), (b) partially supported (for user grouping, interface adaptation, adaptive evaluation and full modification of a course on-the-fly), and, finally, (c) no support (for dynamic modification of learning structure and method in run-time, and adaptive information filtering and retrieval). Moreover, IMS-LD while is a standard in the area of CSCL scripting is accompanied with a series of supporting tools 15 and thus can by no means be neglected when introducing a reference architecture for building AICLSs.

In literature we find studies that propose frameworks and/or architectures for building AICLS systems. aLFanet system presented in 30, uses the named Collaborative Learning Framework (CLF) which aims at adaptively supporting CSCL process by controlling how the issues that cannot be predicted at design time should be managed on run time. In the direction of combining standards (i.e. IMS-LD) with other technologies (i.e. Agents, external Tools, Games) we track some work. For instance, in 32 a working system, namely <e-adventure> is tested in a modified version of the official service-oriented implementation of the Copper-core engine. Walker, Rummel and Koedinger (45) present also an architecture for the integration of tutoring and process scaffolds into existing collaborative applications like Cool Modes (44). Most frameworks in the area that attempt to provide adaptive support in CSCL settings stay independent of the domain knowledge aspect of a CSCL system. Nevertheless, some works (e.g. 46) present frameworks specifically oriented to a plug and play capability of a system that is during run-time oriented to a learning domain.

On one hand, there are studies that propose totally new languages for specifying a CSCL script; for instance 35 with LDL language, 36 with LAG language. Contrary, we locate studies claiming that standards can be extended or combined to achieve the desired result (e.g. 32, 33, 34) and that some aspects of adaptivity can be modeled and supported by today's technology. Similarly, some recent research (e.g. 47) has proposed SLD 2.0 an extension to IMS-LD (and also to tools widely used in the community like Collage) that enables to specify several characteristics of the use of tools that mediate collaboration. SLD 2.0 promises to rethink the learning design in the LMSs context while keeping the most essential features of IMS-LD like its capacity to express collaborative learning activities. Nevertheless this research attempt is very fresh, has not yet gained maturity and acceptance in the CSCL community, but deserves a researcher's attention.

We believe that community has strong evidence through success stories and available tools not to reject any standard, but try to extend them and merge the best of available proposals and standards. We also believe that this direction forms a research opportunity and a requirement for both CSCL and AI communities.

Work in 37 is an example where Grid services are proposed to be used in CSCL applications. Moreover, there are research works combining IMS-LD and learning services in an attempt to extend IMS-LD capabilities. The technological frameworks presented in literature are using grid services (16, 17, 18, 19) and more specifically web services. Other works, like (20 & 21) deal with the same issues in a similar manner, but they build additional components upon an engine running an IMS-LD based CSCL script (i.e. Coppercore). A worth mentioning tendency similar to our work is the utilization of external software components called widgets/mashups in collaborative learning scenarios (24, 25). However, all the above studies do not focus on flexibility-adaptivity needed in CSCL systems, but cater mainly for reusing existing tools and providing communication diodes between them and IMS-LD. For instance, they demonstrate how IMS-LD can call a forum or a chat tool. Similar to our philosophy, we track work in relevant literature with the objective to increase reusability of a IMS-LD by offering the designers an alternative to introduce slight variations on the original design, diminishing the need for changes or extensions to the Learning Design definition 22.

Study presented in 29 is similar to our philosophy in the sense that it provides a generic architecture for extending IMS-LD through specific application interfaces. The differences of the above with our work are twofold:

a) Our work has the ambition to provide a more generic architecture (see following section) of AICLS system realizations in the sense that every possible external component could be bidirectionally communicated by an IMS-LD orchestrating a CSCL VLE. In 29 authors present an also generic architecture that tries to integrate every possible external component with bidirectional communication with a service adapter acting as a middle layer. The difference is that in our approach: 1) our architecture is based solely on standards i.e. IMS-LD and Web services for the interconnectivity part, while 29 leaves the requirements for service adapters general and not standards-based, 2) IMS-LD is left intact as we do not introduce any new XML element in IMS-LD in order to perform any adaptation, unlike work presented in [29]. Our approach is based on Adaptation Patterns and from technical point of view supported by the "Mediator Component" (MC) which has all the logic to play the diode through which any external component can communicate with IMS-LD based courses.

b) A whole framework is provided to the teacher and IMS-LD-author to guide her design and implement an IMS-LD with AICLS features (i.e. combine design patterns, adaptations/adaptation patterns, IRMO design methodology and components of MAPIS architecture presented in next section). In 29 authors also present a way to support authors during authoring and teachers during deployment. But their approach pre-supposes that an API exists towards an external component built probably by a person with programming skills. On the contrary our approach integrates the IRMO methodology which is in fact a method for designing adaptation patterns (and not simple adaptations) –the difference is mainly at the level of abstraction of an adaptation. Thus, the designer-teacher is guided into a process which may or may not involve in the end use of external tools in order an adaptive behaviour to be implemented in an IMS-LD based course.

Table 1 Attempts towards standardized AICLS systems

Strategy	Systems (References)	Implementation exists	Description
Enhance standard	24, 25	YES	Using external components and calling them with IMS-LD extension tags called Widgets
Combination of IMS-LD and external tools	29, 18	YES	Leaving IMS-LD intact and providing communication diodes between IMS-LD run engine and external tools
Enhance Run Engine	20, 21	YES	Implementing additions to IMS-LD engine to incorporate external tools
New Standards	35, 36	NO	Proposing totally new standards in the area of building AICLS systems

In Table 1 we summarize findings of our research and review in the area of IMS-LD based attempts to provide support for building AICLSs at all layers of a collaborative task (i.e. pre-task, in-task, post-task) as presented in 38). Table 1 classifies these attempts in four major types based on the strategy these attempts follow. Characteristic examples of such attempts are depicted in 'Systems' column of Table 1 and a small description of the way each specific strategy is implemented in the relevant systems is noticed. We observe that there are some attempts to enhance IMS-LD standard, others that focus on enhancing IMS-LD run engine(s), some that attempt to combine IMS-LD and external tools while others propose totally new standards for providing support to build AICLS systems. Our approach can be considered as a combination of the first two approaches with the focus on facilitating: a) adaptivity (i.e. providing ways to model adaptations and adaptation patterns) and b) flexibility (i.e. providing the means to re-use intelligent tools in the learning process and seamlessly connect them with IMS-LD design). Moreover, in Table 1 we mention whether a strategy and the systems-works found in literature present existing implementations and results (this is denoted with YES) or they are merely proposals in this research area. Concluding our review of IMS-LD extension attempts, once again is stressed that standardization at the level of adaptation of components and services in CSCL environments has not been addressed adequately.

3   AICLS Reference Framework: IRMO and MAPIS

3.1   Requirements from an AICLS Framework

So far we have managed to implement simple adaptive interventions solely based on tools based on IMS-LD standard (see 7). But, when we came up with more complex adaptations like 'Group Heterogeneity' and 'Peer Interaction Scaffolding', we could not implement them simply because: a) in 'Group Heterogeneity' case the Rule part of IRMO methodology for these adaptation could not be implemented in IMS-LD, and b) in 'Peer Interaction Scaffolding' case bidirectional communication with external tools supporting peer interaction (whether synchronous or asynchronous) is necessary. Study in 8 has demonstrated that there are cases where more advanced programming structures (than those offered by IMS-LD) are needed for complex algorithms to be implemented. Such as, in the 'Group Heterogeneity' case, the Rule part of IRMO methodology for this specific AP could not be implemented with IMS-LD.

The problems met so far, pose two clear requirements for a framework/ architecture:

a) Interoperability,

b) Extensibility & Generality (i.e. IMS-LD communication with external intelligent s/w components or agents or services in general).

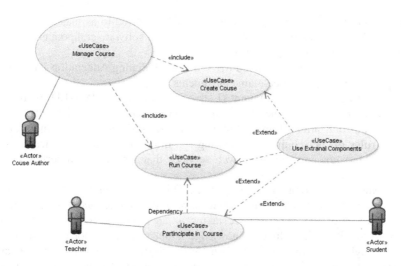

Fig. 1 Use cases supported by IRMO-MAPIS reference framework.

In figure 1 we depict a use case diagram where interested users of our proposed framework are supported to fulfill their purposes. Specifically, the role "Course Author" is supported in managing a course consisting of creating a course and running the course. Creating a course is considered as assisting a person to create an IMS-LD, while running the course implies availability of technological platforms to actually enliven the course under concern. Our framework is also required to support the use case of letting teachers and students participate in a course. This fact depends on the framework's availability of tools to let this happen (i.e. we have an IMS-LD engine with minimum managing capabilities of students, teachers courses and 'runs').

In this study we are proposing a framework for building AICLS systems. The proposed architecture connects IMS-LD with external tools (e.g. a forum in Moodle, a web-conference tool, a chat tool, a group formation tool etc.) and exploiting their 'processing' power in order to provide pedagogical sound and rich virtual learning environments. Thus, in figure 1 the use case called "Use External Components" plays the most vital role of such a framework as it extends practically the whole IMS-LD process from every actor's (i.e. Course Author, Student & Teacher) point of view.

3.2 MAPIS: Architecture for Implementing Adaptations with IMS-LD

The requirements mentioned previously and the conscious choice of re-using s/w components along with standard specifications (i.e. IMS-LD and Web services) formed the background of our reference architecture to be presented in the following (Figure 2). While IMS-LD is a mature standard in the CSCL area, web services which are also basic constituents of our architecture, is a standards-based way of rapidly developing applications 23 –often referred to as mashups or widgets (24, 25) – with the benefits of (a) loose-coupled software components and (b) machine-to-machine interoperability 26.

Our reference architecture called MAPIS (i.e. Mediating Adaptation Patterns & Intelligent Services) has three main constituents, described next.

(1) Learning design environment (LDE): It comprises an IMS-LD editor and IMS-LD engine. When a concrete and specific AP is required, then IMS-LD editor facilitates customization of an AP by modeling -according to IRMO- the properties and rules of the required AP. During run-time, properties and rules are exposed to Mediator component as web services that get or set properties inside IMS-LD engine.

(2) External s/w component/agent (EA): This is in general a s/w component that exposes specific functionality through a web service interface. For instance, in an example case (see 48) a service taking as an input the required number of groups –provided it has already knowledge about students' prior domain knowledge through a questionnaire- performs group formation, stores the results and outputs a message indicating the successful (or not) finalization of the process. This part of the architecture can be whatever tool, even a web conference tool that outputs, for instance, participation indicators through a web service. Notice that the web service definition of a software component is typically published in Web service repository.

(3) Mediator component (MC): This is the core of our architecture incorporating the logic of getting requests from IMS-LD engine (or IMS-LD editor) for searching and calling a functionality offered by a software component through a published web service. This component consists of three logical interfaces to IMS-LD editor, to IMS-LD engine and to a software component. Mediator component has the logic, for example, to receive a call from LDE (e.g. for a group formation service in 48). Then, the software component providing required functionality (i.e. form groups of mild heterogeneity based on prior domain knowledge) is called by MC. The result through the web service is returned to MC. MC finally sets the relevant property representing the mode part of IRMO. Thus, IMS-LD engine can apply a rule accordingly and present adaptive behavior (e.g. selected groups are assigned a specific activity).

The following sequence diagram (see figure 3) illustrates a fraction of the functionality supported by our architecture and at a high level of granularity how components of this architecture are communicating among them in order to support and implement the use case presented in section 3.1. For instance we describe how MAPIS components support participation in a course designed to incorporate in IMS-LD design external to IMS-LD software services. Thus (see figure 3), a published course that resides in a 'run-engine' component can use during run-time an 'external software agent' and even search and locate through the 'external component manager' and use another 'external software agent'. This component, in a following sequence of actions/messages in the sequence diagram (not shown here), gets information from the external tool and sets IMS-LD properties according to IRMO base design to provide adaptively support (i.e. hide or show an activity or a hint) to the group of learners inside IMS-LD run environment.

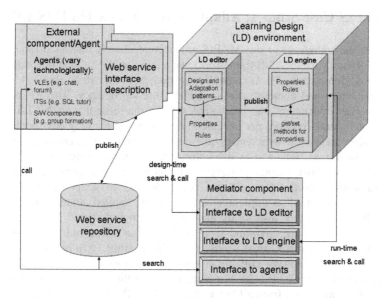

Fig. 2 MAPIS: A reference architecture for implementing ACLS systems with IMS-LD.

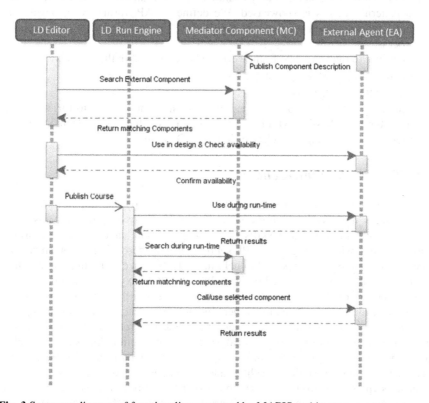

Fig. 3 Sequence diagram of functionality supported by MAPIS architecture components.

4 MAPIS Case Studies: Adaptations Expressed in IMS-LD

To prove the soundness of our MAPIS architecture already discussed (and conse-quently of the IMRO-MAPIS framework), we choose an AP, among others (see Ka-rakostas & Demetriadis 27), as example case study/scenario which employs ideas of our architecture and demonstrates the architecture's efficiency. The AP is called "Trigger Peer Interaction" and is a complex AP in the sense that an AICLS system has to intelligently observe peer interactions in a virtual learning environment and decide when and how to intervene in order to increase interaction among partners.

Thus, in this section (and in the following sub-sections) we present a design case where the above adaptation pattern is integrated in an activity based on a col-laboration script. Notice that APs are independent of the script they are used in. However, when an AP is implemented it has to be applied along with a specific script. In our cases (see also 48 for another AP with an external group formation component) we use the 'pyramid type' collaboration script, but any other script could be equally applied. First we use the Collage editor 16 to produce the initial structure of a pyramid type collaboration script. What we call design pattern (DP) is called elsewhere CLFP (15). For instance, in Collage tool we can implement such a DP as an option from a library of available DPs. We used Collage in order to produce an initial structure of a pyramid script. A miniature of the whole 'adap-tation pattern' process is showcased. We define an AP, formalize it, develop a computer representation of it and through a user interface we enact and evaluate the process.

According to our reference framework we have to work in three levels:

1. Design our adaptation in IRMO terms,
2. Map IRMO to MAPIS components
3. Implement MAPIS components (e.g. publish an agent's ability to form groups, model Input in IMS-LD with Level B properties, mediate IMS-LD and agent in order to provide adaptive behavior in IMS-LD engine etc.)

4.1 Trigger Peer Interactions

Peer interaction (PI) support refers to the actions taken by the system in order to help partners improve their in-group interaction and possibly develop also do-main-general knowledge and skills (for example, argumentation, peer tutoring, and peer reviewing). In general, PI support is considered important since the system interventions are expected to help individuals in acquiring important colla-boration skills (e.g., 49, 50, 51). Typically, an AICLS system offering PI-type support focuses on the collaboration process regardless of the domain (see 38) and provides customized feedback to learners by modeling peer interactions during the activity. To accomplish this, the system tracks the progress of the collaboration and performs a type of interaction analysis (IA). "Interaction Analysis" is a rela-tively new research direction aiming to extract useful indicators by processing data recorded during student-machine and/or student-student interactions. These indicators (presented to users in graphical or literal mode) are expected to help students and/or teachers in order to self-assess their activity (52, 53, 54, 55).

According to 38 review study, such systems are increasingly becoming more capable (automatically or semi-automatically) in identifying both peer interaction patterns and indicators of in-group collaboration. One major observation is the abundance of works trying to model and study the interactions in a CSCL process, whether scripted or not. In the area of interaction analysis, many tools are available for capturing the progress of a collaborative activity 54. However, despite the great interest and some proposals (55, 54) in the area of peer interaction support, no standard way or model or specification has been agreed by the community in order to study the essence of collaboration in a consistent manner and be capable of comparing results.

4.1.1 Designing the Adaptation

The pattern 'trigger peer interactions' (noted as 'Adaptive prompting' adaptation pattern in 27) expresses the simple but general idea that in a group of learners who are interacting through various communication media like speech or text and through virtual learning environments like forums, chats, web-conferencing systems, observation of this interactions can lead to decisions of teacher or system intervention. Interaction analysis techniques are used in the area to denote that such a conversational and interactive group activity produces interaction data indicators that are available to a system or a teacher for possibly further analysis or action taking.

Many indicators, data and combinations among them can be produced and used in order to decide for an adaptive intervention to be performed. In this scenario we simplify the case study by taking into account the indicator called 'Level of Participation' (LoP) which is a percentage indication of posts to a forum of a particular learner within the group of his collaborators. Moreover, in the specific case study we suppose that the IMS-LD-author and teacher wish to use a Moodle (asynchronous) forum as the tool to facilitate peer interactions. An analysis of the adaptation pattern to be designed follows:

1. Name: "Trigger peer interactions" (here based on LoP within a Moodle forum)
2. Key-idea: Trigger (e.g. send appropriate messages to) partners interacting less than others within group. Send, also, messages to learners with high-participation to help/trigger discussions with learners that show low participation levels.
3. Activation conditions: When students have low LoP they are shown a message and information of their LoP and an example of a fruitful collaboration pattern. When students have high LoP they are shown private messages in order to instigate interactions with students exhibiting low LoP.
4. What to model: (a) students' LoP, b) history of students' LoP. To this end, helpful tools might include a tool facilitating learners within group communication and persistence storage for text data and indicators.
5. What to adapt: hints/messages to students. Aim is to scaffold interactivity within group and bring a balance among students based on LoPs of each student, thus implementing in general a beneficial interaction pattern with gains for all participants.

According to the IRMO methodology this AP is described as follows:

1. Input: the outcome of LoP from a forum tool for students,
2. Model: LoPs ,

3. Rule(s): IF LoP is low THEN provide indicator to student and a pattern of beneficial peer interaction, and b) IF LoP is high THEN provide message suggesting to student to help other student(s) with low LoP,
4. Output: Hints and messages according to rule.

Before implementing an adaptation that has been modelled in IRMO terms we have to map IRMO constructs to basic component of MAPIS architecture. Thus, the following table is filled in:

Table 2 IRMO to MAPIS mapping.

IRMO construct	MAPIS component	Description
Input	LDE AND /OR EA	LoP: a) Level B IMS-LD personal property, b) LoP indicator calculated from data and text within a forum (EA).
Model	LDE AND EA	The modelled entities of this adaptation are and are stored respectively as: 1) number of posts per user: EA, 2) number of posts per group: EA, 3) LoP: LDE, EA & MC, 4) Mean of LoP of all participants: EA & MC
Rule	MC AND LDE	A mediator component has all the knowledge of how to contact LDE and EA and when to store information from one to another. Moreover, LDE will decide according to pre-designed rules which messages/activities to trigger
Output	LDE (AND EA if required)	The output is mainly stored in LDE but it can be decided as this is the case to be stored also in EA as a second persistent storage layer

Next, we implement, and then "run" in SLeD (28) IMS-LD run-time engine the adaptive form of the collaboration script and draw conclusions from this attempt. Notice that the main reason for referring to this AP is to demonstrate that there are APs so complex that cannot be implemented with IMS-LD syntax only or external tools only, at least to our knowledge up to now.

4.1.2 Implementing the Adaptation

4.1.2.1 IMS-LD Modeling

The way we elicit and set the property "personal_participation_level" (PLoP) is done using Web Services technology via MC component which propagates these values from Moodle forum (see figure 4) to IMS-LD run-time engine. Notice that all other properties mentioned in Table 3 manipulate values inside EA (i.e. in our case Moodle forum) and MC (i.e. in our case a extra software component that manipulates these values according to rules set by IMS-LD-author).

Table 3 Local Properties modelling a) number_of_learners & b) number_of_groups.

```
- <imsld:loc-property identifier=" number_of_learners ">
        <imsld:title>number_of_learners</imsld:title>
        <imsld:datatype datatype="integer" />
</imsld:loc-property>
- <imsld:loc-property identifier=" number_of_groups ">
        <imsld:title>number_of_groups</imsld:title>
        <imsld:datatype datatype="integer" />
 </imsld:loc-property>
```

Table 4 Local Personal Property modelling personal participation level

```
- <imsld:locpers-property identifier=" group_memebership ">
        <imsld:title>group_memebership</imsld:title>
        <imsld:datatype datatype="integer" />
</imsld:locpers-property>
- <imsld:locpers-property identifier="personal_knowledge_level">
        <imsld:title>personal_knowledge_level</imsld:title>
        <imsld:datatype datatype="integer" />
</imsld:locpers-property>
- <imsld:locpers-property identifier=" average_group_knowledge_level ">
        <imsld:title>average_group_knowledge_level</imsld:title>
        <imsld:datatype datatype="real" />
</imsld:locpers-property>
```

In order to model groups two local properties in IMS-LD terms are introduced: Number of learners and number of groups are necessary local properties for the script (see Table 3). These properties are assumed not to be set until runtime. The property PLoP that classify the learner as more or less "participating" in the collaboration process (i.e. exhibiting lower or higher level of peer interaction) is of type Integer (for reasons of simplicity we hypothesize that learners exhibit either high or low level of participation denoted by "1" and "0" values respectively). Moreover, EA (i.e. Moodle forum) with the use of an extra tool that presents to learners their participation levels analyzing their social network (56), stores: 1) number of posts per user, 2) number of posts per group, 3) PLoP, 4) Mean of LoP of all participants in a group (MGPL). Moreover MC manipulates PLoP and MGPL and according to rules classifies each learner within a group as having

'Low' (value 0) or 'High' (value 1) PPLs. All these properties are the model part of IRMO methodology for the adaptation under consideration.

In table 4, a local personal property in IMS-LD terms (i.e. personal participation level (PLoP)) was mentioned as necessary in order to denote that a learner when working within his group is characterized in real-time as of 'Low' or 'High' participation Levels. In 'Peer Interaction Support' case this property is also required to be set during execution of the script. Setting this property in IMS-LD triggers the Output part of IRMO methodology for this adaptation.

It is pointed out, that no new XML element is introduced in our approach. The integration is performed via the MC which is based on WSs and has all the logic to perform what is missing from both IMS-LD and the external component.

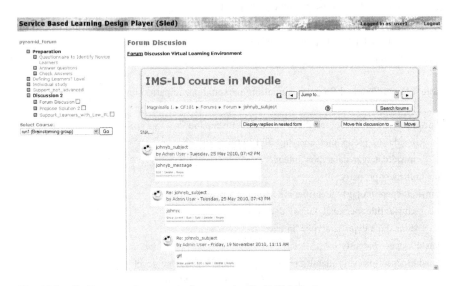

Fig. 4 Moodle Forum tool as seen by learnner inside IMS-LD player.

4.1.2.2   EA: Moodle Forum Component

The Moodle forum service is a software component that exists as an activity inside a Moodle course and was required to be re-used in an IMS-LD based course. Although Moodle provides means and interfaces for statistics manipulation inside activities like forum discussions, in the specific case we chose to incorporate visualization of simple peer interaction analysis (IA) indicators such as number of total posts and posts per user. A tool called SNAPP (56) provided the ability to show to the learners in real-time their collaboration network with graphics and statistics of their on-going discussion (see 0). Every post in the Moodle forum is considered as a single interaction. Each participating student with a message post triggers the system to update its information and update LoP. This is also graphically represented in SNAPP tool. The aim of the current work is not to provide many indicators or discuss which one is the best. Instead, the focus is on providing proof of concept of the proposed framework and demonstrates how easily the MAPIS architecture can utilize tools for CSCL settings, tools already available to teachers. Technically, users from the two servers (Coppercore and Moodle) are handled manually, i.e. the teacher creates the same user account with same credentials in both systems. Although it is out of the scope of our work, we have dealt with this issue and come to conclude that open LDAP is a viable solution in a real system in production state.

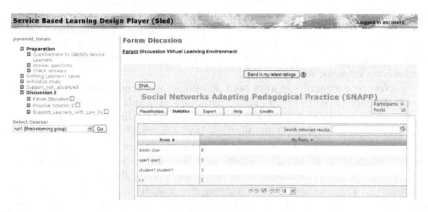

Fig. 5 Peer intaraction analysis (IA) indicators shown to learners.

The important difference of this adaptation when compared to other adaptations is the complexity of the Rules to be modeled and implemented. Such kind of complexity cannot be dealt with IMS-LD, nor the functionality of a Moodle forum service can be provided by IMS-LD –this is out of the scope of IMS-LD actually. Notice that the Moodle forum service works as an independent software component which has its own persistence layer (i.e. tables in a database storing values). Moreover, Moodle forum service has been enhanced with more methods for handling its logic and persistence layer. For instance, it has a published interface (as a web service) that allows setting PLoPs manually. For space reasons we do not analyze and deal with this part of the Moodle forum service as it is too technical.

Technically, it is denoted that the external tools need not be modified in order to be integrated in the learning flow of the proposed architecture. However, if an external tool is to be integrated it should provide an interface via WSs in order to be bidirectionally accessible, meaning that properties and information inside the tool must be able to be retrieved or set.

4.1.2.3 Implementing the Mediator Component

The functionality of EA Moodle forum service is required by the IMS-LD script. In order to be usable, this functionality was exposed and published as a web service that takes as input a user identity within a group and returns values of PLoP and MGPL to MC. MC then applies Rule part of IRMO, classifies the learner as of 'Low' or 'High' PLoP and propagates result back to IMS-LD run-time engine setting the property personal_participation_level. IMS-LD according to Rule specification of IRMO shows the relevant hints/messages to learners. Categories 'Low' and 'High' are hard coded and indicative of what the system could use as categories. Ideally an AP could provide an interface where the specific categories is a (table) parameter and introduced by AP designer.

We have built a component called MC inside the MAPIS architecture. This, as already discussed in our reference architecture, has three components: a) Interface to IMS-LD editor, b) Interface to IMS-LD engine, c) Interface to Agent (service) & Repository. So far, we have prototypically implemented and tested the 'Interface to IMS-LD engine' and 'Interface to Agent (service) & Repository'. 'Interface to IMS-LD editor' is planned to be implemented as an add-on to an IMS-LD editor after extensive evaluation of our approach. Both of the interfaces implemented are given as a choice (i.e. url link) to a teacher of a course in Moodle. This course is seamlessly interconnected with our IMS-LD course implementing our AP along the implemented script DP.

Our prototype system supports both automatic and semi-automatic classification of students as of 'Low' or 'High' participatory levels within Moodle forum discussions. For example, "Update Coppercore IMS-LD" links in Moodle call 'Interface to IMS-LD engine'. This interface was developed both in Java and PHP just to prove that calling this interface can be: a) easily implemented and b) is, in general, independent of the technology used. The interface calls the Coppercore Web Service 'CoppercoreAdmin' and more specifically the web method called 'setProperty'. This method allows changing dynamically a property within a running IMS-LD script provided that the calling interface supplies as input some parameters (i.e. Unit of Learning Identification, User Identification, Run Identification, Property Identification, value of property to be set to).

After this, the property 'personal_participation_level' of IMS-LD is set for all learners. Thus, an adaptation can be performed according to a rule already defined in IMS-LD during design-time. For instance, learners with 'Low' personal LoP within a group of Moodle forum discussion members can be shown an activity especially designed for them (see figure 6).

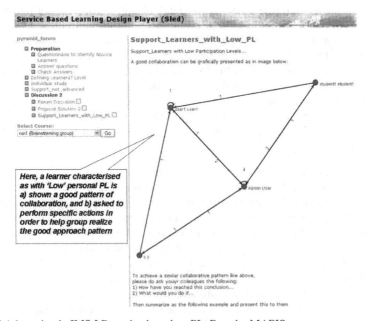

Fig. 6 Adaptation in IMS-LD engine based on PLoP set by MAPIS.

5 Discussion and Future Work

In this work we have presented our latest efforts toward linking the "adaptation pattern" (AP) approach with the IMS-LD modeling language in order to build powerful and interoperable tools to support the design of flexible collaborative learning environments. Flexibility is used as a synonym of adaptability, where a flexible course delivers different material or activities depending on user/group characteristics.

We have argued that the AP approach employs powerful pedagogical ideas and support teachers in understanding what could be successfully adapted in situations of group learning. This is in contrast to simply providing teachers with the capability of altering various parameters of the CSCL system (a technologically driven design which usually does not help teachers to grasp the pedagogical value of their possible actions).

Moreover, we have dealt with some important technical issues during our efforts to tackle the hindrances of implementing IMS-LD compliant APs. We have presented a framework for building an AICLS standards-based consisting of two major "guidelines": a) IRMO methodology for designing adaptations, and b) MAPIS architecture for implementing adaptations.

To achieve this flexible-adaptive use of services (or learning services) in different phases and cycles we chose to represent each re-usable object as a local (or can be global for some cases) property in the IMS-LD description or a simple URL that calls it. The property is defined in IMS-LD, because while the initial state of the property can be set in the IMS-LD document, the external MC can manipulate the content during the learning process via setting the property it is associated with. Moreover, a personal property will be usually required here, because every student involved in the learning process might possess an individual output of the artifact (e.g. the participation level of a student within an external chat tool).

We emphasize the fact that all interaction of the learner is done directly with the learning environment (i.e. inside IMS-LD-engine) and only indirectly with the learning services. On the other hand, no further software installation (i.e. apart from the original learning environments) is needed on the student's learning environment side. Thus, a true virtual and intelligently adaptive learning space is formed for a student and the teacher monitoring the whole CSCL process.

Our approach is not intrusive to the user, but it makes the learning processes and the choice of partners for collaboration even more flexible. It is in-line with other studies, as for instance 21 where an approach is presented for providing a client that is capable of dynamically launching learning environments. This approach makes use of grid technologies to transfer artifacts between the different learners.

Our approach allows re-using a particular script expressed in IMS-LD with various virtual learning environments (or agents providing learning services). This is the case of IMS-LD-editor in LDE choosing to use a different forum than initially designed to. Of course, the prerequisite is that the alternative forum supports similar functionality exposed through Web APIs (i.e. web service interfaces). A possible future research direction is the support through the MAPIS architecture of dynamic search, discovery and selection of a service suitable for the adaptation to be realized. Put briefly, this functionality is important for a running IMS-LD script, as it will provide more flexibility. For instance, in a case where a forum service is not available due to network problems, another available service satisfying the pedagogic requirements of the IMS-LD's designed activity (e.g. a chat) could be discovered, selected and used by a MAPIS 'registry service'. Another relevant example is to build an IMS-LD author that integrates a generic forum to be called at a specific phase of an IMS-LD based course design. The specific forum, providing outputs concerning the participation of each student, is located and called at run-time.

Another future research direction on adaptive CSCL systems emerges because of the fact that the artifacts within the learning process -such as a group formation tool used by the teacher- can change over time. So, they have to be available for the LDE and the teacher that designs a CSCL process through an IMS-LD editor. It is a technical challenge for the community to build systems that support the whole process of designing and running an IMS-LD based course that utilize external components. Ideally, a teacher/course creator at design-time should be able to integrate the desired predefined adaptations during design. Then, IMS-LD editor creates the necessary properties in IMS-LD to facilitate the model part of IRMO representation of these adaptations. Then, if needed (i.e. the adaptation is complex enough) IMS-LD editor calls IC to search for an agent/service that is capable of providing the required result as output (e.g. a value of group membership for each student in a group) after suitable input information and process (e.g. number of groups and prior domain knowledge for each student). Teacher, finally, selects in IMS-LD editor the agent/service that is to be called during run-time. One could even expect that IMS-LD could change the appearance and behavior of the service according to some input (e.g. properties changing appearance of a forum called by IMS-LD).

As a next step we are already working in designing a complete course in IMS-LD and Moodle providing adaptive support at three distinct levels in a script:

1. Pre-task adaptation: for example a questionnaire in Moodle sets prior domain knowledge properties in IMS-LD and according to student classification (as advanced or novice students) relevant learning activities are activated.
2. In-task adaptation: providing hints and careful interventions in discussion within a Moodle forum according to participation levels monitored in Moodle and set into IMS-LD.
3. Post-task adaptation: Assessment of the CSCL process from the students can provide ratings for the hints introduced during in-task adaptation. The system should not use hints rated as not helpful in a next run.

After designing of such a course-prototypical system we target at extensively evaluating it as this will provide valuable feedback for the proposed framework.

An open research question for the community is how can a system –guided by a teacher or designer- incorporate in its own adaptation life cycles assessment-evaluation derived from the students and the teachers. The ultimate target for an AICLS system should be its evolution according to feedback from teachers and learners. Eventually, every next time a course is running the system adaptive operation considers also historical data from its use, whether automatically or semi-automatically derived.

References

1. Karakostas, A., Demetriadis, S.: Adaptation Patterns in Systems for Scripted Collaboration Designing an Adaptive Collaboration Scripting System The DeACS method. In: Scripted vs. Free CS collaboration. Alternatives and Paths for Adaptable and Flexible CS Scripted Collaboration Workshop in CSCL 2009, Rhodes, pp. 43–47 (2009)
2. Brusilovsky, P., Peylo, C.: Adaptive and intelligent Web-based educational systems. IJAIED, Special Issue on Adaptive and Intelligent Web-based Educational Systems 13(2-4), 159–172 (2003)
3. Ronen, M., Kohen-Vacs, D.: Designing and Applying Adaptation Patterns Embedded in the Script. In: International Conference on Intelligent Networking and Collaborative Systems, INCOS 2009, pp. 306–310 (2009)
4. Furugori, N., Sato, H., Ogata, H., Ochi, Y., Yano, Y.: COALE: Collaborative and Adaptive Learning Environment. In: Proceedings of CSCL 2002, pp. 493–494 (2002)
5. Walker, E., Rummel, N., Koedinger, K.R.: Beyond explicit feedback: new directions in adaptive collaborative learning support. In: Proceedings of the 9th International Conference on CSCL 2009, pp. 552–556 (2009)
6. Miao, Y., Hoppe, U.: Adapting Process-Oriented Learning Design to Group Characteristics. In: Looi, C., McCalla, G., Bredeweg, B., Breuker, J. (eds.) Proceedings of Artificial Intelligence in Education, pp. 475–482. IOS Press (2005)
7. Magnisalis, I., Demetriadis, S.: Modeling Adaptation Patterns in the Context of Collaborative Learning: Case Studies of IMS-LD Based Implementation. In: Daradoumis, T., Caballé, S., Juan, A.A., Xhafa, F. (eds.) Technology-Enhanced Systems and Tools for Collaborative Learning Scaffolding. SCI, vol. 350, pp. 279–310. Springer, Heidelberg (2011)
8. Magnisalis, I., Demetriadis, S.: Modeling Adaptation Patterns with IMS-LD Specification: A Case Study as a Proof of Concept Implementation. In: Proceedings of INCOS 2009, pp. 295–300 (2009)
9. Halm, J., Olivier, B., Farooq, U., Hoadley, C.: Collaboration in Learning Design Using Peer-to-peer technologies. In: Koper, R., Tattersall, C. (eds.) Learning Design: A Handbook on Modelling and Delivering Networked Education and Training, pp. 203–213. Springer, Heidelberg (2005)
10. Koper, E.J.R., Olivier, B.: Representing the Learning Design of Units of Learning. JETS 7(3), 97–111 (2004)
11. IMS-LD: IMS Global Learning Consortium: Learning Design Specification (2003),
    http://www.imsglobal.org/specifications.html
    (accessed June 11, 2010)
12. Paramythis, A.: Adaptive Support for Collaborative Learning with IMS Learning Design: Are We There Yet? In: Proceedings of the Adaptive Collaboration Support Workshop, held in Conjunction with the Adaptive Hypermedia 2008 Conference, Hannover, Germany, July 29, pp. 17–29 (2008)
13. Towle, B., Halm, M.: Designing Adaptive Learning Environments with Learning Design. In: Koper, R., Tattersall, C. (eds.) Learning Design: A Handbook on Modelling and Delivering Networked Education and Training, pp. 216–226. Springer, Heidelberg (2005)

14. Burgos, D., Tattersall, C., Koper, E.J.R.: Representing adaptive and adaptable Units of Learning. How to model personalized eLearning in IMS Learning Design. In: Fernández Manjon, B., Sanchez Perez, J.M., Gómez Pulido, J.A., Vega Rodriguez, M.A., Bravo, J. (eds.) Computers and Education: E-learning - from Theory to Practice. Kluwer, Germany (2006)

15. Hernandez-Leo, D., Asensio-Perez, J.I., Dimitriadis, Y.: Computational Representation of Collaborative Learning Flow Patterns using IMS Learning Design. JETS 8(4), 75–89 (2005)

16. Collage: Collaborative learning design editor – Collage (2009), http://ulises.tel.uva.es/collage/ (accessed June 11, 2010)

17. Caballe, S., Xhafa, F., Daradoumis, T., Marques, J.M.: Towards a generic platform for developing CSCL applications using Grid infrastructure. In: Proceedings of the 2004 IEEE International Symposium on Cluster Computing and the Grid, CCGRID, April 19-22, pp. 200–207. IEEE Computer Society, Washington, DC (2004)

18. Bote-Lorenzo, M.L., Hernández-Leo, D., Dimitriadis, Y., Asensio-Pérez, J.I., Gómez-Sánchez, E., Vega-Gorgojo, G., Vaquero-González, L.M.: Towards reusability and tailorability in collaborative learning systems using IMS-LD and Grid Services. Advanced Technology for Learning 1(3), 129–138 (2004)

19. Bote-Lorenzo, M.L., Gomez-Sanchez, E., Vega-Gorgojo, G., Dimitriadis, Y.A., Asensio-Perez, J.I., Jorrin-Abellan, I.M.: Gridcole: A tailorable grid service based system that supports scripted collaborative learning. Computers and Education 51(1), 155–172 (2008)

20. Harrer, A., Malzahn, N., Wichmann, A.: The remote control approach - An architecture for adaptive scripting across collaborative learning environments. JUCS 14(1), 148–173 (2008)

21. Harrer, A., Lucarz, A., Malzahn, N.: Dynamic and Flexible Learning in Distributed and Collaborative Scenarios Using Grid Technologies. In: Haake, J.M., Ochoa, S.F., Cechich, A. (eds.) CRIWG 2007. LNCS, vol. 4715, pp. 239–246. Springer, Heidelberg (2007)

22. Zarraonandia, T., Dodero, J.M., Fernández, C.: Crosscutting Runtime Adaptations of IMS-LD Execution. Educational Technology & Society 9(1), 123–137 (2006)

23. Web services Architecture, W3C, http://www.w3.org/TR/ws-arch/ (accessed June 11, 2010)

24. Sharples, P., Griffiths, D., Scott, W.: Using Widgets to Provide Portable Services for IMS Learning Design. In: Koper, R., Stefanov, K., Dicheva, D. (eds.) Proceedings of the 5th International TENCompetence Open Workshop "Stimulating Personal Development and Knowledge Sharing", pp. 57–60. TENCompetence Workshop, Sofia (2008)

25. Wilson, S., Sharples, P., Griffiths, D.: Extending IMS Learning Design services using Widgets: Initial findings and proposed architecture. In: Proceedings of the 3rd TENCompetence Open Workshop on Current Research on IMS Learning Design and Lifelong Competence Development Infrastructures (2007), http://dspace.ou.nl/handle/1820/963 (accessed February 15, 2012)

26. Web Services Description Language (WSDL) Version 2.0 Part 1: Core Language, W3C, http://www.w3.org/TR/2003/WD-wsdl20-20031110/ (accessed October 27, 2010)

27. Karakostas, A., Demetriadis, S.: Adaptation Patterns as a Conceptual Tool for Designing the Adaptive Operation of CSCL Systems. Educational Technology Research & Development, ETRD 2, 1–23 (2010)

28. SLeD, Service Based Learning Design Player (2005),
    `http://sled.open.ac.uk/` (accessed October 27, 2012)
29. de-la-Fuente-Valentin, L., Pardo, A., Delgado, C.K.: Generic service integration in
    adaptive learning experiences using IMS learning design. Computers & Education 57,
    1160–1170 (2011)
30. Bayón, A., Santos, O.C., Couchet, J., Boticario, J.G.: An architecture for adaptive col-
    laboration support guided by learning design. In: International Conference on Intelli-
    gent Networking and Collaborative Systems, INCOS 2009, pp. 323–328 (2009)
31. Miao, Y., Burgos, D., Griffiths, D., Koper, R.: Representation of Coordination Me-
    chanisms in IMS-LD. In: Lockyer, L., Bennet, S., Agostinho, S., Harper, B. (eds.)
    Handbook of Research on Learning Design and Learning Objects: Issues, Applications
    and Technologies. Idea Group Inc., Wollongong (2008),
    `http://hdl.handle.net/1820/930` (accessed October 27, 2012)
32. Moreno-ger, P., Burgos, D., Martínez-ortiz, I., Luis, J., Fernández-manjón, B.: Educa-
    tional game design for online education. Computers in Human Behavior 24, 2530–
    2540 (2008)
33. Specht, M., Burgos, D.: Modeling Adaptive Educational Methods with IMS Learning
    Design. Journal of Interactive Media in Education, 1–13 (2007)
34. Ghali, F., Cristea, A.I., Hendrix, M.: Augmenting e-Learning Standards with Adapta-
    tion. Computers and Advanced Technology in Education 614, 79 (2008)
35. Ferraris, C., Martel, C., Vignollet, L.: Helping teachers in designing CSCL scenarios:
    a methodology based on the LDL language. In: Chinn, C.A., Erkens, G., Puntambekar,
    S. (eds.) Proceedings of the 8th Iternational Conference on Computer Supported Col-
    laborative Learning, New Brunswick, New Jersey, USA, July 16-21, pp. 193–195. In-
    ternational Society of the Learning Sciences (2007)
36. Paramythis, A., Cristea, A.: Towards Adaptation Languages for Adaptive Collabora-
    tive Learning Support. In: IGACLE_Workshop_ECTEL (2008)
37. Vaquero-Gonzalez, L.M., Hernandez-Leo, D., Simmross-Wattenberg, F., Bote-
    Lorenzo, M.L., Asensio-Perez, J.I., Dimitriadis, Y.A., Gomez-Sanchez, E., Vega-
    Gorgojo, G.: The opportunity of grid services for CSCL-application development. In:
    13th Euromicro Conference on Parallel, Distributed and Network-Based Processing,
    pp. 4–11 (2005)
38. Magnisalis, I., Demetriadis, S., Karakostas, A.: Adaptive and Intelligent Systems for
    Collaborative Learning Support: A Review of the Field. IEEE Transactions on Learn-
    ing Technologics 4(1), 5–20 (2011)
39. Karakostas, A., Demetriadis, S.: Systems for Adaptive Collaboration Scripting: Archi-
    tecture and Design. In: Adaptive Collaboration Support Workshop in 5th International
    Conference on Adaptive Hypermedia and Adaptive Web-Based Systems, pp. 7–12
    (2008), `http://www.ah2008.org/index.php?section=62` (accessed Oc-
    tober 27, 2012)
40. Demetriadis, S., Magnisalis, I., Karakostas, A.: Adaptation Patterns in Systems for
    Collaborative Learning and the Role of the Learning Design Specification. In:
    Scripted vs. Free CS Collaboration: Alternatives and Paths for Adaptable and Flexible
    CS Scripted Collaboration Workshop in CSCL 2009, Rhodes, pp. 43–47 (2009),
    `http://mlab.csd.auth.gr/cscl2009/sfc-`
    `workshop.htm#Proceedings` (accessed October 27, 2012)
41. Hewitt, J.: Toward an understanding of how threads die in asynchronous computer
    conferences. JLS 7(4), 567–589 (2005)

42. Kobbe, L., Weinberger, A., Dillenbourg, P., Harrer, A., Hämäläinen, R., Häkkinen, P., Fischer, F.: Specifying computer-supported collaboration scripts. IJCSCL 2(2), 211–224 (2007)
43. Dillenbourg, P., Tchounikine, P.: Flexibility in macro-scripts for computer-supported collaborative learning. JCAL 23(1), 1–13 (2007)
44. Cool Modes, http://www.collide.info/index.php/Cool_Modes (accessed at April 12, 2010)
45. Walker, E., Rummel, N., Koedinger, K.R.: CTRL: A research framework for providing adaptive collaborative learning support. User Modeling and User-Adapted Interaction, 387–431 (2009)
46. Harrer, A., Ziebarth, S., Giemza, A., Hoppe, U.: A Framework to Support Monitoring and Moderation of e-Discussions with Heterogeneous Discussion Tools. In: ICALT 2008, pp. 41–45 (2008)
47. Durand, G., Belliveau, L., Craig, B.: Simple Learning Design 2.0. In: 2010 IEEE 10th International Conference on Advanced Learning Technologies (ICALT), pp. 549–551 (2010)
48. Magnisalis, I., Demetriadis, S., Pomportsis, A.: Implementing Adaptive Techniques in Systems for Collaborative Learning by Extending IMS-LD Capabilities. In: International Conference on Intelligent Networking and Collaborative Systems, pp. 70–77 (2010)
49. Anaya, A.R., Boticario, J.G.: A Data Mining Approach to Reveal Representative Collaboration Indicators in Open Collaboration Frameworks. In: Proc. Educational Data Mining Conf., pp. 210–219 (2009)
50. Anaya, A.R., Boticario, J.G.: Reveal the Collaboration in a Open Learning Environment. In: Mira, J., Ferrández, J.M., Álvarez, J.R., de la Paz, F., Toledo, F.J. (eds.) IWINAC 2009. LNCS, vol. 5601, pp. 464–475. Springer, Heidelberg (2009)
51. Mørch, A.I., Dolonen, J.A., Nævdal, J.E.: An Evolutionary Approach to Prototyping Pedagogical Agents: From Simulation to Integrated System. J. Network and Computer Applications 29(2/3), 177–199 (2006)
52. Dimitracopoulou, A., Petrou, A., Martinez, A., Marcos, J.A., Kollias, V., Jermann, P., Harrer, A., Dimitriadis, Y., Bollen, N.: State of the Art of Interaction Analysis for Metacognitive Support & Diagnosis, IA JEIRP Deliverable D.31.1.1. Kaleidoscope NoE (2005), http://www.noe-kaleidoscope.org
53. Avouris, N., Margaritis, F.: A Tool to Support Interaction and Collaboration Analysis of Learning Activity. In: Proc. Int'l Conf. Computer Supported Collaborative Learning (2002)
54. Martinez, A., Harrer, A., Barros, B.: Library of Interaction Analysis Tools, Deliverable D.31.2 of the JEIRP IA Kaleido-Scope, Publications (2005), http://www.rhodes.aegean.gr/ltee/kaleidoscope-ia/
55. Martinez-Mones, A., Dimitriadis, Y., Harrer, A.: Interaction-Aware Design for Learning Applications Reflections from the CSCL Field, Proc. Eighth IEEE Int'l Conf. Advanced Learning Technologies (ICALT '08), 539-541 (2008)
56. Social Networks Adapting Pedagogical Practice: SNAPP, http://research.uow.edu.au/learningnetworks/seeing/snapp/index.html (accessed April 1, 2011)
57. De Bra, P., Calvi, L.: AHA: a Generic Adaptive Hypermedia System. In: Proc. of the 2nd Workshop on Adaptive Hypertext and Hypermedia, pp. 5–12 (1998)
58. Chen, W.: Supporting teachers' intervention in collaborative knowledge building. Journal of Network and Computer Applications 29, 200–215 (2006)

Prototype Tools for the Flexible Design of CSCL Activities Based on the Adaptation Pattern Perspective

Anastasios Karakostas, Zaharoula Papamitsiou, and Stavros Demetriadis

Department of Informatics, Aristotle University of Thessaloniki,
PO BOX 114, 54124, Thessaloniki, Greece

Abstract. This work presents the design and preliminary evaluation data regarding two prototype tools (namely, FlexCoLab and PPR), which have been designed according to the prescriptions of the adaptation pattern perspective for promoting a flexible design of CSCL activities. An adaptation pattern is described as a pedagogically useful and well-targeted adjustment process that can be initiated by the teacher or the CSCL system, in order to foster an improved learning setting when specific conditions occur during the collaborative learning activity. Both tools aim to support teachers in the process of developing flexible designs of online collaborative activities by reusing and customizing adaptation patterns, according to the requirements of a particular learning situation. "Advance the advanced" and "Support group of novices" are two adaptation patterns implemented by FlexCoLab, while "Advance the advanced", "Heterogeneity group formation based on prior domain knowledge", "Support groups of novices" and "Lack of confidence" are the adaptation patterns supported by PPR. In the chapter we present the theoretical background of adaptation patterns, the design specifications of the two systems and student evaluation data from implementing an in-school collaborative learning activity supported by PPR.

1 Introduction

Collaborative learning has been proved rather important for students both for social and cognitive reasons [1]. However, collaborating students might fail to engage in productive learning interactions when left without teachers' consistent support and scaffolding [2, 3]. Several approaches have been proposed on how to support the students' collaborative learning activities. Scripted collaboration – taking its origin from the scripted cooperation approach [4] – is the idea that collaboration can be guided and orchestrated by didactic scenarios, aiming to engage students in fruitful learning interactions. By implementing a script, the main mechanisms that promote collaborative learning, such as externalization, articulation, argumentation and negotiation of multiple perspectives [5, 6], can be put forward and integrated in the team learning activity. Lately, the considerable interest that the scripting approach has gained within the computer-supported collaborative learning (CSCL) community, has motivated efforts for the formalization of

T. Daradoumis et al. (Eds.): Intelligent Adaptation & Personalization Techniques, SCI 408, pp. 109–130.
springerlink.com
© Springer-Verlag Berlin Heidelberg 2012

collaboration scripts [7] and the development of computer-based environments for supporting scripted collaborative learning [8, 9].

Nevertheless, scaffolding collaborating students with a script is not a straightforward process. A script, typically, is conceived by an instructor as a helpful tool that will engage student teams in meaningful learning; however it may also become an unnecessary constraint in case advanced learners experience script guidance as an impediment to their self-organization. Adjusting the script level of granularity and flexibility emerges as an important issue that affects the outcome of scripted collaboration and it has been already discussed extensively in the literature [10, 11, 12, 13]. Overall, it is all boiled down to the fact that although a teacher can be flexible enough and adjust various collaboration parameters "on the fly" (i.e. during script run-time) CSCL systems for scripted collaboration are far from exhibiting a comparable level of flexibility.

In order to provide more flexible collaboration support, various research groups focus on the adaptive operation of CSCL systems (e.g. [14]). In general, adaptive collaboration support techniques aim to model the major aspects of the collaborative activity (such as domain, collaborative activity workflow, problem solution, student/group profile, peer interactions) and activate learner/group assistance interventions when needed and in the form needed [15]. This approach, therefore, seeks to exploit the potential of two traditions: the building of Adaptive Educational Systems (AES) and the Computer-Supported Collaborative Learning systems (CSCL systems) with emphasis on script design and implementation.

However, suggesting that adaptation methods could be of help in CSCL systems is but a vague idea. How could we build such systems taking into account the richness and complexity of human interaction in collaborative settings? Toward answering this question, we introduce the concept of "adaptation pattern". We argue that when experienced teachers identify a shortcoming in group learning activity then they enact some type of alternative version of the activity scenario which helps overcome the shortcoming. In the CSCL context therefore, an adaptation pattern can be described as an adaptation process that can be initiated by the system when specific conditions are met during implementation of the collaboration script and offer personalized support or otherwise adjust the various aspects of the script. Our working hypothesis is that these patterns can be formalized and embedded to CSCL systems in a way that the script flexibility is improved in situations where lack of flexibility may result to detrimental learning outcomes for the collaborating learners.

The main goal of this chapter is to present two prototype tools (FlexCoLab and PPR), which implement a number of adaptation patterns. Using these tools, a teacher/designer can customize and enact adaptation patterns during the collaborative activity, promoting flexible design of CSCL activities.In the following, after presenting the main efforts on dynamic collaboration support, the paper focuses on the theoretical background of adaptation patterns, the DeACS methodology for identifying these patterns, and an analysis of the FlexCoLab and PPR. Furthermore preliminary evaluation data from a collaborative activity supported by the PPR system are also presented.

2 Supporting Collaboration

2.1 Collaboration Scripts and the Need for Flexibility

Researchers have systematically emphasized that freely collaborating students might fail to engage in productive learning interactions when left without teachers' consistent support and scaffolding (e.g. [16, 1, 3]).

Collaborative learning is a complex process where it is very difficult – if not impossible – for the instructor to consider all interacting parameters in order to foster productive learning experiences for each individual separately [17]. Instead, it has been suggested that the instructor guides the learners' interactions within the group, by implementing an appropriate collaboration script [18]. Collaboration scripts are didactic scenarios that aim to improve and support the collaborative learning process by specifying the way in which learners interact with one another. A script provides specific instructions "for small groups of learners on what activities need to be executed, when they need to be executed, and by whom they need to be executed in order to foster individual knowledge acquisition" [20, p.195]. In this way one increases the probability of productive student-student and student-teacher learning interactions. Indeed, scripting collaborative learning has been reported to result in improved learning outcomes [20, 21].

Every script can be seen as a sequence of phases with five major attributes each [7]: (a) the kind of task that has to be performed in the specific phase, (b) the composition of the group, (c) the way that task is distributed among group members, (d) the way of interaction and communication among group members and (e) script's time duration [10]. It is important also to emphasize that one can dissociate between two levels of scripting: (a) the macro level and (b) the micro level [19]. The macro level refers to the organization and the structure of the collaborative activity, for example how to organize the group's task or the specific collaborations between group members. The collaboration scripts that are focused on this macro level of collaboration are called accordingly "macro scripts" [10]. In contrast the micro level refers to scripts that provide support for specific activities. The "micro scripts", therefore, (as opposed to the macro scripts) emphasize the activities of individual learners.

As Suthers [23, p. 255] points out "a goal of computer supported collaborative learning (CSCL) systems… is to improve the effectiveness of collaborative learning as an instructional format: i.e., to support peer interactions in a manner that increases learning gains". Following this perspective, a considerable interest for the scripting approach has emerged lately within the CSCL community. This interest has motivated efforts for the formalization of collaboration scripts [7] and the development of computer-based environments for the authoring and operationalization of CSCL scripts [9]. CSCL scripts are described as computer-based representations of collaboration scripts where the role of the computer-based system is both to structure the students' learning activity [12] and to support users with communication services.

An innovative tool for the design and deployment of web-based scripted collaborative activities is the "Learning Activity Management System" (LAMS) [24].

LAMS provides a simple yet highly intuitive graphical user interface that allows teachers to drag and drop activity tools into the workspace and use connecting arrows to organize the activities into a sequential workflow. Based on teacher design LAMS generates and deploys the activity workflow to guide student groups in their collaboration. Following this design/deployment approach current instructional technology offers tools of the "script editor and player" type. Script editors such as Reload Editor [25] or Collage Editor [26] allow users (teachers) to design the sequence of the activities that must be performed by the participants of a collaborative learning scenario as well as the tools and documents that can be employed in each activity. Script players, in turn, like Reload Player [25] and SLeD [27], support the realization of scripted collaborative learning design and guide users (student groups) in performing all the related collaborative tasks. The development of such editors and players is usually based on the CopperCore [28] learning flow engine. CopperCore is the first open source Learning Design engine capable to process all three levels of IMS-LD. IMS Learning Design (LD) [29] is an educational modeling language that enables the formalized description of learning scenarios. IMS-LD can describe a wide variety of pedagogical models including group work and collaborative learning activities. Although it supports the entire IMS-LD specification, CopperCore is not a tool for the end-user. In practice, CopperCore engine is used for developing end-user tools (like the ones described above) that interpret IMS-LD representations of learning scenarios.

Nevertheless, CSCL scripting has been also criticized for its loss of flexibility (difficulty of modifying a script in run time according to the needs of the instructional situation) [10], and the danger of "over-scripting" collaborative activity (the pitfall of overemphasizing script imposed interactions and constraining natural collaboration) [30]. Although a teacher can be flexible enough and adjust various collaboration parameters during script run-time, CSCL systems for scripted collaboration are far from exhibiting a comparable level of flexibility [31]. None of the previously mentioned systems offers the possibility to manage unforeseen situations during collaboration. In practice, this means that once the script is enacted it is not possible to adjust anything during script runtime. This opens the question of how to build CSCL systems that adaptively provide learner and group support, something that is expected to be highly beneficial both for learners and instructors in collaborative learning situations.

2.2 Adaptive Collaboration Support

Several researchers have already tried to support the collaborative activity by using adaptive supportive mechanisms [15]. These efforts extend the tradition of Adaptive Educational Systems (AES) toward the area of CSCL. Adaptive systems are model-based systems that provide flexible support to the learners but their objective is not the same. They focus on adapting various operational characteristics of the system (e.g. content presentation) based on learners' needs and/or characteristics. Thus, adaptive educational systems behave differently for different students/groups considering characteristics of the individual/group model [31].

So far, research has indicated that adaptive interventions can be a very useful scaffolding method in individual learning settings [32, 33].

CSCL systems have already embodied techniques from the AES area into four major and distinct directions: a) group formation, refers to methods for dividing a large group of students into teams [34], b) peer matching, focusing on providing either a suitable partner to an individual learner in order to accomplish a task or the appropriate information to identify the most suitable partner [35], c) peer interaction support, refers to a situation when the system provides support to group members to improve their collaboration (for example, help students make better contributions or efficiently attend and understand other group members' contributions), and, finally, d) decision support, which is the support provided by the system to partners in order to take a decision as a group [36].

However, despite promising evidence, it is a fact that the adaptive collaboration support systems are at an early stage. Walker et al. [14] emphasize that few adaptive collaborative learning support (ACLS) systems have been implemented (and even less evaluated) and one major reason for this is the difficulty to effectively model the partners' dialog and provide feedback.

2.3 Research Motivation

Overall, the issue of flexibly adjusting the collaborative learning activity, much like a skilled tutor would do, is of high priority in current CSCL research agenda. Significant open research questions in the area include: (a) How to flexibly adjust collaboration scripts so that students of various profiles maintain a productive peer interaction during the activity; (b) How to avoid issues of overscripting, i.e. excessively guiding the activity and deemphasize the role of natural free collaboration; (c) How to develop technology tools that facilitate both the design and the deployment of flexible and adaptive collaborative activity computer-based representations.

To address these issues we have introduced the "adaptation pattern" approach which suggests that teachers' common practice to adjust collaborative activities when certain conditions occur, could be formalized and embedded in CSCL systems as a tool for increasing system flexibility. In the following we present both the theoretical underpinnings of the approach and two web-based systems designed according to this perspective.

3 Adaptation Patterns

3.1 The Nature of Adaptation Patterns

The concept of "pattern" emerged in the context of architectural design referring to the outline of a solution, which can be successfully applied to recurring problems of the same typology [38, 39]. Since then, the pattern approach was employed in several other domains. In software engineering, patterns provide a description of successful solutions to common software problems.

Extending the general pattern-based approach, an adaptation pattern [40] is introduced as an outline of some type of alteration of the initial design, which an informed instructor would apply when some specific conditions (that is, a "recurring problem of specific typology") occur during the enactment of a collaboration procedure. Such a pattern epitomizes when and how an adaptation strategy should be implemented to provide improved conditions in a collaborative learning situation. For example, in case an advanced and a novice learner happen to work together, this might call for teacher's intervention and support. Perhaps the teacher guides the advanced learner on how to help the novice and, also, adapts the task for the advanced learner to become more challenging and interesting for him/her. Ideally, one would expect an ACLS system to model such situations and offer adaptive support to students, exhibiting similar to teacher's behavior. An adaptation pattern, therefore, is a process (or a set of processes) that models effective teachers' interventions to change the initial design and better support collaborating peers in a specific situation. It takes into account the learner, the group and the learning activity and adjusts certain aspects of the collaborative setting in order to maximize student engagement, interaction, satisfaction and, consequently, the learning (cognitive and metacognitive) outcomes. Furthermore, a computer-based representation of this process can provide the basis for establishing the adaptive behavior of the system when designing computer supported collaborative activities.

3.2   The DeACS Methodology

However, the question emerges of how to identify pedagogically useful adaptation patterns possible to occur in a collaborative activity. So far, to answer this question, we have proposed the DeACS (Designing Adaptive Collaboration Support systems) method, a qualitative method which prescribes an iterative implementation of the collaborative activity [40]. During the activity iterations, the teachers/designers have the opportunity to a) implement and evaluate the impact of some known adaptation patterns, and, b) identify new possible adaptation patterns beneficial for the group partners, from observing students' behavior and expressed needs. The ultimate goal of the DeACS methodology is to help teachers/designers identify the design specifications of valuable adaptation patterns, thus making a step toward formalizing and implementing the patterns in an ACLS system.

The DeACS method identifies three major processes: (a) a topdown process: integration of adaptation patterns under evaluation in the ideal script (the form of the script that the teacher initially wishes to put into practice), (b) a bottom-up process: identification of adaptation patterns that emerge from students' expressed needs for help, support, adjustments, etc., during script runtime, (c) an evaluation process aiming to assess the impact of the adaptation patterns in the previous two processes. If the evaluation of patterns reveals beneficial impact on student learning then these patterns can become part of the computerized script representation embedded to the collaboration support system.

Moreover, an adaptation pattern could typify two types of adaptivity: "static" and "dynamic". An adaptation pattern could typify "static" type of adaptivity, if it models only one variable (e.g. students' prior knowledge) when starting the activity without further considering peer interaction. In contrast, other adaptation patterns may demonstrate "dynamic" adaptivity considering certain aspects of ongoing peer interaction.

3.3 Adaptation Pattern Analysis

In order to implement and exemplify the DeACS method we conducted two studies [40]. The first one engaged 12 postgraduate students in a scripted collaborative activity of pyramid-type. In the second, 36 undergraduate students formed dyads and worked collaboratively in the LAMS environment, where they were adaptively supported by domain specific prompts. The analysis of the learning experiences provided the basis for identifying a number of possible adaptation patterns [40]. For the purpose of this chapter, in the following we present two of the adaptation patterns that FlexCoLab and PPR systems implement.

3.3.1 "Advance the Advanced"

The key-idea of this pattern is the adjustment of the level of task difficulty for the group with one (or more) advanced participant(s) in order to offer a challenging learning experience to him/her too. This pattern is activated when the ACLS system identifies one or more advanced student participants in a group (regardless of the group size).

In order to implement this pattern a system needs to model: (a) learners' prior domain knowledge, (b) group synthesis, (c) learning material and/or aspects of the task (various characteristics, for example level of task difficulty). This pattern main objective is to adapt the difficulty level of the task for the advanced learner, by providing, for example, more challenging material for study or engaging the student in a more demanding form of activity.

3.3.2 "Support Group of Novices"

Research-based evidence indicates that it is undesirable to form homogenous groups [1], especially with all partners being domain novices. Nevertheless sometimes it is unavoidable to form such groups. When this happens, this pattern objective is to provide necessary support to partners in this type of groups.

Therefore, the pattern is activated when the group members are domain knowledge novices and to become operational the pattern needs to model: (a) individual learners' domain knowledge, (b) group synthesis, (c) supportive material and/or instructors' interventions (e.g. prompts).

Support offered to the group when this pattern is enacted, may include: (a) providing more detailed guidelines on the task and/or the collaboration procedure (i.e. a more detailed form of the micro-script which would include detailed role descriptin, detailed presentation of the task goals, etc.), (b) adapt the difficulty level of the task for novices, (c) adapt the level of the learning material (i.e. provide introductory material).

4   The FlexCoLAB System

FlexCoLab system provides a web-based environment for designing, managing and deploying collaborating learning activities. Moreover FlexCoLab provides a simple user interface that allows teachers to easily design scripted collaborative activities. Based on teacher's design, FlexCoLab generates and deploys the activity workflow to guide student groups in their collaboration. The main innovation of FlexCoLab, compared to other CSCL environments, is that it caters for flexibility during collaboration by implementing adaptation patterns. The version of the system that is presented in this paper is able to handle groups of two (dyads) and enact two specific adaptation patterns: "advance the advanced" and "support group of novices" which already have analyzed above.

4.1   Authoring Environment

4.1.1   Designing a Collaboration Activity

FlexCoLab provides an authoring environment to support teacher in designing a collaboration activity. In the following we present the design process facilitated by FlexCoLab. The process is not strictly sequential.

- General description of the activity: the teacher provides a general description of the whole collaborative activity.
- Collaboration activity phases: the teacher defines the collaboration activity phases and more specifically the start and finish dates, the study guidelines for each phase and the collaboration tool that the students will use to collaborate during each phase (e.g. chat or forum).
- Files: the teacher uploads all the files (e.g. pdf, source files) that will be necessary during the activity.
- Define the questionnaires: This is an important aspect of FlexCoLab. In Flex-CoLab teacher can create closed-type questionnaires that will allow him/her to identify the student domain knowledge level. As described before, adaptation patterns are based on student/group models that include their domain knowledge. The questionnaires will help the system to develop the necessary student knowledge models relevant to each pattern.
- Define student types: In this phase the teacher determines how the system should interpret the questionnaire results to identify student level of prior knowledge. For example, the teacher could define that if a student succeeds in answering more than 60% of the questionnaire then she will be characterized as "advanced".
- Content to students: With FlexCoLab the teacher can define which content (learning material) files will be provided to which groups and in which phase.
- Adaptation patterns: Finally the teacher determines which adaptation patterns will be enacted and their specific options.

4.1.2 Adaptation Patterns Authoring Environment

In general, the teacher/designer of the collaboration activity in FlexCoLab is able to specify which adaptation patterns are going to be enacted in order to support students during the collaboration. Furthermore, the teacher can select the aspects of the collaborative activity that the adaptation pattern will affect. In other words, the aspects in the design of the collaborative activity that will be adapted when the pattern will be triggered.

Generally the adaptation patterns that FlexCoLab implements affect:

- the task guidelines of a specific phase and
- the learning resources that will be available to students

In case an adaptation pattern is not triggered then the task guidelines and the learning resources that will be presented to the group members would be the default ones that the teacher has already defined during the design of the collaborative activity.

In the following we present in detail the steps that a teacher has to do in order to enact and modify an adaptation pattern. The adaptation pattern that is presented in Fig. 1 is the "Advance the advanced" and the target of adaptation is the description guidelines of a specific phase.

Step1: with a radio button option the teacher can select the characteristics of the collaboration that are going to change when the pattern will be triggered. In the example below (Fig. 1, section A) the teacher has selected to adapt the description of the activity in a specific phase.

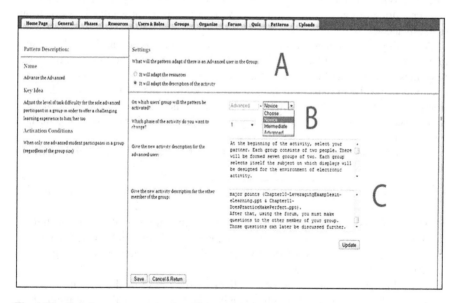

Fig. 1 FlexCoLab environment: adaptation patterns settings

Step2: the teacher sets the conditions under which the pattern will be triggered. For example, referring to the pattern "advance the advanced" the pattern triggering condition could be when a dyad with one advanced and one novice student is formed. The teacher can easily set the group profile that activates the pattern by using a simple drop down menu (Fig. 1, section B).

Step3: the next step depends on teacher's selection in step 1. If the teacher has chosen to adapt the learning resources available to the students, the next step is to determine the level of the learning material that will be given to each group member. The difficulty level of the material has already been defined by the user when the files were uploaded to the system during the design procedure of the collaboration activity. The teacher is able to select educational material with different or the same level for each participant.

Similarly, if the teacher has chosen during step 1 to adapt the phase guidelines, the next step is to choose in which phase of the activity the guidelines will change. Then the teacher has to re-write the instructions for each member separately. In the example, which is presented in Fig. 1 (section C), the new instructions have been added both for the group members (the advance and the novice one).

4.2   Student Environment

As students log into the FlexCoLab, the system informs them about the available collaborative activities that they could participate. The teacher has already created these activities, and also has assigned which groups are going to participate in which activity. At the beginning of the collaborative activity, the student has to answer a domain knowledge questionnaire to get a prior knowledge classification (e.g. advanced).

According to teachers' design, the student could answer additional domain questionnaires in next phases of the collaborative activity. The answers on these new questionnaires will provide to the student a new classification. In other words a student could be classified as "advanced" regarding the domain in a collaboration phase and "novice" regarding other issues in a next phase during the same activity. The teacher has already declared these classifications during the design phase of the activity.

The student environment is divided into four distinct and separated areas as they are presented in Fig. 2:

- A: in this area the general description of the whole collaborative activity is presented.
- B: this area presents the educational resources that are available to each student individually. The format of these resources varies from a link to web page or a pdf file. This is the first of the two areas that can be altered according to the adaptation patterns options.
- C: in this area the instructions/guidelines of the phase of the activity are presented. This is the second area that can be adapted based on the enactment of an adaptation pattern. The teacher has defined these guidelines during the design phase and their alternates during the adaptation pattern adjustment phase.

- D: this area presents the collaboration tool (e.g. chat or forum) that the students are going to use during that phase.

Finally in FlexCoLab student environment, the student is able to identify in which phase of the activity is at any moment. Depending on the teacher's design, the student is able to move to a previous or to a next phase. Moreover, from the tab button Group Area the student can upload a file to the system (e.g. a deliverable for a phase). Summarizing, the adaptation patterns that the FlexCoLab implements are focused on the information presented in sections B and C. According to the pattern, each group member could have different information than the other.

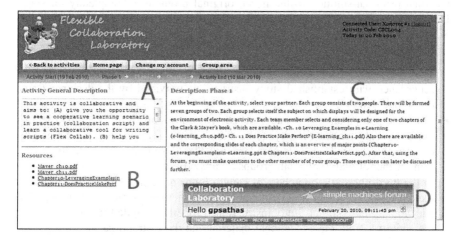

Fig. 2 FlexCoLab student environment

5 PPR (Pyramid Peer Review)

PPR (Pyramid Peer Review) is a proposed implementation of a web-based environment for composing, customizing and reproducing scripted collaborative educational activities, regardless to the knowledge domain. It is a collaborative learning information system for managing educational content and providing adaptive support to the learners' interactions based on a predefined sequential workflow of activities. PPR introduces a computational approach of the core idea of adaptation patterns to improve the learning outcome of the participants according both to the user and group needs, by triggering specific adaptive interventions whenever specific learning conditions occur.

5.1 The Core Idea

The environment supports "creating" (editing) and "running" (playing) educational scenarios of four phases with distinct characteristics in each one of them. The script underlying the implementation is the Pyramid Script of two phases.

In literature the term "Pyramid" generally refers to a pattern of collaboration script (Script Schema) involving gradually increasing number of participants. The Pyramid Script begins with an individual task (Pyramid peak) – intending to foster individual knowledge acquisition – and proceeds by extending the size of the collaboration group (base of the Pyramid) – aiming to form common understanding among the members of the group. However, unlike the typical Pyramid script, the model implemented impedes two additional assessment phases. So, a self-assessment phase is inserted after the individual study phase, and a peer evaluation phase is introduced after the submission of the deliverables from small groups. Finally, the phase at which the whole class constitutes a uniform group submitting the final proposal-solution to the initial original problem is completely missing. Due to these changes, the model developed is characterized as a "hybrid Pyramid Script".

More specifically:

- The first phase of the script includes self-study of the educational material in accordance with established study instructions provided by the designer / author of the activity.
- Then, the participants complete a closed-type questionnaire in order their learning level (only for this activity) to be diagnosed, to form the user model and to classify them based on their performance. This designation shall be the criterion for the formation of small groups (dyads / triples) for the next phase.
- In the third phase, the members of small groups are emerged together to complete a task and compose a deliverable. Each member of the small group has educational material and study instructions properly compiled (by the author of the activity). Both the educational material and the instructions are compatible with the learning profile of the student and the profile of the group. At the end of the phase, each team submits the deliverable to be evaluated by both the teacher / designer of the activity, and the other participants.
- The forth and last phase of the script implements the evaluation of the deliverables under specific criteria set by the author of the activity and according to prescribed evaluation instructions available to all participants. Each student (individually) reviews at least one of the deliverables of other groups and documents the assessment.

Phases 2 and 3 cannot be omitted or replaced, while phases one and four are not compulsory. However, instead of phase one, the teacher may share the educational material using any other CMS.

5.2 Design Analysis– The Design Axes

5.2.1 Design Axis 1: The Adaptation Patterns

Four (4) adaptation patterns have implemented in PPR, that fit more with the core idea of PPR and are more likely to create initiation conditions during the execution of the script. These adaptation patterns are: (a) Heterogeneity group formation based on prior domain knowledge, (b) Lack of confidence, (c) Advance the advanced, and (d) Support groups of novices [40].

5.2.2 Design Axis 2: PPR Interface Design

PPR consists of two separate interfaces to support the corresponding functions: the authoring / editor interface (for the syntax, customization and monitoring section used by the author / designer of the activity) and the student / player interface (for running the activity by the participating students).

The editor (author) interface
The author's environment (Fig. 3) was developed to facilitate the possible managing requirements of the teacher designing the activity, in a functionally consistent, simple and logically semantic-abstract manner. These three dimensions define the intuition of the interface. To achieve that goal, the objects included in the interface and the apparent symbolisms were selected, organized and categorized so as to describe the corresponding functionalities. Two major classifications were considered:

- First classification (horizontal classification): separation between available activities (and the permitted management actions on them) and new activities (complete definition of all the individual characteristics, such as general description of the activity, start date and end date of each phase, description of the sub-tasks, communication tools, etc.). The author can set up, preview, edit (change further parameters before begin date of a phase), monitor and/or participate and support (during runtime), and/or evaluate (when completed) an activity.
- Second classification (vertical classification): identification of necessary and sufficient individual characteristics of each activity and each phase. For each activity the author a) provides descriptive characteristics in the form of general information, b) registers the educational material (files and internet resources) – which will be distributed along the phases of the script – c) constructs the question pool to select from when configuring the self-assessment questionnaire, d) defines the instructions and assigns them to the corresponding educational level category and the group model, and finally e) manages the participants.

Furthermore, the *group moderator* facilitates the discussions, keeps the group on task, assures work is done by all members, and makes sure all have opportunity to participate and learn. This role is challenging for advanced learners in order to improve higher level skills during collaboration. The *group recorder* takes notes of the group discussion and prepares a written conclusion. This role is assigned to the less advanced members of the heterogeneous group so as to improve their monitoring skills and empower their learning level, or to the mutual in a homogeneous group. Finally, the *group summarizer* restates the group's conclusion or answer. Every member of the group can act as a summarizer during the third phase, using the available wiki tool. The description of the responsibilities of each role is included within suitable instructions that are assigned to learners to complement the task. Through the chosen roles to be assigned to the members of the groups, every student can participate and contribute in a unique and suitable way to accomplish the given tasks.

Although the editor interface consists of forms and menus the author can use in random order, there are limitations related to the duration of the phases: the author cannot change the parameters of an ongoing or completed phase.

The player (student) interface
The main features of the student's interface (Fig. 4) retain the same graphic symbolisms and data layout as the teacher's interface. Similarly, this environment includes relevant classifications for activities management: as the student logs in the system, she can register in available activities, which have not started yet or she can choose an ongoing activity. The system itself informs the student about her current activity with notifications related to her participation (e.g. whether she has downloaded the educational material and instructions or not).

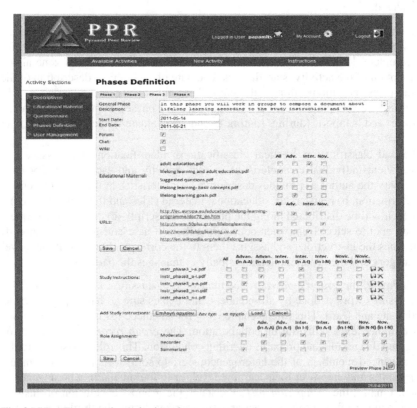

Fig. 3 PPR – The page for assigning phase parameters

However, the feature that differentiates this interface and implements the adaptation patterns from the aspect of the learner is that it adjusts the content presented to the user according to a) the ongoing phase of the scenario, and b) the user model and group model during the third phase. So, the on-screen content shown to a novice user in a group of novices differs from the respective on-screen content shown to a novice user in a group with an intermediate user and from the on-screen content shown to the intermediate user. And basically, these are the rules set by the author during customizing the activity. Thus, each participant sees (each time she logs in) only the information necessary to execute each phase (Fig. 4).

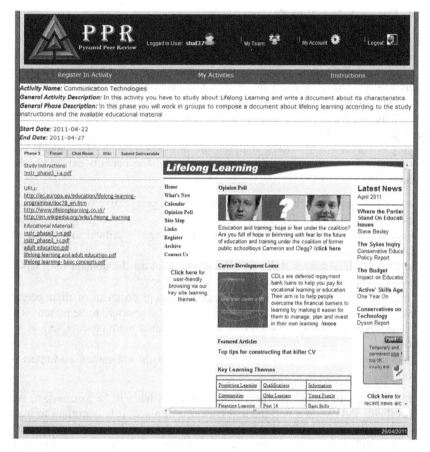

Fig. 4 PPR students environment

5.2.3 Design Axis 3: Distributed Cognition

Human interaction as an enabler of individual cognition is further strengthened through the shared aspect of social learning. Therefore, to achieve teaching support, it was appropriate to attach tools to "bind" the user's action in the environment itself. For this reason, PPR supports:

- Built-in archive opening (in .pdf format) and loading websites within the application window, and
- Built-in tools for synchronous and asynchronous communication (chat and forum respectively. Their use is optional)

In addition, during the third phase, the system provides access to a wiki environment for collaborative synthesis of the deliverable. Due to high technological requirements, the wiki engine was not built from scratch. Instead, PPR makes use of the existing wikispaces system (available at http://wikispaces.com).

5.2.4 Design Axis 4: Peer Review

Peer review enables learners to evaluate the work of other participants and to receive additional feedback on their work.McConnell [41] argues that peer reviewing offers to students the opportunity for a constructive and collaborative learning experience, by engaging them in an active learning exercise. The literature abounds with relevant studies indicating that the method is popular among educators inspired mainly by the constructivist and socio-constructivist paradigms for learning (e.g. [42, 43, 44]). Peer review is primarily expected to support higher-level learning skills such as synthesis, analysis, and evaluation [45] as the students have the opportunity to analyze and evaluate peer work. The benefits expected when implementing the method are:

- Students approach the learning situation from the perspective of the instructor / teacher.
- Students practice and refine their ability to engage in critical evaluation and understanding.
- Students with advances skills or educational level on the domain may have a role in supporting their peers.
- Students with a lower performance seem to accept criticism of their peers – even when it is acute – and face the teacher as a "supporter" to help them. This makes the relationship between teacher and students more productive and less critical.
- Students seem to improve their self-assessment skills when involved in peer reviewing.

Critical issues raise during the peer review are: a) what will be assessed, b) who will evaluate and c) when will the assessment take place. It is important to maintain the anonymity of both authors evaluated the deliverables and their respective reviewers, and compare peer assessments with those of the teacher / author of the activity mainly for reasons of the reliability of the evaluation.

Three types of "participants" are identified in any assessment process with peer involvement in educational activities: a) the author of the deliverable submitted for reviewing, b) the peer reviewer, and c) the instructor / designer of the activity, who also participates as an evaluator. Therefore, during designing the system, the requirements for all these three categories were considered.

5.3 System Architecture

PPR was developed according to the three-tier architecture for adaptive systems for scripted collaboration based on adaptation patterns [46]. It includes: a) the storage layer, b) the layer of group management and c) the runtime layer.In the case of PPR, a database was designed to host the storage layer. The database stores: a) the activities and their descriptive features both general and individual per phase (e.g. start and end date, educational resources, study instructions, review criteria, etc.), b) the user model and the group model, c) questions and answers of the self-assessment questionnaire, d) group deliverables and evaluations, and in general features of the script to be represented in runtime.Regarding to the second layer, PPR implements a clustering algorithm to format the group composition. It runs during the second phase of the script. Finally, the third layer consists of the two separate interfaces described previously.

5.4 The Case Study

This case study was conducted in a secondary education school in Greece with 17 participants and the domain of the activity was "Communication Technologies".

The aim was to test and evaluate PPR a) from the perspective of students – users, and b) from the perspective of teacher – designer. During the preparation time, the educational goals were identified and organized. Additionally, targeted subtasks for each phase of the collaborative activity were planned and scheduled. The objective of the pilot case study focused on the identification and examination of the following parameters:

- Are the initiation conditions of the adaptation patterns activated during runtime?
- Could an implementation of a collaborative script with adaptive features enacted within PPR enable and encourage students' participation?
- Which is the contribution of the enacted adjustments to the effective development of collaboration among students?
- Do the considered adjustments (according to the user and the group profiles) and the corresponding instructions affect the roles and performance of students with high, moderate or less moderate skills?
- Does PPR promote the development of higher level skills (documented argumentation, development of critical approach of the educational resources, rule-discipline dialogue about a predetermined theme)?

The instructional material was selected and edited carefully in accordance with the requirements of each phase of the activity. The study instructions were set and adjusted for each phase in response to the learning level of the participants and the group models.

5.4.1 Results

In order to evaluate PPR from students' perspective, a questionnaire was drafted and completed by the participants. The questionnaire consisted of questions in a 5-level Likert scale. The questions were grouped in 5 categories: a) identification of the participants' profile, b) overall evaluation of the system, c) assessment of the technical and functional attributes of the system, d) evaluation of the scripted collaborative activity with the adjusted features, and e) the prospects of the systematic inclusion of such activities and relevant environments in educational practice. Complementary, the interview method was used to record the teacher's observations.

The general evaluation by both students and teacher involved in the case study was positive. 9 out of 17 students (53%) agreed that they liked the activity, while only 3 of them (17,6%) were placed negatively. This attitude is reinforced by the fact that 11 to 17 students (64,7%) claimed that the activity helped them under-stand the basic concepts of the educational section. According to the students, the instructions given by the teacher (through the system) to guide them during run-time were clear. However, the general appreciation of the environment ranges from neutral to good. 7 out of 17 students (41,2%) were neither positively nor ne-gatively placed, while only 2 (11,8%) said that they did not like this application. Table 1 summarizes the student answers for the category of questions "General Comments".

As for the characteristics of the environment, 76,5% (13 to 17) of the students agreed that the navigation through the system was easy, almost spontaneous (70,6%), that the keywords (for the corresponding functionalities) were appropri-ate (58.8%) and that the icons and graphics of the interface were appropriate (53%).

Regarding the functionalities of the interface, the students argued that although the access to the educational material and instructions was easy and simple (94,1%), however, the study using the computer was demanding (70,6%). Finally, 7 to 17 students (41,2%) claimed that the integrated communication and collabora-tion tools have helped them remain focused on the activity, and notifications and feedback were useful (70,6%). From the teacher's point of view, she agreed that "it was easy to manage the educational material, easy to construct the self-assessment questionnaire and spontaneous to navigate not strictly sequential. In addition, it ensures the objectivity during group formation, facilitating the provi-sion of targeted support geared to the needs of each learning level, but, at the time, it does not provide any mechanism for writing the study instructions or monitoring the user's behavior".

Table 1 Student answers for the category of questions "General Comments"

I would participate again in relevant activities

	Totally disagree	Disagree	Not disagree nor agree	Agree	Totally agree
Frequency	0	5	1	6	5
Percent	0%	29,4%	5,9%	35,3%	29,4%

I would like to use this system again

	Totally disagree	Disagree	Not disagree nor agree	Agree	Totally agree
Frequency	1	2	5	9	0
Percent	5,9%	11,8%	29,4%	52,9%	0%

The instructions given by the teacher were unambiguous

	Totally disagree	Disagree	Not disagree nor agree	Agree	Totally agree
Frequency	0	0	1	7	9
Percent	0%	0%	5,9%	41,2%	52,9%

The teacher helped my through executing the activity

	Totally disagree	Disagree	Not disagree nor agree	Agree	Totally agree
Frequency	0	1	1	10	5
Percent	0%	5,9%	5,9%	58,8%	29,4%

6 Conclusions and Future Work

This chapter has presented the main functional characteristics of the FlexCoLab and PPR tools, which are meant to support the design and implementation of flexible collaborative activities drawing on the "adaptation patterns" approach. Both tools allow the editing of collaborative activities by customizing adaptation patterns. Adaptation patterns are introduced as an outline of some type of alteration of the initial design, which an informed instructor would apply when some specific conditions occur during the enactment of a collaboration procedure.

FlexCoLab and PPR are web-based environments that can be used both by teachers and students. The teacher is able through an authoring environment to define the phase of the activity, the tasks in each phase, the educational resources and the guidelines for each phase. Moreover the teacher is able to enact and adjust specific adaptation patterns. In the current versions, both tools can implement specific adaptation patterns like "advance the advanced" and "support group of novices". We have also presented some encouraging preliminary evaluation data relevant to using the PPR tool to support collaborative learning activities at school.

Our future work includes further evaluations of both systems to assess their efficiency from both teacher and student perspective. Our main objective is to explore: (a) the potential of the adaptation pattern–based approach to capture the dynamics of the collaborative activity and provide an adequate model for building systems for adaptive collaboration support; (b) the efficiency of FlexCoLab and PPR system interface to help teachers design the adaptive activity with minimum of effort and maximum of pedagogical understanding and merit, and (c) the students' benefits and shortcomings when using these systems for learning through the adaptation of some key aspects of their collaborative learning activity.

References

1. Liu, C.C., Tsai, C.C.: An analysis of peer interaction patterns as discoursed by on-line small group problem-solving activity. Computers & Education 50, 627–639 (2008)
2. Hewitt, J.: Toward an understanding of how threads die in asynchronous computer conferences. The Journal of the Learning Sciences 7, 567–589 (2005)
3. Bell, P.: Promoting students' argument construction and collaborative debate in the science classroom. In: Internet Environments for Science Education, pp. 114–144. LEA, Mahwah (2004)
4. O'Donnell, A.M.: Structuring dyadic interaction through scripted cooperation. In: Cognitive Perspectives on Peer Learning, pp. 179–196. Lawrence Erlbaum Associates, Mahwah (1999)
5. Baker, L.: Technical assistance for writers in the workplace: Some heuristic uses of professional writing techniques in a multiauthor writing process. Technical Communication 43, 19–28 (1996)
6. Veerman, A.L., Andriessen, J.E.B., Kanselaar, G.: Learning through synchronous electronic discussion. Computers and Education 34, 269–290 (2000)
7. Kobbe, L., Weinberger, A., Dillenbourg, P., Harrer, A., Hämäläinen, R., Fischer, F.: Specifying computer-supported collaboration scripts. IJCSCL 2, 211–224 (2007)
8. Turani, A., Calvo, R.: The Potential Use of Collaboration Scripts in Synchronous Collaborative Learning. In: Proceedings of IMCL 2007 Conference (2007)
9. Bote-Lorenzo, M.L., Gomez-Sanchez, E., Vega-Gorgojo, G., Dimitriadis, Y.A., Asensio-Perez, J.I., Jorrin-Abellan, I.M.: Gridcole: A tailorable grid service based system that supports scripted collaborative learning. Computers and Education 51, 155–172 (2008)
10. Dillenbourg, P., Jermann, P.: Designing Integrative Scripts. In: Scripting Computer-Supported Collaborative Learning: Cognitive, Computational and Educational Perspectives, pp. 275–301. Springer, NY (2007)
11. Dillenbourg, P., Tchounikine, P.: Flexibility in macro-scripts for computer-supported collaborative learning. Journal of Computer Assisted Learning 23, 1–13 (2007)
12. Haake, J., Pfister, H.R.: Flexible Scripting in Net-Based Learning Groups. In: Fischer, F., Kollar, I., Mandl, H., Haake, J. (eds.) Scripting Computer-Supported Collaborative Learning, pp. 155–175. Springer, New York (2007)

13. Tchounikine, P.: Operationalizing macro-scripts in CSCL technological settings. International Journal of Computer-Supported Collaborative Learning 3(2), 93–233 (2008)
14. Walker, E., Rummel, N., Koedinger, K.R.: Modeling Helping Behavior in an Intelligent Tutor for Peer Tutoring. In: Proceedings of AIED, pp. 341–348 (2009)
15. Magnisalis, I., Demetriadis, S., Karakostas, A.: Adaptive and Intelligent Systems for Collaborative Learning Support: A Review of the Field. IEEE Transactions on Learning Technologies 4(1), 5–20 (2011)
16. Soller, A., Martinez, A., Jermann, P., Muehlenbrock, M.: From Mirroring to Guiding: A Review of State of the Art Technology for Supporting Collaborative Learning. International Journal of Artificial Intelligence in Education 15, 261–290 (2005)
17. Sandoval, W.A., Millwood, K.A.: The quality of students' use of evidence in written scientific explanations. Cognition and Instruction 23, 23–55 (2005)
18. Dillenbourg, P., Baker, M., Blaye, A., O'Malley, C.: The evolution of research on collaborative learning. In: Spada, E., Reiman, P. (eds.) Learning in Humans and Machine: Towards an Interdisciplinary Learning Science, pp. 189–211. Elsevier, Oxford (1996)
19. O'Donnell, A.M., Dansereau, D.F.: Scripted cooperation in student dyads: A method for analyzing and enhancing academic learning and performance. In: Interaction in Cooperative Groups: The Theoretical Anatomy of Group Learning, pp. 120–141. Cambridge University Press (1992)
20. Weinberger, A., Stegmann, K., Fischer, F., Mandl, H.: Scripting argumentative knowledge construction in computer-supported collaborative learning environments. In: Scripting Computer-Supported Collaborative Learning: Cognitive, Computational and Educational Per-spectives, pp. 191–211. Springer, NY (2007)
21. Rummel, N., Spada, H.: Can people learn computer-mediated collaboration by following a script? In: Fischer, F., Kollar, I., Mandl, H., Haake, J. (eds.) Scripting Computer-Supported Collaborative Learning, pp. 39–55. Springer, New York (2007)
22. Kollar, I., Fischer, F., Slotta, J.D.: Internal and external collaboration scripts in web-based science learning at schools. In: Computer Supported Collaborative Learning 2005: The Next 10 Year!, pp. 331–340. LEA, Mahwah (2005)
23. Suthers, D.: Representational Support for Collaborative Inquiry. In: Proceedings of the Thirty-Second Annual Hawaii International Conference on System Sciences (1999)
24. LAMS International: LAMS, Learning activity management system (2009), http://www.lamsinternational.com/ (accessed February 2, 2010)
25. Reload: Reload Learning Design Editor (2005), http://www.reload.ac.uk/ldeditor.html (accessed April 11, 2011)
26. Hernández-Leo, D., Villasclaras-Fernández, E.D., Asensio-Pérez, J., Dimitriadis, Y., Jorrín-Abellán, I.M., Ruiz-Requies, I., Rubia-Avi, B.: COLLAGE: A collaborative Learning Design editor based on patterns. Educational Technology & Society 9(1), 58–71 (2006)
27. SLeD: Service Based Learning Design Player (2005), http://sled.open.ac.uk/ (accessed April 11, 2011)
28. Coopercore: The IMS LD Engine (2008), http://coppercore.sourceforge.net/ (accessed April 11, 2011)
29. IMS: IMS Learning Design v1.0 Final Specification (2003), http://www.imsglobal.org/learningdesign (accessed February 2, 2010)
30. Dillenbourg, P.: Over-scripting CSCL: the risks of blending collaborative learning with instructional design. In: Three Worlds of CSCL. Can We Support CSCL, pp. 61–91. Open Uni-versiteit, Nederland (2002)

31. Dillenbourg, P., Jermann, P.: Designing Integrative Scripts. In: Scripting Computer-Supported Collaborative Learning: Cognitive, Computational and Educational Perspectives, pp. 275–301. Springer, NY (2007)

32. Brusilovsky, P., Millán, E.: User Models for Adaptive Hypermedia and Adaptive Educational Systems. In: Brusilovsky, P., Kobsa, A., Nejdl, W. (eds.) Adaptive Web 2007. LNCS, vol. 4321, pp. 3–53. Springer, Heidelberg (2007)

33. Masthoff, J.: Design and Evaluation of a Navigation Agent with a Mixed Locus of Control. In: Cerri, S.A., Gouardéres, G., Paraguaçu, F. (eds.) ITS 2002. LNCS, vol. 2363, pp. 982–991. Springer, Heidelberg (2002)

34. Davidovic, A., Warren, J., Trichina, E.: Learning benefits of structural example-based adaptive tutoring systems. IEEE Transactions on Education 46(2), 241–251 (2003)

35. Read, T., Barros, B., Bárcena, E., Pancorbo, J.: Coalescing Individual and Collaborative Learning to Model User Linguistic Competences. UMUAI 16, 349–376 (2006)

36. Bull, S., Britland, M.: Group Interaction Prompted by a Simple Assessed Open Learner Model that can be Optionally Released to Peers. In: Brusilovsky, P., Papanikolaou, K., Grigoriadou, M. (eds.) Proceedings of Workshop on Personalisation in E-Learning Environments at Individual and Group Level (PING), User Modeling (2007)

37. Introne, J., Alterman, R.: Using Shared Representations to Improve Coordination and Intent Inference. UMUAI 16, 249–280 (2006)

38. Alexander, C.: The Timeless Way of Building. Oxford University Press, New York (1979)

39. Alexander, C., Ishikawa, S., Silverstein, M.: A Pattern Language: Towns, Buildings, Construction. The Oxford University Press, New York (1977)

40. Karakostas, A., Demetriadis, S.: Adaptation Patterns as a Conceptual Tool for Designing the Adaptive Operation of CSCL Systems. Educational Technology Research and Development 59(3), 327–349 (2011)

41. McConnell, J.: Active and cooperative learning. In: Analysis of Algorithms: An Active Learning Approach. Jones & Bartlett Pub. (2001)

42. Topping, K.: Peer assessment between students in colleges and universities. Review of Educational Research 68, 249–276 (1998)

43. Falchikov, N.: Learning together: Peer tutoring in higher education. Routledge Falmer, London (2001)

44. Liu, C.C., Tsai, C.M.: Peer assessment through web-based knowledge acquisition: tools to support conceptual awareness. Innovations in Education and Teaching International 42, 43–59 (2005)

45. Anderson, L., Krathwohl, D.: A Taxonomy for learning teaching and assessing: A revision of Bloom's taxonomy of educational objectives. Wesley Longman, New York (2001)

46. Karakostas, A., Demetriadis, S.: Systems for Adaptive Collaboration Scripting: Architecture and Design. In: Adaptive Collaboration Support Proceedings, 5th International Conference on Adaptive Hypermedia and Adaptive Web-Based Systems, Hannover, Germany, 29 July-August 1, pp. 7–11 (2008)

Adapting the Collaborative Strategy 'Students Team Achievement Divisions' in an Information Technology Work Place

Maria Kordaki[1], Thanasis Daradoumis[2], Dimitrios Fragidakis[1], and Maria Grigoriadou[3]

[1] Department of Cultural Technology and Communications, University of the Aegean, Mytilene, Greece
m.kordaki@aegean.gr, D.Fragidakis@nsk.gr
[2] Department of Cultural Technology and Communications, University of the Aegean, Mytilene, Greece / Department of Computer Sciences, Multimedia and Telecommunications, Open University of Catalonia, Barcelona. Spain
adaradoumis@uoc.edu
[3] Department of Computer Science and Telecommunications, National and Kapodistrian University of Athens, Athens, Greece
gregor@di.uoa.gr

Abstract. This paper presents an innovative description and an initial implementation of the "Students Team Achievement Divisions (STAD)" collaboration method (Slavin, 1978), in the form of an online adaptive collaborative design-pattern that has been constructed taking into account adaptation techniques, within the context of an open-source learning design-based environments such as the LAMS system (Dalziel, 2003). This method is described with special reference to the learning of essential aspects of an Information System. The innovative description of the aforementioned collaborative method within the LAMS system is based on the fact that: (a) the tasks assigned to the groups consist of investigation of real world scenarios, and not merely the study of learning material as is usually proposed, (b) adaptive techniques are integrated with the method and (c) for the design of the collaborative learning activity, an intuitive learning design tool such as the LAMS system is used. A research study was also conducted aiming the development of an empirical model to allow the implementation of the aforementioned adaptive STAD collaborative method within the context of an IT work place, namely; the Legal Council of the Hellenic State. In fact, the data gathered from this study were used to build the initial learning profile of the user –that is needed for the implementation of Phase 2 of the previously mentioned adaptive STAD collaborative method- so that to be able to provide him/her personalized training, monitoring, scaffolding and evaluation.

1 Introduction

Research in e-learning shows that involving learners in online collaborative learning activities could provide them with essential challenges to: extend and

T. Daradoumis et al. (Eds.): Intelligent Adaptation & Personalization Techniques, SCI 408, pp. 131–153.
springerlink.com   © Springer-Verlag Berlin Heidelberg 2012

deepen their learning experiences, try new ideas and improve their learning outcomes (Roberts, 2005), motivate active involvement in their learning (Scardamalia & Bereiter, 1996), trigger their cognitive processes (Dillenbourg, 1999), enhance their diversity in terms of the learning concepts in question (Johnson & Johnson, 1999) and develop a sense of community and belonging online (Haythornthwaite, Kazmer, Robins & Shoemaker, 2000). Computer-supported collaborative learning has been recognized as an emerging paradigm of educational technology (Koschmann, 1996).

To encourage teams to achieve effective collaboration, some structuring may be useful (Diggelen & Overdijk, 2009). To this end, it is proposed that using computer-supported collaborative design patterns is essential. A pattern is seen as something that will not be reused directly but can support the informed teacher in building up their own range of tasks, tools or materials, drawing on a collected body of experience (McAndrew, Goodyear & Dalziel, 2006). Specifically, best pedagogical practices can be reflected in the formation of context-free 'design patterns' which could be shared across learning contexts and subject domains and essentially support online learning.

The concept of specific collaborative patterns could be well integrated into 'learning design'-based e-learning environments. A 'learning design' has been defined as the description of the teaching-learning process that takes place in a unit of learning, e.g. a lesson or any other learning event, such as a specific collaboration structure (Koper & Tattersall, 2005). An important aspect implied within this definition is that teaching strategies could be conceptually abstracted from context and content, so that excellent instructional models can be shared and reused across educational contexts and subject domains. A 'learning design' represents the sequence of learning activities that need to be performed by teachers and learners within the context of a unit of learning. Within the context of 'learning design', the role of collaborative design patterns is to explicitly indicate the flow of collaboration activities using specific collaboration structures. To this end, LAMS (Dalziel, 2003), is a well-known open source e-learning system that could effectively support the idea of 'learning design'. Recently, a number of collaboration design patterns have been constructed within LAMS (Kordaki and Siempos, 2009; 2010; Kordaki, Siempos & Daradoumis, 2011).

Modern constructivist learning theories suggest that learning is an active, constructive and subjective activity (Jonassen, 1999). In the context of these theories, 'learning design' attempts have to seriously take into account individual learner differences in knowledge, skills, goals and preferences, and each individual learner's needs. Learners could therefore be supported in understanding the learning concepts in question when e-learning is coupled with appropriately-designed adaptation techniques (Brusilovsky, 1996). Specifically, the use of adaptation techniques within e-learning systems can support each individual learner, taking into account some of their individual characteristics, i.e., learning styles, background knowledge including her/his alternative views and misconceptions about the learning concepts in question as well as their experience, goals and preferences. The architecture of adaptive e-learning systems usually consists of the 'learners' model', the 'subject matter model' - or expert model - and the 'learning model'. The first one consists of the aforementioned individual characteristics for each learner, the subject-matter model contains the aspects of the knowledge domain deemed appropriate for students' learning and the learning model consists of the instructional methods –including adaptation strategies- proposed as appropriate for the learning of the subject matter, e.g. the use of specific collaboration strategies.

Taking into account all the above, we have attempted to transform the "Students Team Achievement Divisions" (STAD) collaborative method (Slavin, 1978) into an adaptive collaborative design pattern within the context of LAMS (Kordaki and Grigoriadou, 2010) to construct a sequence of learning activities for the learning of essential issues in Information Systems (IS) aiming at: (a) the realization of the importance and the reasons of using IS in big industries and organizations, (b) the familiarization of students with the practical uses of Computer Science (CS) and especially of the uses of IS, and (c) the realization of the wide impact of CS in the human life and especially in the work place. An initial proposal of the aforementioned sequence of online, adaptive and collaborative learning activities for the learning of IS using the STAD method within LAMS has been made in an IT work place, namely the Legal Council of the Hellenic State. Such an empirical research study has not yet been reported.

In the context of an IT work place, learners-employees may learn better and more effectively through an exploratory learning methodology and through collaboration in small groups, exchanging their personal experiences, helping each other, and learning/obtaining new knowledge together through experimentation, exploration, discovery, problem solving and critical thinking. To achieve effective collaboration we need to employ the STAD collaborative learning strategy with adaptive capabilities in order to provide learners with personalized training, monitoring, scaffolding and evaluation. Our ultimate aim is to develop a research methodology and an empirical model that will allow the implementation of an adaptive and intuitive learning system that will be based on the STAD collaborative learning strategy and which will be used to provide personalized training to learners-employees who are users of an IS and wish to obtain more knowledge about particular aspects of it.

The essential features of LAMS are briefly presented in the following section of this paper, followed by a brief description of the STAD collaboration method. Then, a sequence of online, adaptive, collaborative learning activities using STAD-within-LAMS with special reference to the aforementioned issues of learning about IS is presented. Next, a research study is reported which addresses the way the 2nd phase of the STAD method should be implemented within a work place. Subsequently, the design of this sequence and of the results emerged from the aforementioned empirical study are discussed. Finally, conclusions and future research plans are drawn.

2 Background

2.1 LAMS

LAMS (Learning Activity Management System; http://www.lamsfoundation. org/) is an open-source tool for designing, managing and delivering online collaborative learning activities. In fact, LAMS is a revolutionary environment that can support easy and intuitive learning design – appropriate for the learning of concepts within any subject domain - especially for professionals with no programming experience and knowledge, as is the case with most teachers at primary and secondary level (Cameron, 2007). Teachers are also provided with the ability to 'Preview' the learning activity sequences through the lens of a learner and make suitable adjustments after reflection (Cameron, 2006). In addition, LAMS provides teachers with the chance to overview the entire sequence of learning activities on the computer screen and make appropriate revisions (Cameron, 2007). Furthermore, there are also possibilities for improvement of a sequence even while it is running online in real-time. It is also worth noting that, in the context of LAMS, the role of the teacher is not reduced to the role of a traditional behaviorist practitioner (Skinner, 1968) who necessarily uses 'learning designs' ready-made by expert designers: in fact, LAMS provides teachers with possibilities to transform ready-made sequences of learning activities according to both their own personal views of learning and their students' individual learning characteristics. Collaboration could also be easily supported by using the possibilities of fine-grained grouping and branching. Within LAMS there are also possibilities for adapting a sequence of learning activities according to students' previous knowledge, their preferences and specific learning styles, by using appropriately-designed questionnaires in combination with suitable grouping and branching. Efficient learning design patterns could also be accessed by teachers using the Activity Planner integrated within LAMS. Various generic 'blank' learning sequences representing efficient collaboration learning strategies are also available from/for members of the LAMS community (Kordaki and Siempos, 2009; 2010), http://www.lamscommunity.org/lamscentral/). To this end, the community of learners built around LAMS could prove encouraging to teachers and designers of learning activities by providing them with opportunities to exchange experience and knowledge as well as their own sequences of learning activities.

LAMS also offers designers of educational activities specific tools that support grouping and conditional branching. In fact, the grouping can be random or based on either learner's or author's choice. Additionally, students can be directed to different sequences of activities depending on the group they belong to (grouped branching) or based on what the learner has contributed to a specific activity (tool output branching). LAMS can make branching decisions based on criteria such as the number of correct answers in a questionnaire, or the certain words that a learner has or has not typed into a chat, forum or survey activity. In any case, the author of the learning activity can assign students manually to any branch he likes. Nevertheless, Dalziel (2003) has commented on the absence of tools supporting broader ranges of collaborative tasks and, despite the availability of all the tools mentioned above, sequences of learning activities for the STAD collaboration method within LAMS – using adaptation techniques - for the learning of specific IS concepts have not yet reported.

The said sequence of collaborative activities was implemented using specific tools provided by LAMS[1]. These tools are demonstrated in its interface and are briefly presented below:

The *Assessment tool* allows authors to create a series of questions with a high degree of flexibility in total weighting. The *Chat Activity* runs a synchronous discussion for learners while the *Scribe Activity* is used for collating the chat group's views on questions posed by the teacher. The *Forum Activity* provides an asynchronous discussion environment for learners, with discussion threads initially created by the teacher and the *Scribe Activity* is also used for collating Forum Postings into a written report. The *Mindmap activity* allows teachers and learners to create, edit and view mindmaps in the LAMS environment. The *Multiple Choice activity* allows teachers to create simple automated assessment questions, including multiple choice and true/false questions. The *Noticeboard Activity* provides a simple way to supply learners with information and content. The activity can display text, images, links and other HTML content. The *Question and Answer Activity* allows teachers to pose a question or questions to learners individually and, after they have entered their response, to see the responses of all their peers presented on a single answer screen. The *Share Resources* tool allows teachers to add content to a sequence, such as URL hyperlinks, zipped websites, individual files and even complete learning objects. The *Submit Files Activity* allows learners to submit one or more files to the LAMS server for review by a teacher. The *Survey Tool* presents learners with a number of questions and collects their responses. However, unlike Multiple Choice, there are no right or wrong answers. The *Wiki Tool* allows authors to create content pages that can link to each other and, optionally, allow learners to make collaborative edits to the content provided.

2.2   Student – Teams – Achievement - Divisions (STAD)

STAD (Slavin, 1978) is considered to be one of the basic approaches to introducing learners to cooperative learning. The use of this method is thought of as an effective and efficient way to teach well-defined educational subjects. The teams are heterogeneous, made up of learners of diverse academic achievement, race, and nationality. The rewarding of the best teams motivates the better students in a team to encourage the other members to achieve their mutual goal.

Goals: 1) to motivate students to encourage and help each other, 2) to accelerate student achievement, 3) to facilitate gains in self esteem, liking of class, 4) to improve behaviour.
Process: 1) Personal assessment, 2) Assignment presentations, 3) Team collaboration, 4) Collaborative writing of reports, 5) Team assessment, 6) Praise for best reports.

---

[1] http://wiki.lamsfoundation.org/%20display/lamsdocs/Home

The design of an adaptive STAD method, as it is proposed by our approach, is described in detail in Section 3.

2.3 Adaptation

A four-stage process has been proposed for the design of an adaptive system (Brusilovsky, 2003): (i) design of the 'knowledge-base', including a hierarchy of learning goals and specific learning topics, (ii) design of the 'learner's model', including her/his individual learning characteristics and preferences, (iii) design of the 'media space', including various materials and topics which are interconnected with the topics included in the previously mentioned knowledge-base, and (iv) design of the 'adaptation model', including the rules for the selection of appropriate topics - from both the knowledge-base and the media space - taking into account each learner's individual characteristics as they emerge from the relevant 'learner model'.

As regards the construction of the 'learner model', the learner's profile in terms of her/his knowledge-background and experience, goals, preferences and learning style must be investigated. To this end, the learners' background and experience with regard to the knowledge in question is useful to explore, because learners with different backgrounds need different treatment from the system. In fact, the learner's knowledge has to be diagnosed before they can be characterized as novices, intermediate or experts; their goals and preferences should also be examined. The basic learning style of each individual learner also plays a fundamental role in finding ways to support them in their learning. In terms of learners' learning styles, various classifications have been proposed. Some important classifications separate learners as field-dependent (F/D) and field-independent (F/I; Witkin, Moore, Goodenough & Cox, 1977), some view learners as holistic-analytic and verbalizers-imagers (Riding & Cheema, 1991), while others sort learners into activists, pragmatists, reflectors and theorists (Honey & Mumford, 1992). These individual characteristics are usually explored through the completion of a questionnaire immediately after their entrance into the e-learning system.

A system can be adapted in various ways to support learners in their learning, namely: (a) adaptive presentation, and adaptive navigation techniques (Kay & Kummerfeld, 1997; De Bra & Calvi, 1998) - these techniques are usually proposed for the design of adaptive hypermedia educational materials, where sequences of web pages are created and the adaptation could be implied at both content level and link level, (b) adaptive curriculum sequencing, where sequences of educational materials are formed and proposed to the learner by the system according to her/his individual characteristics (Brusilovsky & Pesin, 1994), (c) problem solving support - here, two, three modes of support have been reported: (i) intelligent solution analysis, where the ideal solution to a problem is compared with the solution provided by the learner and appropriate feedback is given by the system, after the problem-solving process has been completed, (ii) interactive problem solving support - here, the system monitors the learner's problem-solving path and provides appropriate feedback during the problem-solving process, and

(iii) example-based problem solving (Brusilovsky, 1996), where an appropriate repository of examples is provided by the system, to support each learner's problem-solving actions, and (d) collaboration support, where the system can use the learners' personal characteristics to support the creation of appropriate groups for collaboration and communication to deal successfully with suitable learning activities (Brusilovsky, 1999).

Adaptive techniques are also useful for the design of tests used for the assessment of learners' knowledge throughout the course of a learning experiment. Such tests are generated by the system according to each individual learner's knowledge. For example, when a learner successfully answers a set of questions - appropriate for the assessment of a piece of knowledge - then the system provides questions aiming to assess another, probably more complicated, piece of knowledge. On the contrary, when a learner does not succeed in answering a set of questions, the system provides her/him with easier questions and various kinds of help.

In the next section, we will describe all the phases of the STAD method in detail, how some of these phases can include adaptivity aspects and how they have been designed within the LAMS system.

3 Design of an Adaptive STAD Online Learning Activity in LAMS

The design of the online STAD learning activity consists of the following phases: 1) Introduction to the learning activity, 2) Personal assessment 3) Adaptive group Formation, 4) Adaptive individual learning and team collaboration, 5) Adaptive group report preparation, 6) Group report presentation, 7) Team assessment, 8) Adaptive individual assessment, and 9) Praise for best group reports. The implementation of these phases within the context of LAMS is diagrammatically represented - as a 'design pattern' - in Fig. 1. The proposed activity can be used in environments of synchronous and asynchronous collaborative learning. The aim of this section is not to describe the details of each phase which is a long and under construction endeveour. Instead, we provide a brief description of each phase and describe phase 2 in detail, which constitutes an important step to achieve the adaptivity of the STAD method.

3.1 Phase 1: Introduction to the Learning Activity

The main aim of this educational activity is the encouragement of learners through their interaction within a collaborative learning environment, to explore fundamental issues concerning CS and especially of IS. Additionally, through the learners' efforts to fulfill educational objectives, some secondary skills are developed, e.g. a) the practice of word processing and presentation software, and b) the practice of web searching techniques. The learning activity also aims to highlight the value of collaborative learning as a modern method of teaching.

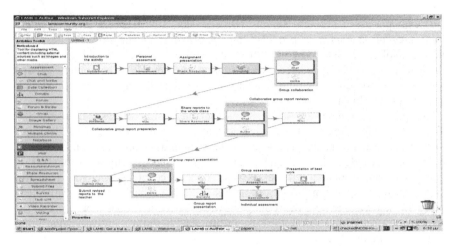

Fig. 1 Implementation of the adaptive STAD as a design pattern within LAMS

To fulfill the aforementioned learning aims, students have to collect diverse data types by visiting significant areas of life where IS are used, such as: (a) a financial organization, e.g. a bank, (b) a private company, (c) a public organization, (d) a health care organization, e.g. a hospital, e) a university lab specializing in IS, and (f) an IS development company.

Here, it is worth noting that the usual process of data collection in school environments is the study of given literature or web searching. In fact, pragmatists can learn through practical activities – e.g. in a financial organization - where they can observe how IS are applied; activists can gain knowledge by being involved in interactive learning activities (e.g. in a private company), reflectors can learn through various IS examples – e.g. reflecting on their experience within a university IS lab - and theorists can absorb knowledge through exploring theoretical materials available in a university CS department. Other appropriate types of content could also be provided for online study.

By visiting the aforementioned real life areas, students should be able to collect data on the following issues regarding each specific IS in use: (a) needs covered and solutions given, (b) specifications, (c) cost, (d) benefits, (e) infrastructure, in terms of types and number of computers as well as kind of networks used, (f) updates performed, (g) validity, (h) maintenance and support, (I) possible/necessary future improvements, and (j) the hiring and firing needed after the use of the IS at hand.

Students should ask the users of each specific 'IS' about such interesting issues as: (a) their background of knowledge about computers, and (b) how their jobs - in terms of tasks, health and socialization - have been changed due to the evolution of the specific IS in their work place. By visiting a company that develops IS, students should ask questions regarding: (a) stages of the development of an IS, (b) initial specifications, (c) programming languages used, (d) debugging, (e) documentation, (f) operator-training, and (g) support. In this phase of the STAD activity, students are informed - using a Notice board - about the whole context of the activity, including its aims and the specific issues of IS that have to be studied during this activity as well as the various places where they could collect appropriate data. Students should exchange ideas on the whole procedure of the activity using a whole-class Forum/ Chat-room.

3.2 Phase 2: Personal Assessment

The proposed design utilizes the 'Assessment' tool for the investigation of students' main learning styles and their basic preferences (in terms of areas of life about which they like to collect information on IS). Our work in this chapter focuses on implementing an approach that is able to determine students' preferences, and most importantly, to explore a variety of learning styles and assign to each student his/her primary and secondary learning styles that best characterize them. This approach is described in detail in Section 4.

We pay a particular attention to an effective and accurate identification of students' learning styles, since this is both an important prerequisite to build a robust student learning profile and a criterion to assign students to working groups. Certainly, another important aspect of the student learning profile and group formation is student's preferences, though in this chapter we deal only with some basic preferences, since it constitutes a separate long issue.

The construction of a rich and well-structured student profile is crucial for achieving adequate adaptations in several phases of the application of the STAD method, so other aspects that our approach wants to incorporate in the student profile are student's specific preferences of the subject matter, student's knowledge and experience, student's interests and abilities, and other aspects that concern everyday working tasks, socialization habits and health issues. All these issues constitute interesting research topics and are currently being investigated by our work.

3.3 Phase 3: Adaptive Group Formation

Students have to be grouped in such a way so that they form heterogeneous groups of 5-6 members which are best adapted to the students' needs and goals. To achieve the best adaptation to these prerequisites, we use a variant of the *collaboration support* adaptive method to instruct the system to consider two basic criteria: assign students to a group with the same primary learning styles but with different preferences as these are depicted in each student learning profile constructed in the previous phase. As such, on the one hand, students feel comfortable in a group that all its members have a similar way of learning so the feeling is that they will collaborate and learn better. On the other hand, the fact that the group respects their personal preferences, this satisfies their personal needs and goals as well, since work and learning will focus on these particular needs while the diversity of preferences enhances discussion, interactivity, willingness to learn from one another, all in all it promotes collaboration!

The formation of such groups is supported by the system, using the grouping tool in combination with the branching tool. In case of dispute, the students who prefer to be in another group could ask the educator to assign them accordingly.

3.4   Phase 4: Adaptive Individual Learning and Group Collaboration

When each group is formed, its members should visit the specific areas of life mentioned in Phase 1 'Introduction to the learning activity', where IS are used, to collect specific data. The system can advise the students of each group on the selection of a specific area by using the data of the student profile referring to their particular preferences and primary learning styles (Papadimitriou, Grigoriadou & Gyftodimos, 2008).

As such, each group member sets a specific goal to accomplish and the system instructs him/her how to carry it through in the most adequate and personalized way. To do so, the system is based on specific aspects of the student's learning profile - such as the student's specific preferences of the subject matter, student's knowledge and experience, student's interests and abilities, and other aspects that concern everyday working tasks, socialization habits and health issues - in order to design the learning path which is most adapted to the student's learning profile.

In other words, using these aspects, the system applies a variant of the *interactive problem solving support* adaptive method to direct, supervise and monitor student's work and learning in the area of his/her preference and to provide appropriate feedback during the learning path process. In particular, the system can recommend the student in which topics should focus his/her study, what material to search (providing also material that the system itself has available), which tasks should carry out and how, which skills and competencies needs to work out or develop in order to succeed in his/her workplace, which socialization habits needs to change, and finally which health issues needs to take into account.

Then, when students come back to their group they should join their forces to achieve the goal of the proposed learning activity through participation in group work. To this end, students in each group should first share the information they collected and the knowledge acquired on the specific area of life they visited during their personal work. The system is assisting students to form small subgroups of two or three members by identifying common points in their profile, so that they perform a focused work on specific topics of the activity. Doing so, collaborative work and learning can be adapted using the *adaptive curriculum sequencing* technique, according to which the system can propose adequate sequences of complementary educational materials and tasks to each subgroup of learners.

Finally, all members get together in their group and exchange their expertise and experiences through a debate (using a forum or a chat room) which is monitored by a coordinator member assigned by the group itself, aiming at completing the proposed learning activity. Tools like a Mind or Concept Mapping tool may be used to identify key ideas and organize appropriate data categories.

3.5   Phase 5: Adaptive Collaborative Group Report Preparation

In this phase, students have to work together in their group to prepare a report by using a wiki tool. Then, they have to send this report to other review groups to receive appropriate comments. The selection of the adequate review groups to send the report is guided by the system that uses a variant of the *collaboration support* adaptive method to support the selection of appropriate groups for collaboration and communication, according to the groups' working profile. Doing so, it is more probable that the selected groups can provide suitable and constructive comments to the report. Subsequently, each group should assess the quality of the evaluation comments of the review groups on their own report and can decide whether to take these comments into account and to reply to the review groups accordingly (Gouli, Gogoulou & Grigoriadou, 2008). Finally, groups should submit their corrected reports to their supervisors.

3.6   Phase 6: Group Report Presentation

Here, it would be useful to provide students with some recommendations as to how to prepare and deliver a good presentation. Some useful guidelines for the former are: (a) the presentation must begin with the main idea of the subject, (b) only the key points of the subject should be presented, (c) on every slide, only 4-5 key points should be presented, (d) a uniform style of presentation must be followed (unnecessary effects must be avoided since these distract the learner from the key concepts), (e) the duration of each presentation should be around 10 minutes (for synchronous presentation) since there is always the danger that the students may get bored. There will be additional time for further discussions. Online presentations could be performed by each group, using a whole-class chat or forum or a videoconferencing tool.

3.7   Phase 7: Team Assessment

During the online presentation, the supervisor can initiate a 'question and answer' session to encourage students to assess the work and the presentation of each team. Each successful answer is annotated in the group's working profile, which finally provides a means to classify groups according to their performance.

3.8   Phase 8: Adaptive Individual Assessment

In this phase, each student should be given an adaptive quiz that aims at assessing the student individually and thus conclude the learning activity. The quiz is adapted according to two main criteria: the student's learning profile which has been updated during the whole learning process and the performance success of the student's group that was built up in the previous phase. Thus, the supervisor can use the "assessment" tool in combination with branching techniques - based on tool output branching capabilities of LAMS - to design suitable questions for students of different levels of knowledge, different learning goals, etc. The question types can be of multiple choice, true-false and open types.

Questions are generated one-by-one, so that the creation of the next question depends on the success of the previous answer. As such, in case of successful answers, the system keeps on presenting more questions until all the topics are covered, whereas in case of a failure the system provides the student various kinds of help.

3.9 Phase 9: Praise for Best Group Reports

In this final phase, the best group work - as assessed by their colleagues and the supervisor - will be published with honors.

4 Implementing the 2nd Phase of the Adaptive Collaborative Method STAD: Determining Learners-Employees' Learning Styles within the Context of an IT Work Place

4.1 Aims of the Approach

All the phases that constitute an adaptive STAD method have been described in the previous section. This section focuses on the implementation of the 2nd phase of STAD. This is an important step in our effort to build a complete student learning profile that is crucial for achieving suitable adaptations for the full implementation of STAD. As such, we initially need to extract knowledge about two main issues that concern the learner-employee profile: first, the learner-employees' main and secondary learning styles, and second the learners' preferences as regards the Information System (IS) of a specific working environment that the learner would prefer to experience and learn about. The aim of the current study focuses more in identifying and classifying employees' learning styles in an IS-based working environment, though some basic employee's preferences and interests are also explored.

This work took place in the central department of the Legal Council of the Hellenic State (LCHS). Employees belong to several departmental sections and are all users of the IS. To find the employees' learning styles, preferences and interests regarding IS aspects, we designed two specific questionnaires, both of them with closed-type questions, following a five-point Likert scale. Both questionnaires were given to the employees and those that were answered back were then evaluated and commented subsequently. The following sections describe the methodology and results obtained for the case of learning styles. More specifically, Section 4.2 provides the context where our research work took place, and explicitly sets the research objectives pursued. Next Section describes the methodology employed and the user aspects that were considered and evaluated. Section 4.4 presents the results of the survey, using adequate graphs and tables, as well as the explanation and interpretation of these results in detail. Then, in Section 4.5 we draw the final conclusions.

Both learning styles and preferences were used as the criteria to construct the learning teams. Each learning team consists of 5-6 members who have the same learning styles but different preferences. These results constitute the input data to build a complete learner profile module, which forms an important component of our adaptive, collaborative learning/training model. The final Section outlines the work done and makes reference to on-going and further work.

4.2 The Framework of the Legal Council of the Hellenic State

The Legal Council of the Hellenic State (LCHS) is a single supreme authority of the State and directly attached to the Minister of Economy and Finance. The duties of the Legal Council of the Hellenic State are the following:

- The legal advocacy of the interests of the Hellenic State and its legal support generally
- The support of the Greek Republic to the European Court of Human Rights and the European Court of Justice
- The support of the Administration Services on the Community law
- The guidance of the administration services with legal opinions
- The recognition of the claims against the government, the compromise of the disputes with the government and the adjustment of the debts of the debtor to the state
- To issue legal opinions for the guidance of the public administration
- The advocacy of the civil servants in criminal courts
- The advocacy of the public administration to the courts
- The processing of the legislation which are assigned to the Legal Council of the Hellenic State

The departments which participated in this work were the following: the personnel department, the accounting department, the secretariats, the registration department, the department of scanning and document processing, and the IT department. They are depicted in Table 1.

Table 1 Description of the departments of LCHS participated in the research experiment

ID	DEPARTMENT'S DESCRIPTION	EMPLOYEES	E-Q	E-Q/ EMPLOYEES %
1	The Personnel Department	14	12	85.71
2	The Accounting Department	12	9	75.00
3	The Secretariats	22	17	77.27
4	The registration department	3	2	66.67
5	The department of scanning & document processing	13	12	92.31
6	The IT department	3	2	66.67
	AMOUNTS	67	54	**80.6**

The above table presents the following Fields/Results: The department id and description are presented in the first two columns. The third column contains the number of the employees which are in the department, while the fourth column "E-Q" contains the valid processed questionnaires. Finally the last column estimates the percentage between "E-Q" / "EMPLOYEES".

4.3 Research Tools and Methodology

To build a well-grounded methodology for identifying valid user learning styles and preferences and further build a solid user profile, we were based on existing research concerning learning styles and preferences (Riding & Cheema, 1991; Honey & Mumford, 1992; Witkin, Moore, Goodenough & Cox, 1977; Liu & Ginther, 1999; Wyss, 2002). We identified seven most referenced and used learning styles, as they described below.

On the one hand, to extract information about possible learning styles that characterize the employees of an IS and, on the other hand, to find out their particular preferences about an IS and identify their interests about which components and uses of an IS they would like to learn about, a specific questionnaire survey was designed, including two different questionnaires. Both of them included closed-type questions using a five-point Likert scale. The final aim was to achieve an initial learner profile which can be used to implement the rest phases of the STAD method in an adaptive manner.

In sum, the two questionnaires evaluated the following 10 user aspects, as follows:

1. Some **personal characteristics** like their gender, education, age, etc.
2. The user learning style as a *field independent - field dependent* person
3. The user learning style as an *Imager* person
4. The user learning style as a *Verbalizer* person
5. The user learning style as an *Activist/Pragmatist* person
6. The user learning style as a *Reflector* person
7. The user learning style as a *Holistic/Analytic* person
8. The user learning style as a *Theorist* person
9. **User preferences** about the IS of a particular organization he/she would prefer to visit and the specific aspects of this IS that he/she wants to explore further.
10. **User interests** about particular components and uses of the IS that he would like to learn, improve and apply in his/her own working environment.

The learning styles that were considered are briefly described below:

1. **Field independent - field dependent:** The Field Independent learner excels in classroom learning that involves analysis, attention to details, and mastering of exercises, drills, and other focused activities. The Field Dependent learner, by contrast, seems to achieve a higher degree of success in everyday language situations beyond the constraints of the

classroom; he/she deals with tasks requiring interpersonal communication skills (Witkin, Moore, Goodenough & Cox, 1977; Wyss, 2002).

2. **Imager:** The user learns more when he carefully observes charts, graphs, timelines, films and presentations (Riding & Cheema, 1991).

3. **Verbalizer:** Verbalizers learn better with text-based methods of instruction and they prefer to process information by verbal-logical means (Riding & Cheema, 1991; Liu & Ginther, 1999)

4. **Activist / Pragmatist:** They involve themselves fully and without bias in new experiences and their philosophy can be summarized as 'I'll try anything once' (Honey & Mumford, 1992).

5. **Reflector:** These students prefer to think before acting (Honey & Mumford, 1992).

6. **Holistic / Analytic:** They prefer logical approaches and they want to learn when the material is a step by step during the learning progress (Honey & Mumford, 1992).

7. **Theorist:** Theorists learn best when they are put in complex situations where they have to use their skills and knowledge (Honey & Mumford, 1992).

Based on the above descriptions, the questions used for the investigation of the employees' learning styles are listed below while the questionnaires investigating the learners-employees' preferences regarding IS will be presented along with the results emerged from the empirical study in the next section.

Table 2 Questionnaire used for the investigation of LCHS learners-employees learning styles

Questionnaire used for the investigation of employees' learning styles	
s/n	Questions
	1. Field independent-field dependent
1.1.1.	I face some problem of concentration in an environment with noise and confusion.
1.1.2.	I like to analyze the structures of a problem.
1.1.3.	I feel that I have to understand every single word of what I read or hear.
1.1.4.	I believe that the comprehension of a program is more effective in the labour environment.
1.1.5.	I prefer to work alone despite with other persons.
1.1.6.	The reception of feedback in my labour environment by other persons really does not have repercussions in my learning.
	2. Imager
1.2.1.	Regularly I ask persons with regard to their basic admissions.
1.2.2.	I am charmed more by original, non-conventional ideas rather than practical ideas.
1.2.3.	I tend to be fascinated by technical analysis, planning and prevention (eg network analysis, flow diagrams, branching programs, contingency planning).
1.2.4.	I like being able to correlate current operations with a big picture long term.
1.2.5.	I find stifling the formal process of the definition of the specific objectives and drawings.
1.2.6.	I always feel boredom from methodical and detailed work.

Table 2 (*continued*)

	3. Verbalizer
1.3.1.	I like to learn an application by studying mainly the user manual and afterwards to apply it.
1.3.2.	I consider that i need to understand the logic of the job-process in a department and afterwards to involve in its information system.
1.3.3.	In discussions I like to be communicative and persuasive with my interlocutors.
1.3.4.	I like to draw diagrams in my answers.
	4. Activist / Pragmatist
1.4.1.	I prefer to have a lot of information sources – as many data to study as better.
1.4.2.	I suggest in meetings practical and realistic ideas.
1.4.3.	I tend to discuss certain things with people rather than taking part in vague discussions.
1.4.4.	I prefer people who approach things realistically rather than theoretically.
1.4.5.	I tend to reject irrational, spontaneous ideas as impractical.
	5. Reflector
1.5.1.	I watch carefully every detail and afterwards i reach to a conclusion.
1.5.2.	I don't produce ideas spontaneously.
1.5.3.	I am careful to don't make hasty conclusions.
1.5.4.	I believe that decisions based on detailed analysis of all information is safer than those based on intuition.
1.5.5.	I prefer to stand back from a situation and to study all perspectives.
	6. Holistic / Analytic
1.6.1.	When I hear about a new idea or approach I immediately begin to study the practical part of the application.
1.6.2.	I am careful for the interpretation of available data and i get used to avoid hasty conclusions
1.6.3.	I prefer to come to a decision carefully considering several alternatives.
1.6.4.	I believe that the problems are faced with rational, reasonable thoughts.
1.6.5.	It is best to think carefully before we act.
1.6.6.	I tend to be tough with people when they find difficult to adopt a sensible approach.
	7. Theorist
1.7.1.	I am a passionate for the exploration of the basic assumptions, principles and theories where these are behind things and events.
1.7.2.	I like the meetings to follow a methodical flow, a particular program, etc.
1.7.3.	I avoid subjective and ambiguous issues.

The aforementioned questionnaires were given in civil servants to the LCHS, which work exclusively to departments of information technology. They were sent to 67 employees of this public organization, which they agreed to participate in this study. 54 valid questionnaires were sent back, thus returning a response rate of 80.6%. It is much longer than 20%, which is the average of such studies, as it emerges from earlier research.

Here it is worth noting that a meeting was scheduled with a sample of 5 respondents thus ensuring that all participants comprehended all the questions adequately. A five point Likert scale was used in order to select the answers of the respondents. The structuring of the questionnaire and the wording of the questions was finalized after testing the content validity. On average, it took 20-25 minutes for each respondent to answer the questions of the questionnaire.

The tools that were used for the completion of this research and the export of various conclusions were SPSS version 13.00 and Excel version 2007.

4.4 Presentation-Annotation of the Aggregate Results of the Survey

4.4.1 Comments on Aggregate Results Related to the Employees' Learning Styles

Each questionnaire was separately processed so that to detect the learning style of each person. The survey indicated that employees may have two or three possible learning styles. The aim was to find the primary and secondary (second and third learning style) of all the participating employees in the Legal Council of the Hellenic State. The results are described in the following Table 3:

Table 3 Primary and secondary learning styles of the employees of the IT department of the LCHS

		Primary learning style %	Second learning style %	Third learning style %
1	Field independent-field dependent	1,85	1,85	7,41
2	Imager	0,00	1,85	5,56
3	Verbalizer	24,07	22,22	7,41
4	Activist / Pragmatist	33,33	31,48	11,11
5	Reflector	3,70	11,11	25,93
6	Holistic / Analytic	27,78	22,22	9,26
7	Theorist	9,26	9,26	9,26

Primary and secondary learning styles: The results according to the percentages of the primary and secondary learning styles show that 33.33% of the users (primary learning style) and 31.48% of the users (secondary learning style) are *Activist / Pragmatist*. These results are very important for the organization of the learning/working groups and the Legal Council of the Hellenic State itself, as concerns the way learning and work could be focused more efficiently. In other words, the learning activities that will be planned to be performed by the learners/employees should be adapted to the way Activist/Pragmatist users learn better. Such users may learn better from activities, like:

- New experiences and challenges from which to learn
- Excitement, change and variety
- Techniques that can be directly applicable to their own work
- Techniques for doing things with obvious practical advantages

And what *Activist / Pragmatist* users don't like:

- to have a passive role (lectures, instructions, reading)
- to be observers
- to be required to assimilate, analyse and interpret lots of 'messy' data

Other learning styles that have high ratio are Verbalizer and Holistic/Analytic. Consequently, we summarize the results as follows:

Employees' primary learning styles: As regards the employees' primary learning styles, the results are: Activist/Pragmatist (33.33%), Holistic/Analytic (27.78%), and Verbalizer (24.07%). SUM=85.18%.

Employees' secondary learning styles: As regards the employees' secondary learning styles, the results are: Activist/Pragmatist (31.48%), Holistic/Analytic (22.22%), and Verbalizer (22.22%): SUM=75.92%.

Employees' third learning styles: The results according to the percentages of the third learning style show that 25.93% of the users are Reflectors. These results are also very important for the organization of the learning/working groups and the Legal Council of the Hellenic State itself, since one has to focus on what Reflector users learn best from activities, like:

- are allowed or encouraged to watch / think / ponder on activities
- can carry out careful, detailed research
- can reach a decision without pressure and tight deadlines.

And what *Reflector* users don't like:

- feel 'forced' into the limelight
- must act without time for planning
- given insufficient data on which to base a conclusion

It is also significant to know that 41 persons out of 54 have at least 3 learning styles, being the rate 75.93, whereas 13 persons out of 54 have 2 learning styles, with a rate of 24.07.

4.4.2 Comments on Aggregate Results Related to the Employees Preferences

The percentage on what kind of information system the employees would prefer to visit is as follows:

- Ecological organization (1.85%)
- Financial Organization (12.96%),
- Health Care Organization (11.11%)
- IS Development Company (16.67%)
- Private Company (16.67%)
- Public Organization (12.96%)
- Tourist Agency (1.85%)
- University Lab Specializing in IS (25.93%).

Here, one can observe that the choice with the highest ratio is the "University Lab Specializing in IS". That is, most of the employees (25.93%) prefer to visit a "University Lab Specializing in IS" and learn some aspects about their system and their tasks. The variety of these aspects, which constitute the employees' interests, is investigated further through the second questionnaire.

4.4.3 Employees' Interests about Specific Topics of an IS

Here, it is important to examine the percentages of the proposed questions that were evaluated by the employees with the grades "Agree" and "Strongly Agree" as well as 'Disagree' and "Strongly Disagree'. In Table 4, these percentages present the employees' desires as regards what they would like/not like to mostly learn about an IS.

Table 4 Employees' interests of the IT department of the LCHS about specific topics of IS

Employees of the IT department of the LCHS and their interest about specific topics of IS		
Code	Questions	"Agree" & "Strongly Agree" %
Q2.2.1	I would like to know what kind of service needs are covered	94.44
Q2.2.2	I would like to know how the IS faces up any possible application problems and how it resolves them	72.22
Q2.2.3	I'm interested to know the requirements of the IS	62.96
Q2.2.4	I would like to know the cost of the IS in relation to its potential	59.26
Q2.2.5	I'm interested to know the benefits gained by using the current IS	96.29
Q2.2.6	I want to be informed about the infrastructure of the IS, especially about computer hardware, network, Internet and Servers	68.52
Q2.2.7	I would like to learn about the frequency that new software releases are installed	38.88
Q2.2.8	I would like to learn how information validity is provided by the IS	79.63
Q2.2.9	I would like to be informed about the level of maintenance and support as regards the applications and networks used	51.85
Q2.2.10	I would like to know there is a provision for IS prevention from natural disasters	66.66
Q2.2.11	I would like to know the users' opinion about possible / necessary future improvements	74.07
Q2.2.12	I would like to learn about the usability of the IS	90.74
Q2.2.13	I would like to be informed whether there is interoperability of services / applications with other workplaces	72.22
Q2.2.14	I would like to find out if e-Gov services are provided such as G2C or G2B or G2G	57.41

It is significant to comment the questions that have a ratio up to 70%. We observe that the personnel would mostly like to learn about the following aspects of the preferred IS which they want to visit:

- The kind of service needs which are covered
- The way that an IS faces up application problems and how it resolves them.

- The information validity that is provided by an IS
- Future improvements for an IS
- The usability of an IS

There are also questions which have a low interest ratio, fewer than 60%. The aspects that they refer to are the following:

- The cost of the IS in relation to its potential.
- The frequency that new software releases are installed
- The maintenance and support level that concern the networks and the applications used in the IS.
- The provision of services like G2C or G2B or G2G

5 Summary and Future Work

This paper proposed an innovative approach to provide adaptivity to the STAD collaborative method, through specific adaptation techniques which are applied in some of the STAD phases, within the context of online learning-design based learning. In fact, we described an adaptive online collaborative design pattern of the STAD method that has been formed within LAMS, a well-known open source learning-design based system.

The innovative description of the adaptive STAD collaborative method is based on the fact that: (a) the activity takes place in an online context, (b) the tasks assigned to groups emphasize investigation of real world situations, (c) adaptive techniques are integrated, and (d) an intuitive learning design tool such as LAMS is used for the design of the adaptive collaborative learning activity. The design of the aforementioned pattern has been implemented and illustrated through a specific adaptive collaborative STAD- learning activity that concerns the training of public organization employees in essential issues of Information Systems (IS).To this end, the construction of the users' learning profile which can be used as an important component to provide the adaptivity of the STAD learning activity was viewed as significant. In fact, three aspects of student learning profile were examined here: students' learning styles, preferences and interests. To extract and classify this information, two questionnaires were given to the employees of the IT department of the Legal Council of the Hellenic State, aiming at determining their primary and secondary learning styles, their preferences for a specific IS of a real world organization they would like to visit, as well as the issues of the IS they would be interested to investigate and acquire more knowledge about.

The analysis of the data showed that the most dominant primary and secondary learning styles of these employees are 'Activist'/'Pragmatist', 'Holistic'/ 'Analytic', and 'Verbalizer' while their third learning style is 'Reflector'. Most learners-employees also preferred to visit a University Lab specialized in IS to gain extra knowledge about IS. Finally, the issues of IS that were addressed as most important for most learners-employees are related to: (a) the kind of service needs which are covered, (b) the way that an IS faces up application problems and how it resolves them, (c) the information validity that is provided by an IS, (d) future improvements of IS, and (e) usability of IS.

The results of this study can be used for adaptive team formation, whereas a complete employees' profile can be used to support the rest of the STAD phases, especially adaptive individual learning and team collaboration, group report preparation, group report presentation, team assessment, and adaptive individual assessment. In the adaptive team formation, each learning team consists of 5-6 members who have the same learning styles but different preferences. Future work is still under progress for the complete implementation of all phases of the adaptive STAD design pattern as well as for its evaluation in a real world situation.

References

1. Brusilovsky, P., Pesin, L.: An intelligent learning environment for CDS/ISIS users. In: Levonen, J.J., Tukianinen, M.T. (eds.) Proc. of the Interdisciplinary Workshop on Complex Learning in Computer Environments (CLCE 1994), Joensuu, Finland, May 16-19, pp. 29–33 (1994)
2. Brusilovsky, P.: Methods and techniques of adaptive hypermedia. User Modeling and User-Adapted Interaction 6(2-3), 87–129 (1996)
3. Brusilovsky, P.: Adaptive and Intelligent Technologies for Web-based Education. Künstliche Intelligenz (4), 19–25 (1999),
   http://www2.sis.pitt.edu/~peterb/papers/KIreview.html
4. Brusilovsky, P.: Developing adaptive educational hypermedia systems: from design models to authoring tools. In: Murray, T., Blessing, S., Ainsworth, S. (eds.) Authoring Tools for Advanced Technology Learning Environment, pp. 377–409. Kluwer Academic Publishers, Dordrecht (2003)
5. Cameron, L.: Using LAMS to facilitate an effective program of ICT instruction. In: Proceedings of the 2007 European LAMS Conference: Designing the Future of Learning, pp. 39–49 (2007)
6. Cameron, L.: Picture this: My Lesson. How LAMS is being used with pre-service teachers to develop effective classroom activities. In: Proceedings of the 1st International LAMS Conference 2006: Designing the Future of Learning, Sydney, pp. 25–34 (2006)
7. Dalziel, J.: Implementing Learning Design: The Learning Activity Management System (LAMS). In: Proceedings of ASCILITE Interact, Integrate, Impact, Adelaide, December 7-10, pp. 593–596 (2003)
8. De Bra, P., Calvi, L.: AHA! An open Adaptive Hypermedia Architecture. The New Review of Hypermedia and Multimedia, 115–139 (1998)
9. Diggelen, W.V., Overdijk, M.: Grounded design: Design patterns as the link between theory and practice. Computers in Human Behavior (2009), doi:10.1016/j.chb.2009.01.005
10. Dillenbourg, P.: Collaborative learning: Cognitive and computational approaches. Pergamon, Oxford (1999)
11. Gouli, E., Gogoulou, A., Grigoriadou, M.: Supporting Self-, Peer- and Collaborative-Assessment in E-Learning: the case of the PECASSE environment. Journal of Interactive Learning Research 19(4), 615–647 (2008)

12. Haythornthwaite, C., Kazmer, M.M., Robins, J., Shoemaker, S.: Community development among distance learners: temporal and technological dimensions. Journal of Computer-Mediated Communication 6(1) (2000), http://www.ascusc.org/jcmc/vol6/issue1/haythornthwaite.html

13. Honey, P., Mumford, A.: The manual of learning styles. Peter Honey, Maidenhead (1992), http://www.studyskills.soton.ac.uk/studytips/learn_styles.html

14. Johnson, D.W., Johnson, R.T.: Learning together and alone: Cooperative, competitive, and individualistic learning, 3rd edn. Allyn and Bacon, Boston (1999)

15. Jonassen, D.H.: Designing constructivist learning environments. Instructional Design Theories and Models 2, 215–239 (1999)

16. Kay, J., Kummerfeld, B.: User Models for Customized Hypertext. In: Nicholas, C., Mayfield, J. (eds.) Intelligent Hypertext. LNCS, vol. 1326, pp. 47–69. Springer, Heidelberg (1997)

17. Koper, R., Tattersall, C.: Learning Design: A handbook on modeling and delivering networked education and training. Springer, Berlin (2005)

18. Kordaki, M., Siempos, H.: Encouraging collaboration within learning design-based open source e-learning systems. In: Dron, J., Bastiaens, T., Xin, C. (eds.) Proceedings of World Conference on E-Learning in Corporate, Government, Healthcare & Higher Education (E-Learn 2009), Vancouver, Canada, USA, October, 26-30, pp. 1716–1723. AACE, Chesapeake (2009)

19. Kordaki, M., Siempos, H.: The JiGSAW Collaborative Method within the online computer science classroom. In: 2nd International Conference on Computer Supported Education, Valencia, Spain, April 7-10, pp. 65–74 (2010)

20. Kordaki, M., Grigoriadou, M.: A Collaborative and Adaptive Design Pattern for the 'Students Team Achievement Divisions' Method: An implementation within Learning Design-Based e-learning Systems. In: IWASCL-2010 WORKSHOP (Held in Conjunction with the 2nd International Conference on Intelligent Networking And Collaborative Systems), Thessaloniki, Greece, November 24-26 (2010)

21. Kordaki, M., Siempos, H., Daradoumis, T.: Collaborative learning design within open source e-learning systems: lessons learned from an empirical study. In: Magoulas, G. (ed.) E-Infrastructures and Technologies for Lifelong Learning: Next Generation Environments. IDEA-Group Publishing (2011)

22. Koschmann, T.: CSCL: Theory and practice of an emerging paradigm. LEA, Mahwah (1996)

23. Liu, Y., Ginther, D.: Cognitive styles and distance education. The Journal of Distance Learning Administration 2(3), Article 005 (1999), http://www.westga.edu/~distance/ojdla/fall23/liu23.html (retrieved May 5, 2011)

24. McAndrew, P., Goodyear, P., Dalziel, J.: Patterns designs and activities: unifying descriptions of learning structures. International Journal of Learning Technology 2(2-3), 216–242 (2006)

25. Papadimitriou, A., Grigoriadou, M., Gyftodimos, G.: Adaptive Group Formation and Interactive Problem Solving Support in the Adaptive Educational Hypermedia System MATHEMA. In: Proceedings of the, World Conference on Educational Multimedia, Hypermedia & Telecommunications (ED-MEDIA 2008), Vienna, Austria, June 30-July 4, pp. 2182–2191 (2008)

26. Riding, R., Cheema, I.: Cognitive styles - an overview and integration. Educational Psychology 11(3-4), 193–215 (1991)

27. Roberts, T.S.: Computer-supported collaborative learning in higher education: An introduction. In: Roberts, T.S. (ed.) Computer-Supported Collaborative Learning in Higher Education, pp. 1–18. Idea Group Publishing, Hershey (2005)
28. Scardamalia, M., Bereiter, C.: support for knowledge-building communities. In: Koschmann, T. (ed.) CSCL: Theory and Practice of an Emerging Paradigm, pp. 249–268. Erlbaum, Mahwah (1996)
29. Skinner, B.F.: The Technology of Teaching. Appleton, NY (1968)
30. Slavin, R.E.: Student teams and achievement divisions. Journal of Research and Development in Education 12, 39–49 (1978)
31. Witkin, H.A., Moore, C.A., Goodenough, D.R., Cox, P.W.: Field-dependent and field-independent cognitive styles and their educational implications. Review of Educational Research 47(1), 1–64 (1977)
32. Wyss, R.: Field Independent/Dependent Learning Styles and L2 Acquisition. The weekly column (2002), http://www.eltnewsletter.com/back/June2002/art1022002.html (accessed on May 5, 2010)

Part II

Interactive and Intelligent Learning Systems

Examining the Interrelation between the Interaction Analysis and Adaptation Research Fields within Communication-Based Collaborative Learning Activities: Convergence, Divergence or Complementarity?

Tharrenos Bratitsis

Early Childhood Education Department, University of Western Macedonia, Florina, Greece
bratitsis@uowm.gr

Abstract. Communication is an integral part of any Computer Supported Colla-borative Learning (CSCL) research approach. Furthermore the issues of Adapta-tion and Interaction Analysis have been intensively researched during the past years, under the scope of web-based educational approaches. Both research fields seem to share similar or complementary techniques, aims and outcomes. This chapter attempts to examine at what extend these two fields are complementary to one another or whether they can converge (or diverge) in the future, focusing on Computer Mediated Communication tools, especially Asynchronous Discussion Platforms. The existing work on applying IA methods in communication-based CSCL approaches is examined and correlated with the main constituents of the re-search on adaptive systems. Issues related to flexibility, adaptability and interope-rability are also discussed upon, in an attempt to distinguish the future trends of the IA research field and its relation to Adaptation, outlining their conceptual si-milarities and examining the possibilities of developmental interconnections among the two fields.

1 Introduction

In all cases of Computer Supported Collaborative Learning (CSCL), following learning theories, such as constructivism and the sociocultural theory, or even modern approaches, such as Learning Communities, participants' interaction and the need to support and enhance it is highlighted. In this vein, Computer Mediated Communication (CMC) tools are widely used in formal or informal educational contexts, applying principles of constructivism, emphasizing in social interaction during learning activities (Gunawardena et al, 1997). Tools, such as discussion fo-rae, chats, blogs and wikis, even social networking services are utilized within col-laborative learning activities. Towards this direction, supporting mechanisms in the form of adaptive tools addressed directly to the users should be researched (Bratitsis & Dimitracopoulou, 2010). Computer-Based Interaction Analysis (IA) is a research field, focusing on automated analysis of interactions among users, in

T. Daradoumis et al. (Eds.): Intelligent Adaptation & Personalization Techniques, SCI 408, pp. 157–178.
springerlink.com

various collaborative situations (Dimitracopoulou, 2009). The core aim is the implementation of supporting tools for all the involved actors (students, teachers, moderators and/or researchers). On the other hand, the design and implementation of collaborative learning systems able to adapt to the collaborators' profiles and needs is a significant issue for the research community. User modeling, Activity Patterns and Collaboration Scripting are some of the related research topics.

This chapter attempts to connect the research findings of the IA field with the research on systems' adaptability in CSCL settings, mainly focusing on communication based activities. The chapter is structured as follows: initially the IA field is presented, followed by an overview of the state of the art, focusing on analysis tools for communication based, Technology Enhanced Learning (TEL) activities. Then, the topics of collaborative systems' adaptation, flexibility and interoperability are raised. The concluding discussion attempts to distinguish the possible future trends of the IA research field in correlation with those of system adaptation, under the scope of CSCL, examining whether the two research fields can converge towards a common goal, diverge of just be complementary within the TEL research area.

2 Computer-Based Interaction Analysis (IA)

Computer-based Interaction Analysis (IA) can be defined as the set of automatic or semi-automatic processes that aim at understanding the computer mediated activity, drawing on data obtained from the participants' activities. This understanding can serve in order to support human or artificial actors to partially undertake control of the activity, contributing to awareness, (self)assessment or even (self)regulation. The IA field focuses mainly on collaborative activities, within a learning context. An IA process consists in recording, filtering and processing data regarding system usage and user activity, thus producing analysis indicators. The latter may concern: a) the process or the 'quality' of the considered 'cognitive system' being formed by an individual or a group of collaborating students, b) the features or the quality of the interaction product; or c) the mode, the process or the quality of the collaboration, when acting within the frame of a social context formation via the technology based learning environment (Dimitracopoulou, 2009).

The IA results are presented to the participants, as well as the observers of the (learning) activities in an appropriate format (graphical, numerical or literal), interpretable by them. The core aim is to offer the means directly to the human actors, so as to be aware of and regulate their behavior and/or actions, either as individuals or as cognitive groups. In fact, the IA tools support the users in three major levels: awareness, metacognition and evaluation. The objective is the optimization of the learning activity through: a) refined students' participation via reflection, self-assessment and self-regulation, b) better activity design, regulation, coordination and evaluation by the teachers. Moreover, the researchers are provided with means to analyze the complex phenomena that occur within collaborative learning activities.

Reviewing the literature, two main directions exist. The first (Research Direction 1 – RD1) is that of systems which based on the IA output and considering the profiles and the cognitive processes of individuals or collaborating groups, adapt the learning environment to their own needs and preferences or provide guiding messages, thus facilitating participation and collaboration. In this case, the system makes the decisions. The second direction (Research Direction 2 – RD2) is that of providing information directly to human actors, so as to assist them in selfregulating their decisions, actions or behavior, supporting them in a level of awareness and metacognition. In this case the human actors have the locus of control on the collaborative activity. On one hand collaborating students need supporting information, describing their own and their collaborators' actions, in order to (self)evaluate, in an operational way, both the collaborative/learning process and the quality of the overall activity. On the other hand, teachers need supporting information in order to decide upon teaching strategies, perform corrective interventions, or even formative evaluation of their educational actions. Examples of these two directions, focusing on communication based collaborative learning activities are provided in the next section (State of the Art).

The need to support participants' awareness and metacognition is enhanced by the intensive interest to use technology based learning environments, especially collaborative ones, in everyday educational practice, where the need to (self)evaluate both the learning process and the quality of the activity in an operational way (Dimitracopoulou, 2009). During collaborative learning activities or when students work with exploratory learning environments, very complex interactions take place between two or more students or groups or even students and the teacher-moderator. Thus, on one hand students seem to need information regarding their own actions, as well as their collaborators that could support awareness, metacognition and therefore selfregulation of their learning within the overall activity and process. On the other hand, teachers need information, usually structured and condensed, regarding what is happening, allowing them to decide if and how they should intervene, influencing the quality of the activity outcome and/or the quality of the actual collaboration taking place. This need is greater in real teaching, in class situations and not in structured, laboratory activity settings in which the corresponding research approaches take place. Also, teachers need support in order to perform formative evaluation of their educational action (design, strategy and implementation), thus deciding on how they should implement their teaching approaches in the future.

Actually, most of the existing learning systems present limitations when used in educational settings. Some of them are attributed to the fact that students find it difficult to assume and apply metacognitive knowledge to their actions and internal processes or to interact with their partners. Furthermore, teachers are usually in charge of several students and fail to interpret and confront the difficulties the latter face, due to the huge number of interactions that take place simultaneously. In such cases, teachers are not able, for example, to detect a collaboration breakdown that could lead to frustrating experiences for both the students and the teachers, causing abandonment of the new learning environments and experiences, in favor of the traditional teaching approaches (Dimitracopoulou, 2009).

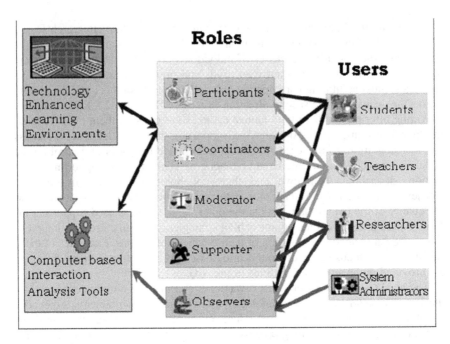

Fig. 1 Informational needs of all actors in TEL approaches

The need to address these problems and meet the informational needs of all the actors involved in TEL approaches (Fig. 1) has lead to the emergence of a new research field, during the past few years, that of Computer-based Interaction Analysis. It is related to the design of technology based learning environments, with the core aim of this field being the studying of participants' interactions and the implementation of tools in order to support regulation, when addressed to the teacher or the moderator, and selfregulation, when addressed directly to the students (Dillenbourg et al. 2002; Reimann 2003; Petrou & Dimitracopoulou, 2003; Jermann 2004; Vassileva et al. 2004; Gerosa et al. 2005; Michozuki et al. 2005, Nakahara et al. 2005, Reyes & Tchounikine 2005; Bratitsis & Dimitracopoulou 2006; Teplovs et al. 2007; Kay et al. 2007; Bratitsis, 2010a).

3 State of the Art

Several collaborative systems integrating IA tools exist in the literature. Jermann (2004) provided tools to dyads of students and observed them in laboratory settings, thus showing that IA tools facilitated students' selfregulation, during synchronous, game-like simple tasks. The students were involved in a problem-solving situation (tuning traffic lights in order to minimize car congestion in a crossroad), having the opportunity to discuss upon their solution strategy through a synchronous chat, while attempting to apply the desired solution at the same time. The provided IA tool assisted them in better regulating their discussion and solution strategy, by designating whether they were just trying out solutions, possibly random ones (Tuning approach), or over-discussing their strategy (Talking approach), at the expense of an actual solution application. The corresponding gauge-like indicator (Fig. 2) calculated a metric for each student, as well as the dyad, varying from "Tuning" to "Talking". In this case, the system provides the students with metacognitive information in order to assist them in regulating their actions and their collaboration, towards a refined solution to the provided problem, thus coming under the RD2 direction, described in the previous section.

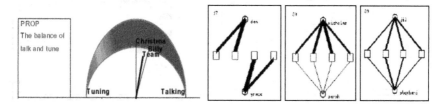

Fig. 2 Interaction Analysis tools – COTRAS system (Jermann, 2004)

Moreover, diagrams show the amount (thickness of connecting lines) of interactions with all the artifacts (traffic lights) of the environment (Fig, 2). This information reveals the division of labor among the students to the teacher or the researcher, who can evaluate it following a Role based, a Task based or a Concurrent evaluation approach. This kind of indicators fall under the RD2 approach, as described in the previous section, in a wider sense, given that the provided information assist the corresponding actor to evaluate and act accordingly. In this case the actor is the teacher who can decide to intervene or evaluate in order to further work with the students following a different didactical approach.

Detailed study on teachers' selfregulation, in matters of applied teaching design and strategies, has been conducted by Petrou (2005), in a context of synchronous modeling activities in school classes. Using the CoRPET tool, teachers were able to monitor various collaboration aspects, such as communication, artifact manipulation and workspace evolvement. By utilizing facilities such as a playback of all the activity in a collaboration and communication platform, along with additional monitoring and evaluation tools, teachers were able to decide upon the effectiveness of their teaching strategy and the design of the applied learning activities.

Focusing on asynchronous collaborative, communication-based activities, one can find systems like the AulaNet (Gerosa et al, 2005), which produces various diagrams, facilitating teachers' tasks (Fig. 3). These diagrams provide condensed statistical information, related to the discussions' metrics (thread depth, messages per logical level, thread width, etc), assisting the teacher to monitor the discussions. Specifically, he/she can decide if the dialogues are evolving following the required manner, based on the expected discussion pattern and thus decide if he/she should intervene, performing his/hers moderating tasks.

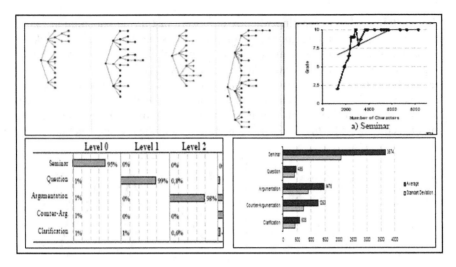

Fig. 3 Statistical visualized information for the teacher – Aulanet (Gerosa et al, 2004)

On the other hand, the MailGroup system (Reyes, 2005) uses Social Network Analysis (SNA) tools, addressed to researchers. By examining structural metrics of the conducted discussions through SNA diagrams, the researchers were able to evaluate the effectiveness of their proposed innovative representation method of an asynchronous discussion, which took into account both the logical and the chronological constituents of the messages' sequence (Fig. 4). Specifically, they used the Eigenvector-centrality approach for producing SNA diagrams and calculated parameters such as Cohesion, which indicates the tendency of information diffusion within a collaborating group, by counting the number of nodes that have to be subtracted from the diagram in order for the latter to be divided into two completely separated parts.

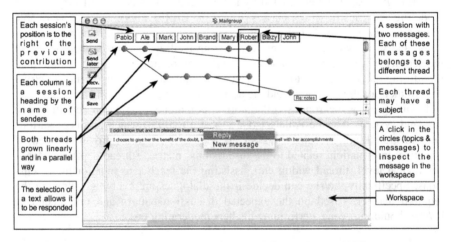

Fig. 4 Representation of synchronous discussion– MailGroup (Reyes, 2005)

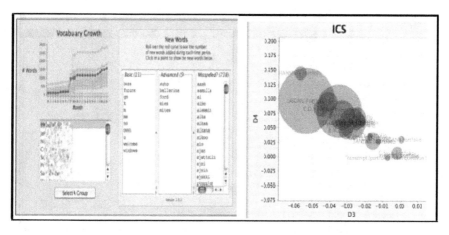

Fig. 5 IA tools for the teacher– Knowledge Forum (Teplovs et al, 2007)

Moreover, the Knowledge Forum system (http://www.knowledgeforum.com/) provides metacognitive tools, assisting students to reflect upon their performance and improve their learning strategies in problem solving situations. The messages are grouped into logical trees, depicting the thinking strategy followed by collaborating actors in order to solve a given problem. By studying these depictions, as well as utilizing other informative diagrams, one can assess his/her thinking strategy or even the one followed by a collaborating group, while solving a problem. The Knowledge Forum has been used by many researchers, who have implemented addon analysis tools, some of which can be used during the learning activity, but they are mainly addressed to the teacher or the researcher. For example, Teplovs et al (2007) provide a set of indicators for the teachers which are directly linked to the Knowledge Forum environment (Fig. 5). These indicators reveal information, such as the "evolution of vocabulary" of the students or visualize "the semantic field of the students' discussion topics".

The Argunaut system (de Groot et al, 2007) supports teachers in understanding when to intervene, in order to assist students (Fig. 6). It provides several offline analysis tools, based on a predefined, by the teacher, scheme of a discussion which derives from the expected outcome of the discussion. These tools mainly facilitate assessment by examining activity patterns. Additionally, on-the-fly visualization tools provide an insight of the discussions' structural elements, such as relation of message types, user-to-user interaction and content related variables.

Other systems provide interesting visualizations, facilitating students' participation. For example the i-Bee system (Michozuki et al, 2005) provides the students with a representation of their synchronous, chat-based discussion, using a set of flowers and bees. The bees correspond to students and the flowers to keywords, the use of which indicates the proper orientation of the ongoing discussion. By the flowers' status (blossomed or closed) and the bees' direction (facing towards

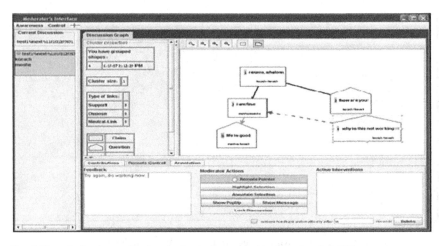

Fig. 6 IA tools for the teacher/moderator – Argunaut (de Groot et al, 2007)

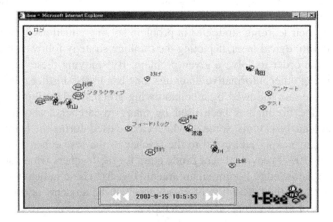

Fig. 7 IA tools for the teacher– iBee (Michozuki et al, 2005)

the flowers or not), students could better orientate their discussion, by using more appropriate vocabulary and thus, staying on topic (Fig. 7).

Also i-Tree system (Nakahara et al, 2005) uses a tree image to represent structural metrics of a discussion forum. The size of the tree, the width of the stem, the number of the branches, leaves and fruit, as well as the color of the sky, depict the discussion evolvement. These images operated as an alerting mechanism, as well as an additional motivation for the students to increase their activity in the discussions (Fig. 8). The tool was addressed to the students. All these tools fall under the RD2 approach, as described in the previous section.

Fig. 8 IA tools for the teacher– iTree (Nakahara et al, 2005)

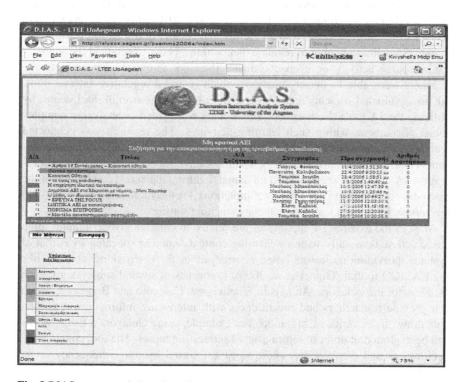

Fig. 9 DIAS system main interface (Bratitsis & Dimitracopoulou, 2008)

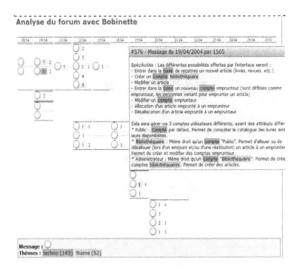

Fig. 10 Bobinette analysis tool – Calico toolkit (Giguet et al, 2009)

The DIAS system is more focused on both asynchronous discussions and the IA field (Bratitsis & Dimitracopoulou, 2008; 2009; 2010). It provides an extensive set of IA indicators, addressed to all the involved actors of discussion learning activities. Students were able to regulate their actions, better understand the scope of the discourse activity or even coordinate their collaboration more effectively. Teachers were able to detect situations which required regulative interventions, but also evaluated students' participation and assessed the overall discussions, by utilizing the IA tools. Furthermore, researchers were aided in analyzing complex social phenomena within such learning activities. The research was conducted with adult learners (Bratitsis & Dimitracopoulou, 2010), as well as 3rd grade students (Bratitsis & Kandroudi, 2010). The innovative aspect of the DIAS system is that the IA indicators are produced on demand, during the actual conduction of the discussion activity (and not after its completion). Furthermore, the effectiveness of the approach has been tested in real class teaching settings.

Most of the existing IA approaches are related to collaborative, communication based activities, usually within a learning context. Another subcategory is that of systems, providing indicators based on analysis of the discussions' content, like the CALICO toolkit (Giguet et al, 2009). It consists of several sub-systems, such as: Showforum, Anagora, Authagora, Themagora, Colagora and Bobinette, which can provide researchers and practitioners with interesting information regarding data drawn from various CMC tools. For example, using Colagora a list of words can be exploited in order to return a set of interesting topics. The tool counts word occurrences and displays the whole word list sorted by usage frequency or by alphabetical order. The Bobinette tool (Fig. 10) provides an alternative visualization of the discussion threads, taking into account the time factor. Furthermore, it produces statistical information about word topics for a forum and for each post (Giguet et al, 2009).

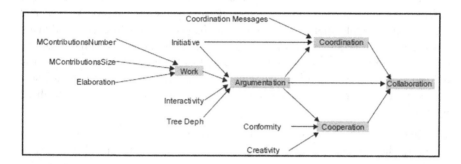

Fig. 11 Analysis scheme of the Degree tool (Barros & Verdejo, 2000)

The Degree system (Barros & Verdejo, 2000) is a representative example of the RD1 approach, as described in the previous section. Integrating fuzzy logic algorithms, this system combines quantitative elements of an asynchronous discussion, such as amount and types of messages or discussion initiations, for producing several indicators (e.g. Collaboration Level). Based on the values of the indicators, an advisory mechanism provides suggestions to the students in order to improve their performance. The suggestions are based on a comparison of the individual indicators with an ideal student model.

Finally, IA tools have been implemented in order to support the collaborating members of a Community of Practice, such as the Kaleidoscope Network of Excellence (Bratitsis et al, 2008), in matters of enhancing social queues and supporting decision making processes.

All the aforementioned examples constitute a representative set of the existing approaches in the IA research field. They are applied to communication based activities, based on an abstract issue, such as a forum topic. Up to now, there are no IA approaches, applied to dynamically alterable communication queues, such as the ones feasible with media annotation systems. In the latter case, a unique communication queue can be initiated in every instance of an annotated video file or every portion of annotated pictorial data. Even in such cases, the communication queues may be split to more than one, separate queues, by distinguishing new initiation points in subsequent annotation targets. For that matter, the set of IA tools integrated into the DIAS system were used to analyze students' participation in blogging systems (Bratitsis, 2010a). The results indicate that the conclusions drawn by several of the produced diagrams can be applied in blog-based activities, facilitating the teacher's evaluation tasks. The latter research is still ongoing.

4 Adaptation in CSCL

Adaptive educational systems aim at facilitating the educational process by providing individualized support to the students (Bruisilovsky, 2001). Adaptability in
an educational system/tool refers to the ability of the tool to dynamically adapt,
based on the student and his/her progress. Personalization is the process of adapting a computer application or environment to the needs of specific users and takes
advantage of the acquired knowledge about them (Gasparini & Lichtnow, 2009).
On the system's side the term used to describe this feature is Adaptation, which is
defined as the application of adjustments to the actual tool so as to meet the student's needs and abilities (Papanikolaou & Grigoriadou, 2005). There are three
main adaptation methods:

- *Adaptability*. This category refers to the adjustment of the educational system
  during the initiation of a working session, based on student characteristics
  which are usually fixed (cannot be altered) and introduced to the system by
  answering to certain questionnaires. This leads to the construction of user models, based on the end-user's preferences, regarding system's usage. Elements of
  such models are, for example, preferred language, learning style, etc.
- *Adaptivity*. This category refers to the automated modification of the learning
  environment during its use by the student, based on the latter's characteristics
  which can be altered during the educational process. Cognitive level is an example of a student characteristic which is altered within/during the educational
  process. These characteristics are traced by the system through the student's interaction with it.
- *Tailorability*. This category refers to adjustments of the educational environment (e.g. user interface) which are applied by the student him/herself.

The main research topics of the adaptation research field are: a) the user characteristics that actually define system adaptation, b) the system's features that are
adaptable, and c) the adaptation techniques and technologies to be used.

All of the adaptive systems store a model for every student, based on which the
adaptation takes place (Bruisilovsky, Karagiannidis & Sampson, 2004). Student
characteristics used for constructing such a model are: goals, cognitive level, cognitive background, navigation experience, preferences, navigation history within
the educational material, test performance, etc.

In any case, the adaptive educational systems adjust the content's presentation
and/or the navigation to a specific student model. A common adaptation method is
that of customizing the User Interface taking into account the student model in
consideration, thus adjusting to the perception better fitting the student's needs.
This technique copes with what Brusilovsky (1998) refers to as *curriculum sequencing*. According to this notion, either the next concept or topic to be taught is
determined or the next task to be carried out. Correlating these approaches, the

technique remains the same. It is also referred to as Instructional planning and aims at providing the student with the most suitable individually planned sequence of knowledge units to learn and sequence of learning tasks (examples, questions, problems, etc). Thus, the goal is to help the student to find an "optimal path" through the learning material (Bruisilovsky, 1999). There are two types of sequencing: active and passive. Active sequencing implies a learning goal (a subset of domain concepts or topics to be mastered). Systems with active sequencing can build the best individual path to achieve the goal. Passive sequencing (which is also called remediation) is a reactive technology and does not require an active learning goal. It is initiated when the student is not able to solve a problem or answer a question (questions) correctly. Its goal is to offer the student a subset of available learning material, which can fill the possible student's knowledge gap of resolve a misconception. It makes sense to distinguish two kinds of curriculum sequencing techniques. High-level sequencing or knowledge sequencing determines the next concept or topic to be taught. Low-level sequencing or task sequencing determines the next learning task (problem, example, test) within current topic to be carried out (Brusilovsky, 1998). In the context of Web-based education, curriculum sequencing technology becomes very important for guiding the student through the hyperspace of available information.

A fundamental goal of most of the educational systems has been, for years, *problem solving support*. The corresponding technologies aim at helping the student in a process of solving an educational problem, but they do so in various ways (Bruisilovsky, 1999). One technology is that of *intelligent analysis of student solutions* which deals with students' final answers, regardless of how they were obtained. Usually a detailed error feedback is provided while the system's internal student model is updated. A more interesting category is that of *Interactive problem solving support*, which assists the student during the process of reaching a solution to a given problem. The level of assistance may vary from a simple notification about "taking a wrong step", to giving a hint/advise or even executing the next step for the student. Both the aforementioned technologies are applicable in web-based educational activities and consist in producing guiding or recommendation queues for the students, supporting them in better understanding the learning activity goals. In the case of problem solving situations, these queues can be related to the produced solution (Bruisilovsky, 1998), as well as the solving process. In the former situation comments and advice are provided regarding the correctness of the provided solution, comparing it to the ideal (or the only correct) one. In the latter situation, intelligent help can be provided to the students throughout all the intermediate steps, towards the final solution, utilizing several techniques, such as the use of agents (Bruisilovsky, 1998).

Adaptive hypermedia is a research area aiming at applying different forms of user models in order to adapt the content and the links of hypermedia pages to the users' needs and/or preferences. Two major technologies in adaptive hypermedia can be distinguished: *adaptive presentation* and *adaptive navigation support*. The content of a web page or the User Interface is correspondingly adjusted, implementing an *adaptable presentation technology* (Bruisilovsky & Peylo, 2003). The goal is to adapt the content of a hypermedia page to the user's goals, knowledge or any other information stored in the user model. In a system which implements adaptive presentation, the pages are not static, but are adaptively generated or assembled from pieces for each user. For example, with several adaptive presentation techniques, expert users receive less detailed but in-depth information, while novice ones receive more additional details and explanation. Adaptive presentation is very important in WWW context where the same "page" has to be suited to very different students.

The goal of the *adaptive navigation support* technology is to support the student in hyperspace orientation and navigation by altering the appearance of visible links. In particular, the system can adaptively sort, annotate, or partially hide the links of a web page in order to make the choice of the next link to proceed easier. Adaptive navigation support can be considered as an extension of curriculum sequencing technology into a hypermedia context. It shares the same goal - to help students find an "optimal path" through the learning material. At the same time, adaptive navigation support is less directive than traditional sequencing: it guides students implicitly and leaves the choice of the next knowledge item to be learned and the next problem to be solved to them.

Finally, *adaptive collaboration support* aims at supporting collaboration using system's knowledge about different users (stored in user models). In these cases, research focuses in group formation (Hoppe, 1995), peer attribution and peer assistance (McGalla et all, 1997), virtual peers and class monitoring (Chan & Baskin, 1990; Oda, et al, 1998).

In all cases, adaptation is implemented by comparing the actual situation to the ideal situation and then instructing the system to act accordingly, thus altering elements which differentiate the actual user interface of the educational system for each individual user. The differentiations comprise in advisory hints and/or queues or even total reconstruction of the user interface layout.

5 Flexibility and Interoperability of IA Tools

CSCL approaches are nowadays widely used in education. Furthermore, the Internet has been gradually transformed into a platform of collaboration in which every user actively participates in the construction of meaningful content. Web 2.0 tools, such as blogs, wikis and social networking services are used also in everyday life, mainly for communicative (direct communication and/or information exchange) purposes. In fact, the term *Education 2.0* has been used in the literature in order the integration of Web 2.0 tools in educational approaches.

In all cases communication, especially in written language form, is a fundamental constituent. Communication among collaborating actors is a prerequisite in order to achieve information exchange, argumentation and expression of their thinking processes, rationalization of their actions and finally common knowledge acquisition (Dillenbourg, 2002). On the other hand, such a level of communication is often difficult to be achieved, as the collaborating actors, especially younger ones, often lack the necessary dexterities (Soller, 2001). For that matter, several techniques have been proposed in order to facilitate argumentation and the development of constructive dialogues (Jermann et al, 2004). For example, *sentence openers* have been widely used (e.g Knowledge Forum), as well as *types of messages* operating as declaratory actions (e.g. DIAS, AulaNet). These approaches assist students in developing argumentation dexterities, such as: the ability to formulate questions, substantiation, negotiation, coordination, etc (Andriessen et al, 2003). According to the literature, this type of dialogue structuring contributes to the development of communication dexterities, but also allows the automated dialogue analysis, as well as the evaluation of the actors' interaction and the development of regulative mechanisms (Andriessen, et al, 2003; Jermann et al, 2004). Nevertheless, the proposed techniques are not always utilized properly by the collaborators, thus further designating the need for additional supporting tools. These can be found in the literature in the form of advisory mechanisms (e.g. Baros & Verdejo, 2000), or IA indicators (e.g. Bratitsis & Dimitracopoulou, 2010).

Futhermore, when designing collaborative learning activities, the development of the collaboration platform and the communication tool is not enough. Strategic design and constant effort to sustain collaboration on a desired level are necessary in order to fully exploit the electronic medium (Hiltz, 1997). On the strategic planning level, the use of *collaboration scripts* has been proposed as a solution (Fischer et al, 2007). Furthermore collaborating actors formulate different cognitive systems, usually having different informational and supporting needs, so as to sustain a high collaboration level (Dimitracopoulou, 2009). Also a teacher has increased informational needs for monitoring learning activities and intervene whenever it is necessary or even evaluating the cognitive processes and/or the learning outcome. Finally a researcher has more complex data analysis needs. In all these cases, similar collaboration and communication tools can be used in diversified manners. Thus, *flexibility* is an important issue. A tool should be developed so as to be used by different types of users and cognitive schemas (individually and collaboratively), different types of activities and collaboration settings, as well as to serve different informational and analysis needs.

One concrete conclusion is that supporting tools seem to be necessary for improved collaboration. On the other hand, one may find numerous communication tools, such as: asynchronous discussion forae, chat tools, blogs, wikis, instant messaging tools, etc. Furthermore, there is a wide variety of available software in order to implement all these types of communication via electronic means. Some of them are Open Source Software, allowing code modification, Free Software or even adhoc solutions, usually as integrated parts of wider systems (e.g. Content Management Systems - CMSs). Although the operational logic of these tools is always the same (for example in a discussion forum, discussants exchange messages asynchronously, which are published in a common web page), there are significant structural differences, mainly due to the underlying technology. For example, in asynchronous discussion forae, a database system is often used for storing and accessing the posted messages. The structure of the database (tables and relations) is usually different in every available forum platform, especially in the Free – Open Source platforms, or even not available at all, especially in the case of forum tools being integrated as a subsystem in wider collaboration platforms, such as CMSs. This is usually due to the fact that most of the available software has not been developed for strictly educational purposes, but have a different target group. Nevertheless, they are used by educators for learning activities. Furthermore, the underlying technology may significantly vary among similar systems (e.g. different programming language, web service or even operating system), increasing the diversity of the available software and thus the difficulty in developing common analysis tools. For such a tools to be utilized for analyzing educational activities, despite of the technological tool used to implement the activities, *interoperability* is a key issue. It can be achieved, for example, by developing analysis tools able to collect activity data from diverse systems, using proper parsing filters and techniques.

6 Discussion

The current chapter attempts to discuss upon the possibility of further converging the IA and the CSCL research fields, additionally involving the Adaptation research field, focusing on communication-based collaborative learning activities. Examining the IA field's literature several approaches can be found, emphasizing in the implementation of supporting analysis tools for the teacher, as well as the students. One of the field's current trends is that of supporting students on a metacognitive level, so as to be able to selfregulate their actions (Dimitracopoulou, 2009). In fact, the research conducted with the DIAS system is consistent with this trend, having a significant differentiation; the research was conducted in real teaching settings (in situ). For example, the findings of the research conducted with the DIAS (Bratitsis & Dimitracopoulou, 2009) and the AulaNet (Gerosa et al, 2005) systems seem to be complementary. Likewise, all the approaches presented in the State of the Art section, describe positive outcomes when utilizing IA, visualized tools. In some cases the research was conducted in laboratory settings and in other cases in real teaching settings. Despite the context of research implementation, the research findings point to the same direction in almost all cases; visualized IA tools facilitate collaborative learning activities. Thus one could argue that this conclusion is concrete enough, so as to consider advancing the corresponding research one step forward.

Once more, focusing on communication based educational activities and taking into account the diversity of the available communication software and the educational approaches, *interoperability* and *flexibility* are the important issues to be examined. On the former issue, it is a matter of implementing collaborative and/or analysis tools, able to inter-communicate. Technologies such as parsers, XSLT filters, etc could be implemented by collaborating researchers, in order to allow data exchange among diverse systems. An ideal solution would be the design and development of autonomous analysis toolkits which could operate as "black boxes". These toolkits could receive input data from the most commonly used communication software (e.g. wordpress blogging system, phpBB forum system, etc), thus overcoming the need of the researchers to develop new, similar tools as adhoc solutions, for every designed research approach. Such attempts have already been made, having unofficially tested the operation of the Calico toolbox with discussions conducted with the DIAS system. Furthermore, parsers were used in order to import and analyze data from commercial blogging software, such as Wordpress (Bratitsis, 2010a) and wiki software, such as MediaWiki (work is still in progress), with the IA tools integrated in the DIAS system. On the other hand, these toolkits should provide distinct sets of IA indicators, for all types of users (students, teachers, researchers) and educational settings (collaboration script(s), group formation, etc). For that matter, sharing of expertise and educational strategies is necessary, not only through the literature, but through researchers' international collaboration. The technology is mature enough to allow such collaboration.

One of the most important contributions of the IA research field is the provision of adequate proof that the utilization of IA indicators by the students in order to self-regulate their actions actually relieves the work load of a moderator, usually the teacher, in many cases (Bratitsis, 2010a; 2010b). Examining the next steps for the IA research field, adaptation techniques seem to be a logical, as well as interesting area for expansion. Adapting the communication tool to the student's needs could further facilitate self regulation, so as to improve the communication outcome by enhancing the prerequisites for a fruitful dialogue.

Specifically, as aforementioned, there are two main research directions in the IA field. The most commonly applied is that of providing information directly to human actors, so as to assist them in selfregulating their decisions, actions or behavior, supporting them in a level of awareness and metacognition, thus facilitating the undertaking of locus of control of the activity (described as Research Direction 2 in the Computer based IA section). The research conducted with the systems which are presented in the State of the Art section and comply with this approach, show that collaboration as a process, as well as an educational approach is clearly facilitated. The proposed analysis methods are mostly disaggregated to calculations based on quantitative activity data with a qualitative interpretation approach applied to the calculation product. In the case of IA tools addressed to the teachers-moderators and/or the researchers, mainly for collaboration meta-analysis and assessment, the applied technique resembles the intelligent analysis of student solutions technology, falling under the problem solving support, as described in the Adaptation in CSCL section. The core aim of the IA research field in these cases is to assist the corresponding actors in assessing the collaborative process, in order to improve their teaching strategies and approaches in the future. Likewise, the intelligent analysis of student solutions technology consists in analyzing students' answers in order to provide feedback, as well updating the student model which is stored internally in an adaptive system. Correlating these technologies, the fundamental approach is the same; after the completion of the educational activity/task, which is mainly problem solving oriented, the product (collaborative or individual, correspondingly) and/or the process of constructing it is evaluated, mainly by comparing it to an ideal one. Then, the adaptive system or the teacher/researcher (also in some cases educational designer), updates the educational approach to be followed in the future. The similarity is obvious, thus enhancing the statement that these two research fields (IA and Adaptation) share common goals and follow cognate techniques and technologies.

Refocusing on the IA research field, the second distinguished research direction is that of utilizing IA analysis methods in order to implement assistance and/or guidance mechanisms for the students. Likewise, the interactive problem solving support technology in the case of adaptive systems, aims at assisting the students during the process of reaching a solution to a given problem. In both cases, the aim is the support of students during a problem solving situation, as individuals (mainly in the adaptive systems) or as collaborating groups (mainly in the IA based systems). The notification mechanisms vary from simple notifications that "something is not correct" to hints and/or advises on how to proceed further within the educational activity. The assistance can be related to the understanding of the activity goal or the path that leads to the proper solution of the problem.

This seems to be the most interesting case of convergence between the IA and the Adaptation research fields, as both of them could benefit from one another. For example, the aforementioned approach is implemented in the Degree system (Baros & Verdejo, 2000). Based on the calculations of the corresponding discussion elements, the recommendation mechanism prompted actions to be undertaken by the students. For example they were asked to initiate more discussion threads or to write more answers to other students' posts. The prompting of the advice was usually explicit and implemented during the user's logging into the system, with a rather strict manner. On the other hand, adaptive systems have been using agents in order to implement these types of actions, following a more indirect and less strict approach. Furthermore, the IA research field studies more complex phenomena that take place within collaborative settings, focusing more on the process than the adaptive systems which focus more on the produced solution. Taking into account that the establishment of collaboration and specifically the engagement in a constructive dialogue is actually studied as a problem solving situation (the problem to be solved is, for example, reaching a commonly accepted argument) the similarities, as well as the mutual benefits for the two research fields under consideration become evident.

Attempting to provide a simple example of convergence, one could consider a common asynchronous discussion forum. It is common for such a system to facilitate minimal alerting mechanisms in order to assist the user to better navigate through the ongoing discussions. For example topics with unread messages are highlighted, the sequence of appearance on the computer screen for the topics is based on the timestamp of the last message in each topic (the most recently altered topics appear first), a list of messages written since the user last joined the system are provided and so on. In the case of dialogue based learning activities, the literature describes the most common problems faced that lead to inconclusive discussions and thus in unsuccessful learning activities (Bratitsis, 2010a). Advisory approaches proposed by the Adaptation research field can be utilized, based on IA results. For example students can be indirectly prompted to collaborate with specific co-students, based on Social Networking Analysis of the ongoing discussions, through an electronic agent. Following a reversed approach, the Adaptation research field aims at assisting the students in better reaching a goal (usually the solution to a given problem) and could benefit from the analysis method of the IA research field. Studying of interactivity patterns through IA indicators can be utilized in order to update student models with more refined information, especially in web based collaborative systems, thus leading to improved adaptation. Additionally, one of the main research axes of the IA field to distinguish the informational needs of all the involved actors in collaborative learning activities, thus producing sets of IA tools to meet those needs. As stated in section 2, different types of users have different needs. Moreover, a user's needs vary when he/she is asked to act as an individual or as a member of a collaborating group or even an entire (learning) community. By better understanding user needs, more efficient user models can be constructed, leading to more effective adaptive approaches.

Finally, examining the issues of flexibility and interoperability, which are more important nowadays, some of the existing tools should also be tested in real teaching settings, under different educational conditions. For that matter, the exchange of data among researchers could be helpful. The validity of their findings should be generalized, via verification in different settings. For example, the IA indicators of the DIAS system seem to function adequately for the teacher, when trying to evaluate students' participation, in blogging systems too (Bratitsis, 2010a). More tests in this direction should be attempted in order to avoid repetitions of similar case studies with similar analysis tools. Of course the design and implementation of several case studies is necessary in order to verify this hypothesis.

As a concluding remark, one could state that the IA research field actually has emerged mainly due to the expansion of the Internet. Web based collaborative tools and applications are the core aim of the corresponding research, although the actual initial point of the field lies earlier in time. Likewise, the research field of Adaptation exists for many years, but has moved to a whole new level of research, due to the Internet, that of Adaptive Hypermedia. The ultimate goal of both fields is to enhance learning, by analyzing student activity and/or interactivity. In the former case this analysis is used mainly as a supporting, visualized feedback mechanism, whereas in the latter case it is used mainly in order to predict user behavior and needs, thus providing the appropriate resources. In both cases, the teacher's tasks are also facilitated, directly or indirectly. All the above gain added value in a collaborative situation, namely CSCL settings, where the interactions are much more complex and adaptation is harder. Consequently a logical next step for both research fields would be an attempt to "join forces", under the scope of CSCL, thus exploiting the analysis conducted mainly for metacognition in the IA field, in order to enhance activity pattern and user modeling recognition for the Adaptation research field. In reverse, the IA field could benefit from the adaptation techniques and technologies in order to utilize additional informative queues, other than presenting visualized indicators. Besides, ideas in that direction have already been formulated (Bratitsis, 2010b).

References

1. Andriessen, J., Baker, M., Suthers, D.: Argumentation, computer support, and the educational context of confronting cognitions. In: Andriessen, J., Baker, M., Suthers, D. (eds.) Arguing to Learn. Confronting Cognitions in Computer-Supported Collaborative Learning Environments, pp. 1–25. Kluwer Academic Publishers (2003)

2. Barros, B., Verdejo, M.F.: Analysing student interaction processes in order to improve collaboration. The DEGREE approach. International Journal of Artificial Intelligence in Education 11, 221–241 (2000)

3. Bratitsis, T.: Using Computer-Based Interaction Analysis Tools For Evaluating Blog Participation In Tertiary Education: An Exploratory Proposal. In: International Conference on ICTs in Education, ICICTE 2010, Corfu, Greece, July 8-10 (2010a)

4. Bratitsis, T.: Thoughts on utilizing Interaction Analysis for adaptable asynchronous discussion platforms in collaborative learning activities. In: International Conference on Intelligent Networking and Collaborative Systems, INCOS 2010, Thessaloniki, Greece, November 24-26 (2010b)

5. Bratitsis, T., Dimitracopoulou, A.: Indicators for measuring quality in asynchronous discussion forums. In: International Conference on Cognition and Exploratory Learning in Digital Era (CELDA 2006). IADIS (International Association for Development of the Information Society) Barcelona (2006)

6. Bratitsis, T., Dimitracopoulou, A.: Interpretation Issues in Monitoring and Analyzing Group Interactions in Asynchronous Discussions. International Journal of e-Collaboration 4(1), 20–40 (2008)

7. Bratitsis, T., Dimitracopoulou, A.: Studying the effect of interaction analysis indicators on students' selfregulation during asynchronous discussion learning activities. In: 8th International Conference on Computer Supported Collaborative Learning, CSCL 2009, Rhodes, Greece, June 8-13 (2009)

8. Bratitsis, T., Dimitracopoulou, A.: Interpretation of Computer Based Interaction Analysis indicators: a significant issue for enhancing collaboration in Technology Based Learning. In: Kock, N. (ed.) Interdisciplinary Perspectives on E-Collaboration: Emerging Trends and Applications, pp. 31–59. IGI-Global, USA (2010)

9. Bratitsis, T., Dimitracopoulou, A., Martínez-Monés, A., Marcos, J.A., Dimitriadis, Y.: Supporting members of a learning community using Interaction Analysis tools: the example of the kaleidoscope noe scientific network. In: Diaz, P., Kinshuk, Aedo, I., Mora, E., (eds.) The 8th IEEE International Conference on Advanced Learning Technologies (ICALT 2008), pp. 809–813. IEEE Computer Society (2008)

10. Bratitsis, T., Kandroudi, M.: Improving 3rd grade students' writing ability via asynchronous discussions: A case study. In: International Conference on Intelligent Networking and Collaborative Systems, INCOS 2010, Thessaloniki, Greece, pp. 24–26 (November 2010)

11. Brusilovsky, P.: Adaptive educational systems on the WWW: A review of available technologies. In: Ayala, G. (ed.) Proceedings of Workshop Current Trends and Applications of Artificial Intelligence in Education, 4th World Congress on Expert Systems, pp. 9–16. ITESM, Mexico City (1998)

12. Brusilovsky, P.: Adaptive and Intelligent Technologies for Web-based Education. In: Rollinger, C., Peylo, C. (eds.) Künstliche Intelligenz, Special Issue on Intelligent Systems and Teleteaching, vol. 4, pp. 19–25 (1999)

13. Brusilovsky, P., Karagiannidis, C., Sampson, D.: Layered evaluation of adaptive learning systems. International Journal of Continuing Engineering Education and Lifelong Learning 14(4/5), 402–421 (2004)

14. Brusilovsky, P., Peylo, C.: Adaptive and Intelligent Web-based Educational Systems. International Journal of Artificial Intelligence in Education 13, 159–172 (2003)

15. Chan, T.W., Baskin, A.B.: Learning companion systems. In: Frasson, C., Gauthier, G. (eds.) Intelligent Tutoring Systems: At the Crossroads of Artificial Intelligence and Education, pp. 6–33. Ablex Publishing, Norwood (1990)

16. De Groot, R., Drachman, R., Hever, R., Schwarz, B., Hoppe, U., Harrer, A., De Laat, A., Wegerif, R., McLaren, B., Baurens, B.: Computer Supported Moderation of E-Discussions: the ARGUNAUT Approach. In: Chinn, C., Erkens, G., Puntambekar, S. (eds.) Proceedings of the International Conference CSCL, Rutgers, The State University of New Jersey, USA, July 21-26. , pp. 165–167 (2007)

17. Dillenbourg, P.: Over-scripting CSCL: The risks of blending collaborative learning with instructional design. In: Kirschner, P.A. (ed.) Three Worlds of CSCL. Can we Support CSCL, pp. 61–91. Heerlen, Open Univ. Nederlands (2002)

18. Dillenbourg, P., Ott, D., Wehrle, T., Bourquin, Y., Jermann, P., Corti, D., Salo, P.: The socio-cognitive functions of community mirrors. In: Fluckiger, F., Jutz, C., Schulz, P., Cantoni, L. (eds.) Proceedings of the 4th International Conference on New Educational Environments, Lugano, May 8-11 (2002)

19. Dimitracopoulou, A.: Computer based Interaction Analysis Supporting Self-regulation: Achievements and Prospects of an Emerging Research Direction. In: Spector, M., Sampson, D., Kinshuk, Isaias, P. (eds.) Special Issue: Cognition and Exploratory Learning in Digital Age, Technology, Instruction, Cognition and Learning TICL, vol. 6(4), pp. 291–314 (2009)

20. Fischer, F., Kollar, I., Mandle, H., Haake, J.: Scripting Computer Supported Collaborative Learning: Cognitive, Computational and Educational Perspectives. Springer (2007)

21. Gasparini, I., Lichtnow, D.: Quality ontology for recommendation in an adaptive educational system International Conference on Intelligent Networking and Collaborative Systems, INCOS 2009, Barcelona, Spain, November 4-5, pp. 329–334 (2009)

22. Gerosa, M.A., Pimentel, G.P., Fuks, H., Lucena, C.: No need to read messages right now: helping mediators to steer educational forums using statistical and visual information. In: Koschmann, T., Chan, T., Suthers, D. (eds.) Proceedings of Computer Supported Collaborative Learning 2005: The Next Ten Years!, May 30-June 4, pp. 160–169. ISLS, LEA, Taipei (2005)

23. Giguet, E., Lucas, N., Blondel, F.-M., Bruillard, É.: Share and explore discussion forum objects on the Calico website. In: Dimitracopoulou, A., O'Malley, C., Suthers, D., Reimann, P. (eds.) Computer Supported Collaborative Learning Practices: CSCL 2009 Community Events Proceedings, pp. 174–176. ISLS (2009)

24. Jermann, P.: Computer Support for Interaction Regulation in Collaborative Problem-Solving. PhD Thesis. University of Geneve, Switzerland (2004)

25. Jermann, P., Soller, A., Lesgold, A.: Computer Software Support for Collaborative Learning. In: What We Know About CSCL in Higher Education, pp. 141–166. Kluwer, Amsterdam (2004)

26. Gunawardena, C., Lowe, C., Anderson, T.: Analysis of global online debate and development of interaction analysis model for examining social construction of knowledge in computer conferencing. Educational Computer Research 17(4), 397–431 (1997)

27. Hiltz, S.R.: Impacts of college level courses via asynchronous learning networks: Some preliminary results. Journal of Asynchronous Learning Networks 1(2) (1997)

28. Hoppe, U.: Use of multiple student modelling to parametrize group learning. In: Greer, J. (ed.) Proc of AIED 1995, 7th World Conference on Artificial Intelligence in Education, Washington DC, USA, pp. 234–249 (1995)

29. Jermann, P.: Computer Support for Interaction Regulation in Collaborative Problem-Solving. PhD Thesis. University of Geneve, Switzerland (2004)

30. Jermann, J., Soller, A., Lesgold, A.: Computer Software Support for CSCL. In: Strijbos, J.W., Kirschner, P.A., Martens, R.L. (eds.) What we know about CSCL and Implementing it in Higher Education, pp. 141–166. Kluwer Academic Publisher (2004)

31. McCalla, G.I., Greer, J.E., Kumar, V.S., Meagher, P., Collins, J.A., Tkatch, R., Parkinson, B.: A peer help system for workplace training. In: Boulay, B.D., Mizoguchi, R. (eds.) Artificial Intelligence in Education: Knowledge and Media in Learning Systems, pp. 183–190. IOS, Amsterdam (1997)

32. Michozuki, T., Kato, H., Hisamatsu, S., Yaehashi, K., Fuzitani, S., Nagata, T., Nakahara, J., Nishomori, T., Suzuki, M.: Promotion of Self-Assessment for Learners in Online Discussion Using the Visualization Software. In: Koschmann, T., Suthers, D., Chan, T.-W. (eds.) Proceedings of Computer Supported Collaborative Learning 2005: The Next Ten Years!, Taipei. ISLS, Taiwan (2005)

33. Nakahara, J., Kazaru, Y., Shinichi, H., Yamauchi, Y.: iTree: Does the mobile phone encourage learners to be more involved in collaborative learning? In: Koschmann, T., Suthers, D., Chan, T.-W. (eds.) Proceedings of Computer Supported Collaborative Learning 2005: The Next Ten Years!, Taipei. ISLS, Taiwan (2005)

34. Oda, T., Satoh, H., Watanabe, S.: Searching deadlocked Web learners by measuring similarity of learning activities. In: Proceedings of Workshop "WWW-Based Tutoring" at 4th International Conference on Intelligent Tutoring Systems (ITS 1998), August 16-19, San Antonio, TX (1998)

35. Papanikolaou, K.A., Grigoriadou, M.: Adaptive educational hypermedia systems in the Internet. In: Retalis, S. (ed.) Advanced Internet Technologies in Education, Kastaniotis, Athens, pp. 204–236 (2005)

36. Petrou, A., Dimitracopoulou, A.: Is synchronous computer mediated collaborative problem solving 'justified' only when by distance? Teachers' point of views and interventions with co-located groups during every day class activities. In: Wasson, B., Ludvigsen, S., Hoppe, U. (eds.) Proceedings of Computer Supported Collaborative Learning 2003: Designing for Change in Networked Learning Environments, Kluwer Academic Publishers (2003)

37. Petrou, A.: Teachers' roles and strategies during usage of technology based collaborative learning environments, in real school conditions. Unpublished doctoral thesis, School of Humanities, University of the Aegean, Rhodes, Greece (2005)

38. Reimann, P.: How to support groups in learning: More than problem solving (keynote talk). In: Aleven, et al. (eds.) Supplementary Proceedings Artificial Intelligence in Education (AIED 2003), pp. 3–16. University of Sydney, Sydney (2003)

39. Reyes, P.: Structural awareness in mediated conversations for collaborative learning environments. PhD Thesis. Université du Maine (2005)

40. Reyes, P., Tchounikine, P.: Mining learning groups' activities in Forum-type tools. In: Koschmann, T., Suthers, D., Chan, T.-W. (eds.) Proceedings of Computer Supported Collaborative Learning 2005: The Next Ten Years!, Taipei. ISLS, Taiwan (2005)

41. Soller, A.L.: Supporting Social Interaction in an Intelligent Collaborative Learning System. International Journal of Artificial Intelligence in Education 12(1), 40–62 (2001)

42. Teplovs, C., Donoahue, Z., Scardamalia, M., Philip, D.: Tools for Concurrent, Embedded, and Transformative Assessment of Knowledge Building Processes and Progress. In: Chinn, C., Erkens, G., Puntambekar, S. (eds.) Proceedings of the International Congress CSCL 2007: Computer Supported Collaborative Learning: Mice, Minds and Society, July 21-26, pp. 720–722. ISLS, Rutgers State University of New Jersey (2007)

43. Vassileva, J., Cheng, R., Sun, L., Han, W.: Designing Mechanisms to Stimulate Contributions in Collaborative Systems for Sharing Course-Related Materials. In: ITS 2004, Workshop on Computational Models of Collaborative Learning, Maceio, Alagoas, Brazil, August 30-September 3 (2004)

Making Adaptations of CSCL Scripts by Analyzing Learners' Online Behavior

Ourania Petropoulou, Georgia Lazakidou, Petros Georgiakakis,
and Symeon Retalis

University of Piraeus, Department of Digital Systems, 80 Karaoli and Dimitriou,
GR-18534 Piraeus, Greece
rpetro@biomed.ntua.gr, glazakidou@gmail.com,
petros@itisart.com.gr, retal@unipi.gr

Abstract. Teachers often use flow design patterns of particular learning strategies in order to define the type of the learning and supportive tasks, their duration, their orchestration as well as the use of learning objects, tool and services needed to support the execution of these tasks. It is quite easy for teachers (even novice learning designers) to create a learning script by applying a flow design pattern. However, they often need to mix and match learning strategies in order to create customized learning scripts that are more appropriate to the learners' preferences and the learning context in general. This task is even more challenging when such adaptations need to be made on-the-fly, i.e. during the learning process and in response to the learner' online behavior. The aim of this paper is to discuss how the learners' interaction data that is collected during an online learning process and analyzed using interaction analysis indicators can be used by a teacher to alter the learning script on the fly. It is shown that with the aid of a learning interaction analysis tool, which is called CoSyLMSAnalytics, a teacher can modify a learning script that is based on a typical Think-Pair-Share strategy in order to offer scaffolds to the learners during the learning process. Also, an example will be shown that will depict how a learning designer can create a variation of a learning scenario that is based on a typical Think Pair Share strategy by using the tool referred above.

1 Introduction

Teachers, acting as learning designers, are called to design effective interactive learning scripts [9]. A learning script describes the type and the flow/sequencing of learning and supportive tasks, the roles, pedagogical rules as well as the learning objects, the tools and the services need to be used for performing the various tasks [18]. There are several learning strategies such as the Predict-Observe-Explain (POE), the Pyramid, the Jigsaw, the Think Pair Share (TPS), Six Hats, etc. which can be used by teachers [34]. Strategies have been documented in a designer friendly way as flow design patterns [13], [14]. The term "flow" is used to

T. Daradoumis et al. (Eds.): Intelligent Adaptation & Personalization Techniques, SCI 408, pp. 179–194.
springerlink.com

portray the coordination and the sequence of tasks during the learning process. Collaborative Learning Flow Patterns (CLFPs) define effective sequences of collaborative learning activities that can be easily reused and communicated to others [12]. A CLFP defines the sequence of the tasks that the strategy dictates as well as other elements needed for the various tasks, such as the duration of a task, the use of a particular tool for a given task, and so on [42]. Teachers can easily create a learning script (lesson plan or scenario) for their course by applying a flow pattern of a specific strategy.

However, very often they need to mix and match strategies thus creating customized scripts which are more appropriate to the learners' preferences, knowledge level, needs and the learning context in general. Also, during the learning process teachers need to adapt their learning script based on learners' performance and online behavior. Making on-the-fly adaptations of a learning script tailored to the specific learning context can result in fewer, less tedious and more engaging tasks [22]. The main question is how teachers can be informed about learners' performance and behavior in an online environment in such a way to easily and quickly adapt their script into the learners' needs. Our way could be the use of interaction analysis indicators. Interaction analysis involves the process of studying and analyzing in-depth the learners' behavior in an online environment and extracting meaningful information that concern learner–learner interactions, learner–instructor interactions, and learner–content interactions.

The aim of this paper is to discuss how the learners' interaction data that are collected during the learning process and analyzed using interaction analysis techniques can be used by teachers to alter a learning script depending on the learning progress. For helping teachers in understanding learners' progress, an interaction analysis tool called CoSyLMSAnalytics was developed [30]. Using the reports about learners' progress which has been extracted from the CoSyLMSAnalytics tool, teachers can quickly proceed in making easily the necessary configurations in the initial learning script. As it will be discussed, teachers not only need interaction analysis tools but also more automatic ways in form of "if then/else" rules i.e. "IF [something occurs] THEN [adapt your design in this way]".

The structure of this chapter is the following. Firstly we describe the role of interaction analysis indicators for monitoring learners' online behavior. Then, using the example of a learning script based on the Think-pair-Share (TPS) strategy, we show how a teacher could make on-the-fly configurations of the CSCL script using the results from real time interaction analysis. The paper ends with some ideas for further research and development on the idea of using interaction analysis indicators for making valuable alternatives of CSCL scripts depending on learning conditions.

2 Interaction Analysis

2.1 Interaction Analysis Indicators

Over the last decade, the international scientific community in the field of e-learning environment (and especially computer supported collaborative learning-CSCL) has shown great interest in the ways to record, analyse and interpret data about interactions that occur during the e-learning process. This is how interaction analysis indicators, methods and tools have been proposed [1], [2], [3], [10], [44], [8], [11], [23], [43]. The life cycle of interaction analysis process consists of two phases (sub-processes). The first phase concerns the collection of data about learning interactions such as messages exchanged among learners, learning objects studied by learners, etc. These data can be retrieved by the logfiles of the e-learning platforms system and can be further annotated depending on their type (e.g. type of messages such as social, question, explanation, etc) for further analysis. The second phase includes the application of methods for the analysis and visualization of the data according to a set of indicators (Interaction Analysis Indicators). The indicators are essentially measurable metrics reflecting in quantitative terms the "volume" and "quality" of interactivity [6], [7], [10], [28].

Thus, a teacher can detect the number and nature of contributions of each learner during the learning process (e.g. number and type of messages exchanged, type and order of learning objects studied, the time for accomplishing a learning task, etc.). These data can be reported either in the form of simple statistical tables, or visualizations such as bar charts, pie charts and other graphs (including sociograms) that can help a teacher draw useful conclusions about the participation of every learner in the learning session (e.g. flow of communication and cooperation among learners in a group). Some of the IA have been proposed by educational researchers [39], [43] or by the teachers themselves [4], [24], [46] or are included in specialized IA tools [5], [32].

The following table (Table 1) contains a set of IA indicators which have been categories according to the classic Moore's model of learning interactions [26].

Table 1 A list of Interaction Analysis Indicators

A. Learning Products	
A1.	The correctness of the learning product (e.g final solution of the problem)
A2.	The completeness of the learning product
A3.	Final mark of the student per activity (e.g final test/final report),
A4.	Total number of correct answers in relation the insufficient or erroneous ones per student (in connection with his/her role)
A5.	Final mark of team per activity
A6.	Final mark of team in the total of activities
A7.	Observation of timetable of activities per student
A8.	Observation of timetable of activities per team
A9.	Duration of completion of an activity (eg in the resolution of problem)

Table 1 (*continued*)

B. Learning Process	
B.1- Interaction: learner-learner & learner-teacher	
B1. 1	Total number of students' messages (write) per activity
B1.2	Total number of team' messages (write) per activity
B1.3	Total number of files shared per student per activity
B1.4	Number and type of students' contributions per activity (e.g. messages argumentation analysis)
B1.5	Comparison of the average of messages (reading) in an assigned activity per student
B1.6	Comparison of the average of messages (reading) in an assigned activity per team
B1.7	Flow of communication (e.g. to one or to many)
B1.8	Total number of discussion threads per student
B1.9	Average of discussion threads per team
B1.10	Degree of central position per student per activity and per phase of instructive scenario
B1.11	Degree of density of social networks developed per team
B1.12	Total number of demands (help –support- encouragement) per student/team towards teacher/schoolmates
B.2 – Interaction: learner-learning resources	
B2. 1	Number of visited resources per student
B2. 2	Number of visited resources per team
B2.3	Popular resources per student
B2.4	Resource's history per student
B2.5	Popular resources per team
B2.6	Total number of visited resources according to his role
B2.7	Total number of visited resources of other roles
B2.8	Number and type of resources that students propose
B2.9	Successive courses of learning resources access per team per activity (theory, example, exercise)
B2.10	Clustering of students (view learning resources)
B2.11	Clustering of teams (view learning resources)

2.2 Interaction Analysis Tools

Various specialized analysis tools have been created that collect data about learners' behavior in an e-learning environment, analyse these data according to IA indicators and produce reports. An overview of interaction analysis tools or educational data mining tools can be found in [33]. Well known IA tools are the following: The system DIAS (Discussion Interaction Analysis System) is an IA tool that mainly emphasis on analyzing learners' interaction in asynchronous computer supported collaborative learning environment using a wide range of IA indicators (over 65 different indicators) [3]. The ColAT tool can be used for qualitative and quantitative IA of synchronous computer supported collaborative learning environment [1].

Degree is an IA tool that helps teachers analyze learners' interaction in asynchronous computer supported collaborative learning environment as well as learners have an idea of their learning behavior [2]. GISMO is another usable tool for collecting, analyzing and visualizing data that concern learners' interaction that occur in a Moodle learning management system [25]. Another one tool is the Student Activity Monitoring using Overview Spreadsheets (SAMOS), which facilitates the automatic generation of weekly monitoring reports derived from data contained in server log files. These reports provide online instructors with visual information regarding students' and groups' activity, thus allowing for a quick and easy classification of students and groups according to their activity level [29].

CoSyLMSAnalytics is the most recently developed IA tool (see Figure 1) which is a free open source software tool. Like the GISMO tool, CoSyLMSAnalytics has been developed for helping teachers analyze learners' behavior that occurs in an e-learning environment that is supported by the Moodle Learning Management System (LMS). CoSyLMSAnalytics can be connected to the Moodle database in order to retrieve that necessary interaction data for a specific course and/or a specific time period that a teacher may want to analyse. CoSyLMSAnalytics analyses the retrieved data and produces reports in the form of graphs and/or tables that a teacher can edit or save.

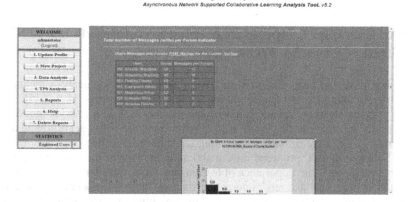

Fig. 1 Screenshot of the CoSyLMSAnalytics Interaction Analysis Tool

More specifically, the CoSyLMSAnalytics interoperates with Moodle LMS in order to:

- Produce usage statistics such as count of visits, average time interval spent on an activity, and present them in various formats such as cross tabs and charts.
- Provide detailed descriptive statistics regarding the interactions via an asynchronous discussion forum.
- Exploit learning activities flow sequential patterns by drawing the exact paths followed by each learner individually or in groups.

- Perform path analysis with the creation of more complex queries that reveal interesting correlations and association rules among students' learning paths.

Data can also be exported in such a format that can be further analysed using the SPSS tool for an in-depth statistical analysis, the NetDraw software for social networks analysis, and the WEKA tool for sophisticated data mining processing.

CoSyLMSAnalytics supports the following IA indicators:

- Flow of Communication
- Total number of messages (write) per Forum/per Group
- User Contribution per Forum/Discussion (Messages Argumentation Analysis)
- Number of Group Contribution per Forum
- Total Number of messages (write) per Student
- Comparison of student messages (write) order by Forum
- Popular resources per student
- Resource's history per student
- Popular Resources per group
- Distribution of actions on learning resources per user
- Clustering of groups (view learning resources)
- Clustering of students (view learning resources)

3 An Example: Adapting a Learning Script Based on a TPS Strategy Using IA Techniques

Recently, there has been a movement toward using interaction analysis techniques for offering adaptive assistance during an e-learning process [19], [35], [20]. Monitoring the learners' online behavior and performance, may provide the teacher with useful information that could help him/her intervene, offer scaffold and adapt the future tasks of the instantiated learning script in order to improve learning [36], [40]. Soller, et al. [38] proposed the collaboration management cycle (see Figure 2) which clearly states the importance of learning analytics tools for gathering, analyzing and interpreting learners' interaction data in order to proceed to corrective actions during a learning scenario (e.g. providing guidance, changing group composition, etc.) in order to create on-the-fly a more effective learning script.

In order to illustrate more clearly teacher's on-the-fly adaption of a learning script based on the results from an interaction analysis, let's take into consideration the following case. In an environmental educational program at an urban school in Athens students of the second grade of lyceum (15-16 years old) were asked to participate into a computer supported collaborative learning scenario about nuclear energy (12/10/2009 till 26/10/2009). Using the Moodle learning management system, students accessed online resources which were categorized into 7 categories (associated with 7 open questions such as the pros and cons of having nuclear power plants), exchanged ideas and resources using an asynchronous web-forum and submitted their reports. The learning script was designed addressing the principles of the Think-Pair-Share (TPS) collaborative strategy.

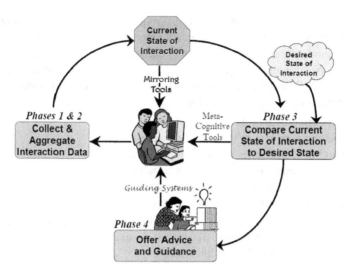

Fig. 2 The Collaboration Management Cycle accordingto [38]

According to the TPS strategy [21], which aims to increase all students' engagement in various learning activities, the learning process is comprised of the following steps:

- Think about the subject of the nuclear energy and individually submit a report that answers to seven (7) questions such as the pros and cons of the use of nuclear power plants worldwide and in their country, the safety regulations for building an new nuclear power plant, examples of activists who are in favor or against the use of nuclear power, etc. During this phase, students had access to online resources suggested by the teacher. Students could also communicate with each other to discuss about these resources, suggest other ones at a web-forum.
- Work in pairs and discuss their thoughts about the nuclear energy. Each pair of students had to electronically submit their assignment which was a report that answers the same seven questions. Each pair had their own workspace to share resources, their individual report of the Think phase as well as the interim versions of the group deliverable. Also, they could communicate asynchronously via their own web-forum. According to the script, the teacher did not want the pairs to communicate (pairs were competing)
- Each pair shares their report with the rest of the students in the classroom, and all students should discuss each question and make a decision about the use of nuclear energy in Greece. For the needs of this phase, students could exchange ideas at a web-forum created for this purpose.

TPS is a valuable collaborative strategy because: a) it asks students to spend more time thinking individually for executing a task and take ownership of the learning process rather than rely solely on the peers' ideas; b) it increases the interactions among peers during the pair phase and c) it requires all learners learn about their peers' ideas during the share phase in an attempt to reach to a consensus.

When instantiating this learning script in classroom, the teacher was monitoring the students' performance using specific interaction analysis indicators integrated into the CoSyLMSAnalytics tool. As shown in Figure 3, these indicators per phase of the learning script were the following:

- The correctness of the learning product
- The completeness of the learning product
- Number of Messages per student per Forum
- Number and type of students' contributions
- Number of visited resources
- Number and type of resources that students propose

Also, the teacher was assessing the students' deliverables at the end of each phase in order to give helpful feedback as well as to evaluate their performance and identify misconceptions or lack of knowledge.

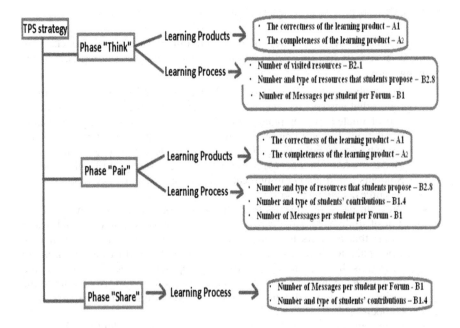

Fig. 3   The CoSyLMSAnalytics interaction analysis indicators for the TPS strategy

When applying the TPS strategy the teacher realized that the students did not have the expected performance per phase. Thus, when students faced difficulties in acquiring the required knowledge and skills during the learning process, adaptations of the learning script have been made.

During the Think Phase

During the Think phase, the teacher using a rubric i) assesses the accuracy of answers for the 7 given questions which were included in the individual learners' deliverables and ii) monitors learners' behavior with respect to whether they studied the given learning resources [31]. An extract of the assessment rubric that a teacher can use is shown in Table 2.

Table 2 An extract of an assessment rubric for Think phase

Criterion	Excellent Performance	Medium Performance	Low Performance
Accuracy of Facts per question (Content)	For almost all the posed questions (more than 6) the supportive facts are reported accurately.	For most of the given questions (4-6) the supportive fact are reported accurately.	For only few questions the supportive facts are reported accurately OR NO facts are reported OR most are inaccurately reported.
Making use of the learning resources (IA indicator: Clustering of students (view learning resources per question)	Learner visited almost all the given learning resources (e.g. more than 8)	Learner visited some of the given learning resources (e.g. 5-8)	Learner did not visit the given learning resources (e.g. less than 5)

Using CoSyLMSAnalytics and more specifically the IA indicator "Clustering of students (view learning resources per question)", a teacher can easily find out whether students accessed or not the given learning resources per question. For example in Figure 4, for a set of six (6) students, the teacher can observe that two (2) of them used of more than 8 learning resources which were related to research question 4, two (2) students visited most of them (7 and 5 learning resources respectively) while two (2) students did not access most of the learning resources (e.g. 3 and 1 learning resources respectively).

At this phase, the teacher can combine the performance that some students had in their individual reporting task (i.e. accuracy of the answers to specific questions) with the IA data. If a student visited all the given resources but failed to give a good answer to a given question, teacher may decide to pair that student with another student who also studies most of the learning resources and performed well thus ensuring apporopriate conditions of support, encouragement and active engagement and active engagement of weak problem solvers during the peer collaborative learning process. If a student had a low performance due laziness the teacher can suggest the student to study the remaining learning resources before proceeding to the second phase(Pair). This idea also accords to Isotani & Mizoguchi[16] which considers not only the number of participants but also their current learning state, in order to make the most appropriate group compositions for collaborative learning activities.

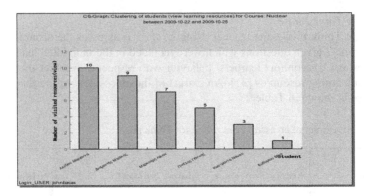

Fig. 4 CoSyLMSAnalytics Interaction Analysis Indicator: Cluster of students' visits of learning resources

Table 3 An extract of an assessment rubric for Pair phase

Criterion	Excellent Performance	Medium Performance	Low Performance
User Contribution per Forum/per Discussion (IA indicator : Messages Argumentation Analysis)	Consistently and actively contributes knowledge, opinions, and skills without prompting or reminding (e.g. more than 3-4 answers and advices)	Contributes information to the group with occasional prompting or reminding (e.g. 1-2 answers and advices)	Contributes information to the group only when prompted

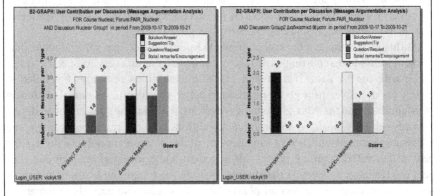

Fig. 5 CoSyLMSAnalytics the Interaction Analysis Indicator: Flow of communication (num of annotated messages)

During the Pair Phase

During the Pair phase students are asked to work as a team and deliver a joint deliverable which contains the team's views on the topic of nuclear energy. At this phase, if the teacher, evaluating the performance of students according to the IA indicator index "Flow of communication in a group", realises that certain groups do not sufficiently collaborate, can alternatively apply the Jigsaw strategy designed to generate more intense interactions among learners. The teacher will assign roles to the learners (roles which will be related to the questions posed). Thus learners will not only have to collaborate in their own group but also with their classmates who are assigned the same role. However, in this way, the competitive mode of the TPS strategy is transformed to cooperation among students.

The teacher can evaluate the students' progress using the rubric that is shown in Table 3 and with the aid of the results from the "Flow of communication in a group" IA indicator that are visualized by the CoSyLMSAnalytics tool. As shown in Figure 5, students in group 1 (group 1) actively participate in collaborative learning by contributing equally to investigate the given set of questions. However, students in group 2 have less and uneven collaborative activities. One of the students sent two (2) messages that indicated only some suggestions for answering on questions while the other group member sent 4 messages in total without suggesting any solid answer to a given question. Thus, the teacher needs to help these students by mixing them with students from other groups. One way, as already mentioned, is the use of the alternative collaborative strategy such as the Jigsaw.

During the Share phase

Equally adaptations of the learning script for the share phase can be made in case the teacher realizes that the students are actively participating to the exchange of ideas for reaching to a consensus about the nuclear energy. The teacher could enforce the brainstorming strategy if some students are lurkers during that phase. Again with the aid of IA indicators of the CoSyLMSAnalytics, the teacher can monitor the students' performance and intervene when needed (e.g. IA indicator: number of messages per student).

Thus, finally the learning script for this case was a variation of the TPS strategy which is shown in Figure 6.

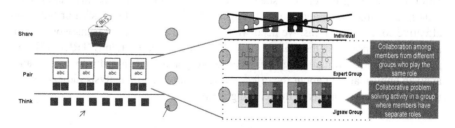

Fig. 6 The adaptation of a TPS learning script when monitoring students' online behavior

Of course, the teacher could have changed the learning script in another way using the same or other IA indicators. The key idea of this exemplar case is to illustrate that IA indicators may help a teacher propose scaffoldings on time during a collaborative learning process. Such scaffoldings can take the form of changes in the sequence of learning tasks that better fit to the needs of the learners for given learning conditions. There are no fixed criteria for choosing an alternative strategy, going for instance from TPS to Jigsaw or to brainstorming strategy. As indicated in idspace project D1.3 deliverable [15] such criteria could be classified into three main categories:

- Higher order criteria such as the type of learning objectives need to be accomplished, the complexity of implementing a strategy as a whole and as per set of activities of the strategy. For example, Hernández-Leo et al. [13] have identified characteristics of collaborative learning strategies that can help practitioners select the most appropriate strategy
- Organisational criteria which will involve decisions about the formation of groups, leadership schema, etc. For example, Katzenbach and Smith [17] in their book called "The Wisdom of Teams", as well as Strijbos et al. [41] suggested that six factors are necessary to effective teamwork. Nemiro [27] presented criteria for selecting the most appropriate leadership structures in virtual teams based on contextual constraints.
- The technological criteria which will concern the use of specific tools, features for the implementation of the strategy into a real specific scenario/case. For example, Shneiderman [37] mentions various factors that should be taken into consideration when choosing the tools for facilitating the collaborative creativity process.

4 Concluding Remarks

Since learners' interactions in an e-learning process can be automatically collected and analyzed using interaction analysis techniques, they should be used by teachers as input for making on-the-fly adaptations of the learning scripts in order to offer effective learning experiences. A step forward will be to create CSCL "coaching systems". As Kumar et al. [19] state in their review of collaboration support systems, "coaching systems help students engaged in computer-mediated collaboration by assessing the current state of student interaction, comparing the current state to a desired state, and then offering assistance to the students". For building effective coaching systems, we need well structured architectural frameworks such as the one proposed by the Collaborative Tutoring Research Lab (CTRL) [45]. Finally, systematic evaluation studies need to be performed in order to collect and analyze empirical data about the effective use of interaction analysis indicators for the on-the-fly adaptations of the learning scripts.

References

[1] Avouris, N., Dimitracopoulou, A., Komis, V.: On evaluation of collaborative problem solving: Methodological issues of interaction analysis. Journal of Computers in Human Behaviour 19(3), 147–167 (2003)

[2] Barros, M., Verjedo, M.: Analysing student interaction processes in order to improve collaboration. The DEGREE approach. International Journal of Artificial Intelligence in Education 11, 221–241 (2000)

[3] Bratitsis, T.: Development of flexible supporting tools for asynchronous discussions, by analyzing interactions among participants, for technology supported learning. Doctoral Dissertation, University of the Aegean, Rhodes, Greece (2007)

[4] Bratitsis, T., Dimitracopoulou, A.: Interpretation of Computer Based Interaction Analysis Indicators: a significant issue for enhancing collaboration in Technology Based Learning. In: Kock, N. (ed.) Interdisciplinary Perspectives on E-Collaboration: Emerging Trends and Applications, Advances in E-Collaboration. Book series, pp. 31–59. IGI-Global, USA (2010)

[5] Bratitsis, T., Dimitracopoulou, A., Martínez-Monés, A., Marcos, A., Dimitriadis, Y.: Supporting members of a learning community using Interaction Analysis tools: The example of the Kaleidoscope NoE scientific network. In: Diaz, P., Kinshuk, Aedo, I., Mora, E. (eds.) The 8th IEEE International Conference on Advanced Learning Technologies, pp. 809–813. IEEE Computer Society (2008)

[6] Bravo, C., Redondo, A., Verdejo, F., Ortega, M.: A framework for process-solution analysis in collaborative learning environments. International Journal of Human-Computer Studies 66(11), 812–832 (2008)

[7] Collazos, C., Guerrero, L., Pino, J., Renzi, S., Klobas, J., Ortega, M., Redondo, M., Bravo, C.: Evaluating collaborative learning processes using system-based measurement. Educational Technology & Society 10(3), 257–274 (2007)

[8] Dettori, G., Persico, D.: Detecting self-regulated learning in online communities by means of Interaction Analysis. IEEE Transactions on Learning Technologies 1(1), 11–19 (2008)

[9] Dillenbourg, P., Järvelä, S., Fisher, F.: The evolution of research on computer-supported collaborative learning: from design to orchestration. In: Balacheff, N., Ludvigsen, S., de Jonh, T., Lazonder, T., Barnes, S. (eds.) Technology-Enhanced Learning. Principles and Products, pp. 3–19. Springer, Heidelberg (2009)

[10] Dimitracopoulou, Martinez, A., Dimitriadis, Y., Morch, A., Ludvigsen, S., Harrer, A., Hoppe, U., Barros, B., Verdejo, F., Hulsof, C., Fessakis, G., Petrou, A., Lund, K., Baker, M., Jermann, P., Dillenbourg, P., Kollias, V., Vosniadou, S.: State of the art on interaction analysis: Interaction analysis indicators. ICALTS Project Deliverable: D.26.1 (2004),
http://www.rhodes.aegean.gr/ltee/kaleidoscope-icalts/Publications/D1StateoftheArtVersion_1_3ICALTS_KalNoE.pdf (retrieved October 27, 2006)

[11] Duque, R., Noguera, N., Bravo, C., Garrido, J., Rodríguez, M.: Construction of interaction observation systems for collaboration analysis in groupware applications. Advances in Engineering Software 40(12), 1242–1250 (2009)

[12] Hernandez-Leo, D., Asensio-Perez, J.I., Dimitriadis, Y.: Computational representation of collaborative learning flow patterns using IMS Learning Design. Educational Technology & Society 8(4), 75–89 (2005)

[13] Hernández-Leo, D., Villasclaras-Fernández, D., Asensio-Pérez, J., Dimitriadis, Y.: Generating CSCL scripts: From a conceptual model of pattern languages to the design of real scripts. In: Goodyear, P., Retalis, S. (eds.) E-learning Design Patterns Book, pp. 49–64. Sense Publishers (2010)

[14] Hernández-Leo, D., Villasclaras-Fernández, E.D., Jorrín-Abellán, I.M., Asensio-Pérez, J.I., Dimitriadis, Y., Ruiz-Requies, I., Rubia-Avi, B.: Collage: A collaborative learning design editor based on patterns. Educational Technology & Society 9(1), 58–71 (2006)

[15] IdSpace project Deliverable D1.3. "Creativity enhancing pedagogical strategies for idSpace learning Design Document - The potential of pattern based learning strategy recommendations for idSpace" (2010), http://www.idspace-project.org/ (retrieved on August 31, 2011)

[16] Isotani, S., Mizoguchi, R.: Using Ontologies for an Effective Design of Collaborative Learning Activities. In: Proceedings of the International Conference on Artificial Intelligence in Education (AIED), Marina del Rey, CA, USA, pp. 578–580 (2007)

[17] Katzenbach, J.R., Smith, D.K.: The Wisdom of Teams: Creating the High-performance Organization. Harvard Business School, Boston (1993)

[18] Kollar, I., Fischer, F., Hesse, F.: Collaboration scripts – a conceptual analysis. Educational Psychology Review 18(2), 159–185 (2006)

[19] Kumar, R., Rosé, C.P., Wang, Y.C., Joshi, M., Robinson, A.: Tutorial dialogue as adaptive collaborative learning support. In: Luckin, R., Koedinger, K., Greer, J. (eds.) Proceedings of the 13th International Conference on Artificial Intelligence in Education, pp. 383–390. IOS Press, Amsterdam (2007)

[20] Lazakidou, G., Retalis, S.: Using computer supported collaborative learning strategies for helping learners acquire selfregulated problem solving skills in Mathematics. Computers and Education 54(1), 3–13 (2010)

[21] Lyman, F.: The Responsive Class Discussion. In: Anderson, A.S. (ed.) Mainstreaming Digest. College of Education, University of Maryland, College Park (1981)

[22] Magnisalis, I., Demetriadis, S.: Modeling Adaptation Patterns with IMS-LD Specification: A Case Study as a Proof of Concept Implementation. In: Proceedings of the 2009 International Conference on Intelligent Networking and Collaborative Systems–INCOS, pp. 295–300. IEEE Computer Society, Washington, DC (2010)

[23] Martinez-Mones, A., Harrer, A., Dimitriadis, Y.: An interaction-aware design process for the integration of interaction analysis into mainstream CSCL practices. In: Puntambekar, S., Erkens, G., Hmelo-Silver, C. (eds.) Analyzing Interactions in CSCL: Methods, Approaches and Issues, pp. 269–292. Springer (2011)

[24] Mazza, R.: Using information visualisation to facilitate instructors in web-based distance learning. (Doctoral dissertation). University of Lugano, Switzerland (2004)

[25] Mazza, R., Botturi, L.: Monitoring an Online Course with the GISMO Tool: A Case Study. Journal of Interactive Learning Research 18(2), 251–265 (2007)

[26] Moore, G.: Three types of interaction. The American Journal of Distance Education 3(2), 1–6 (1989)

[27] Nemiro, J.: Creativity in virtual teams: Key components for success. Jossey-Bass/Pfeiffer, San Francisco (2004)

[28] Pena-Shaff, J., Nicholls, C.: Analyzing student interactions and meaning construction in Computer Bulletin Board (BBS) discussions. Computers and Education 42(3), 243–265 (2004)

[29] Pérez, J., Daradoumis, A., Faulin, T., Xhafa, F.: SAMOS: A model for monitoring students' and groups' activity in collaborative e-Learning. International Journal of Learning Technology (IJLT) 4(1/2), 53–72 (2009)

[30] Petropoulou, O., Altanis, I., Retalis, S., Nicolaou, A., Kannas, C., Vasiliadou, M., Pattis, I.: Building a tool to help teachers analyse learners' interactions in a networked learning environment. Educational Media International 47(3), 231–246 (2010)

[31] Petropoulou, O., Vasilikopoulou, M., Retalis, S.: Enriched Assessment Rubrics: A new medium for enabling teachers easily assess students' performance when participating to complex interactive learning scenarios. Operational Research: An International Journal 11(2), 171–186 (2009)

[32] Romero, C., Ventura, S.: Educational Data Mining: A Review of the State of the Art. IEEE Transactions on Systems, Man and Cybernetics, Part C: Applications and Reviews 40(6), 601–618 (2010)

[33] Romero, C., Ventura, S.: Educational data mining: A survey from 1995 to 2005. Expert Systems with Applications 33(1), 135–146 (2007)

[34] Rowan, K.: Glossary of Instructional Strategies (2010), http://www.beesburg.com/edtools/glossary.html (retrieved May 25, 2011)

[35] Rummel, N., Weinberger, A.: New challenges in CSCL: Towards adaptive script support. In: Kanselaar, G., Jonker, V., Kirschner, P.A., Prins, F. (eds.) International Perspectives of the Learning Sciences: Cre8ing a Learning World, pp. 338–345 International Society of the Learning Sciences (2008)

[36] Shepard, L.: The role of assessment in a learning culture. Educational Researcher 29(7), 4–14 (2000)

[37] Shneiderman, B.: Creativity support tools: Accelerating discovery and innovation. Communications of the ACM 50(12), 20–32 (2007)

[38] Soller, A., Martinez, A., Jermann, P., Muehlenbrock, M.: From mirroring to guiding: a review of state of the art technology for supporting collaborative learning. International Journal on Artificial Intelligence in Education 15(4), 261–290 (2005)

[39] Spada, H., Meier, A., Rummel, N., Hauser, S.: A new method to assess the quality of collaborative process in CSCL. In: Koschmann, T., Suthers, D., Chan, T.W. (eds.) Proceedings of the CSCL 2005, pp. 622–631. Lawrence Erlbaum Associates, Mahwah (2005)

[40] Stiggins, R.: New Assessment Beliefs for a New School Mission. Phi Delta Kappan 86(1), 22–27 (2004)

[41] Strijbos, J.W., Martens, R., Jochems, W., Broers, N.: The effect of functional roles on group efficiency. Using multilevel modeling and content analysis to investigate computer-supported collaboration in small groups. Small Group Research 35(2), 195–229 (2004)

[42] Turani, A., Calvo, R.: Beehive: A software application for synchronous collaborative learning. Campus-Wide Information Systems 23(3), 196–209 (2006)

[43] Villasclaras-Fernández, E.: A design process supported by software authoring tools for the integration of assessment within CSCL scripts (Doctoral dissertation). University of Valladolid, Spain (2010)

[44] Voyiatzaki, E., Margaritis, M., Avouris, N.: Collaborative Interaction Analysis: the teacher's perspective. In: Proceedings of the 6th IEEE International Conference on Advanced Learning Technologies, Kerkrade, Netherlands, July 5-7, pp. 345–349 (2006)

[45] Walker, E., Rummel, N., Koedinger, K.: A Research-Oriented Architecture for Providing Adaptive Collaborative Learning Support. User Modeling and User-Adapted Interaction 19(5), 387–431 (2008)

[46] Zinn, C., Scheuer, O.: Getting to Know Your Student in Distance Learning Contexts. In: Nejdl, W., Tochtermann, K. (eds.) EC-TEL 2006. LNCS, vol. 4227, pp. 437–451. Springer, Heidelberg (2006)

Behavioural Prototyping©: Making Interactive and Intelligent Systems Meaningful for the User

Stephen Benton, Boris Altemeyer, and Bryan Manning

University of Westminster, London, UK
bentons@westminster.ac.uk
boris.altemeyer@my.westminster.ac.uk
bryan.manning@btinternet.com

Abstract. Current studies in the field of assistive design highlight the aim to develop intelligent and interactive services capable of maintaining and augmenting support during broader conditions of decline in personal and infra structural resources (Bos, 2007). These systems are frequently challenged to produce measureable improvements in user learning and function while also responding to the attendant challenges of coherency, interoperability and usability (Manning and Benton, 2010). Such systems are frequently intended to build and transfer augmented coping capacity via the interaction of system- use and user-behaviour, essentially offering a personal learning space and functional resilience. The operating environment for many such systems can be found, for example within both the formal and informal care services. Challenges to the provision of care and the maintenance of personal choice continue to escalate, as does the pressure on systems designed to meet ever changing user needs and aspirations (Benton and Manning, 2008). In the context of health care, well documented resource and demographic challenges (Symonds et al, 2007, Manning et al, 2008), will escalate the need for system utility as a core source of user autonomy. Generally any interface between a system and user will be characterised by redundancy, a deficient use resulting from issues concerning misalignment between; accessibility, learning support and system functionality. Essential user flexibility, acceptance and acquisition of system based protocols will be influenced by the degree of 'personalisation' achieved by the system with a related impact on use (Rigby, 2010). In this context personalisation relates to the degree to which users' perceptual expectations and 'affordances' are aligned with system design and feedback. It is suggested that individual differences of; cognitive style, techno-acceptance and personality may be evaluated to form a Behavioural Prototype©, an evaluation of users' likely interactive expectations and behaviour with the system. This profile would provide a behavioural link between system design and a broader vision of user need, capacity and aspiration.

Keywords: Behavioural Prototyping©, Intelligent Systems, Meaningful, User Needs, Mental Representation, Profiling, Interdisciplinary.

T. Daradoumis et al. (Eds.): Intelligent Adaptation & Personalization Techniques, SCI 408, pp. 195–210.
springerlink.com   © Springer-Verlag Berlin Heidelberg 2012

1   Background

The development of products and services in many sectors still adhere to the standard procedures which base the focus of design on the dominant purchasing appetite and power within a given market. The consequent market focus highlights generational differences which are – according to the product developers, to be observed in the western world – more interested in products which are affordable, highly complex and style/ trend oriented (compare (Carroll & Kincade, 2007; Dempsey, et al., 2000; Fisk & Rogers, 2002)). These main influences on design and market features represent a design mind frame, one that has been driven by established market behaviour, yet this marketplace is changing and will soon accommodate a significant shift in generation demographic, a shift that may prompt a seed change in system design thinking. For example, nations across the world, not withstanding significant differences between industrialised and developing nations, are facing the prospect of nearly 40% of their population being over 65 by 2051 (Bos, 2007). Failure to capture emerging behavioural trends, including associated shifts in user functional expectations, aspirations and cognitive support preferences, are likely to result in significantly increased system redundancy.

An example of such a failure can be found in the current practices used to guide design for the older generation marketplace. For example, it has been noted that older customers, despite being a potentially lucrative group of consumers, are still "one of the most under-appreciated consumer segments" (p.102, Nam, 2006), the fixation on design for younger generations largely endures. Challenges to developing emerging market niches, through adaptive design, may exceed those originating from technological and functional considerations. This mismatch between marketing focus and the monetary/ economic potential of the target group has been identified in the form of an age bias distribution among marketers themselves (Thomas & Wolfe, 1995).

Importantly, the generic issue is about aligning intelligent design with users perceptions of the product (e.g. 'system') such that the perceived utility of the garment; gadget, health product, application or support service prompts and reinforces its uptake, use and exploration.

2   Psychological Perspectives

In the specific context of developments in assistive technology, from a medical and health oriented point of view, there are many different professional perspectives of the groups involved which have to be taken into account. These groups may broadly be divided into academic or scientific development; specialist professional sub groups (e.g. Clinicians, teachers) and user groups.

The development group needs certain information in order to build useful and meaningful devices that can increase the perceived quality of life of the users. This information need may not initially be obvious to either the user or the developer of the technology. On the one hand the developer might normally not see the need for information about personal preferences of the user in a certain area as it is deemed 'less important than the actual functionality' – as defined according to the mental representations and constructs of the developer. For example, designers may be focused on delivering a system that meets the user requirements specified to them based upon a specific set of tasks/ objectives which are then taken to prompt and guide the design of system capability. The relationship between the function and the potential range of user behaviour with this functional capacity may be obscured by professionally constrained sets of delivery priorities. Information available to designers can also reflect a narrow focus, one derived from an analysis of a limited functional range guided by naive user feedback and professional criteria that limit exploration and flexibility of use. In practical terms, the user might either not see the need to provide certain information when asked as it might seem obvious to them or might not even be given the opportunity to give feedback, while different user groups can reflect disparate perceptions which may only align at the cost of potential system capacity. The underlying risk to effective and adaptive design is that a convergence of user and designer 'functional use' biases can produce a constricted user profile with associated limitations on navigation support for the learner/user. Moreover, as the information is reviewed and translated across and between technical specialists behavioural characteristics/profiles of the "users" can be replaced by emphasises on system driven internal coherence rather than overall interactive coherence.

In Figure 1 below, an outline of the complexity of user behavioural profiling is illustrated, highlighting the behavioural profiling of interactions that is needed in order to integrate cognitive, emotional and aspirational aspects of a system- user interface.

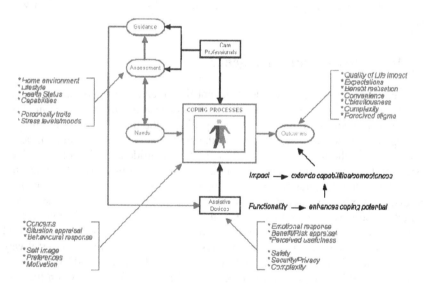

Fig. 1 Designing system support: the system will need to accommodate difficult to quantify behavioural dimensions as an intrinsic part of design, for in this case, technologically supported effective coping.

Any system may have a number of different user groups both primary and secondary, Palmore & Luikart, (1972); Penninx et al., (1999); Savikko et al., (2005). These groups have different ways of interacting with the technology and may have different information needs regarding the technology and this can be divided into two subcategories. In generic terms these categories reflect underlying affordances which shape the quality of system use and 'learnability'. The working definition of an affordance for this paper is any characteristic of the user and system that interacts to shape user behaviour. The first category addresses the information regarding the way in which the device operates and the functions it provides. This is a more passive view as here the technology is seen as a pre-set device which completes certain tasks according to its 'factory settings'. This information is important for the initial and subsequent mental representation of the technology and the in the way it will become part of the user's cognitive routine.

The second category focuses on the interactive and the intelligent part of the technology – learnability and the personalisation and customisation of the systems response to the individual, with regard to the specific needs of target user groups.

At this point the information generated would benefit from a broad evaluative approach, as outlined in Figure 1 and one that was reflective of the physical, cognitive and sensory aspects of system design and users' behaviours. In order to best support usability the system's design concepts have to be assimilated into the "users" behavioural system. Behavioural Prototyping© is intended to collect user's expectations of need and potential benefits across a range of physical, cognitive and sensory aspects to include aspirational use, and furnish designers with ways to reduce redundancy.

3   Interactivity and Intelligence

The user needs an interface in order to be notified of action taken or that suggested by the system, as well as entering information or transmitting information about the users own choices to the system. The quality of the interface for both physical and cognitive dimensions will determine the emotional and judgement confidence experienced by the user in the system, and where confidence is high the product/ system up take, use and sustained exploration is likely to increase. An extension to this is the human interaction at the other end of such systems where medical professionals might be included in the setting-up and programming of the system as well as the long-term and emergency monitoring of the users. The intelligence of such a system can be judged in many different ways and is foremost assessed on the ability of the system to learn and respond to the user in a meaningful way (compare (Ben-Zur, 2002; Dempsey et al., 2000; Fisk & Rogers, 2002;)).

At this point it is important to point out that the relationship between the user and the system is not a simple one-way interaction. This interaction represents the behavioural interface, the point at which the designers' mental representations (affordances) of system use are transformed by actual use into formal operating parameters, as exhibited through user behaviour. In theory proper and effective alignment between the design aim and utility of actual use will be largely influenced by the alignment of designer and user affordances. In brief, a working definition of what makes an intelligent system can be found in the demonstrable alignment between intended and actual use. Following from this the design of a product can be seen as intelligent if it allows an interaction with the user, which fosters and reinforces pre-set learning strategies (cognitive affordances) as well as compensating for possible cognitive impairment in a respectful way.

"Humans and machines interact because of human adaptation, via training and experience, to a machines design, and the machine interacts with humans through the logical possibilities of its initial design and to whatever possibilities for learning through experience and evolving that may be designed into it." (Day, 2011, p. 85)

Day (2011) outlined a general idea of interaction between humans and technology in his critique of information and retrieval systems. In his paper he introduced the idea of quasi- objects and the interaction of subjects with these, turning the subjects into quasi-subject. While being a very valuable approach to the issue of how interaction with technology affects the human agent, seen from a Behavioural Prototyping© perspective it highlights both the potential of user prototyping as well as the pitfalls of poor interaction design. The essence of intelligent design is the degree of accessibility and connectivity that is 'readily' acquired by a wide range of users for a given system. Connectivity will be related to the quality of the Behavioural Prototype© used to design into the system opportunities for the user to learn ways of use and ways to explore further use. It is common that when we meet and work with people for the first time problems of communication, intent and style can present, any new relationship can raise problems to be solved, and the user's relationship to, for example a new technological application, is no different.

3.1 Problem Solving and Learning

It has been pointed out by Anderson (1993) with reference to the work of Newell and Simon (1972) that humans continuously engage in problem solving activities in daily life. The perception of the different problems encountered are those of a 'problem state' (Anderson, 1993), i.e. a situation is perceived as being based or defined by a problem which has to be solved by the agent (user). In this case it is very important to define whether the reference is made to an internal or an external situation (*ibid*), as the situation itself might be a perception or mental representation created by the user, therefore not being directly accessible or modifiable by other agents.

For the problem solving itself the agent considers the different operators which are available (ibid). Operators can be regarded as catalysts which allow the agent to change one situation into another, which – ideally – is more closely aligned with the end-goal of the agent, which motivated him / her to perform the interaction (i.e. solve the problem) in the first place. The combination of different operators and different states, defining the so called 'problem space' (Anderson, 1993) which offers a certain number of possible interactions, which can be described as being :

$$NPI_i = z^s$$

With NPI being the number of possible interactions in a given setting i, z being the number of operators in this setting and s being the number of states the problematic situation can be represented in, including the goal state, where the problem has been solved and is therefore only a latent but inactive variable.

Keeping this formula in mind, any assistive device can only be of benefit to the prospective user if one or more of the following conditions is/are met:

- The number of interactions (use of operators to change the state of the problem situation) using the device must be lower than the number of interactions necessary without the help of the device.
- The operators which are required to use the device in a facilitating manner must be easier to conceptualise than the operators present in the original problem state.
- The operators which are required to use device in a facilitating manner must be easier to 'operate / apply' by the agent than the operators which are given in the original problem state.

These conditions can be regarded as factors for the user / agent to engage with for any assistive device or facilitating piece of equipment in the first place. The dimensions of engagement would be based on the mental representation of the usability of this technology. This mental representation would be directly related to the ease with which errors were processed and corrected by the user. While only one of the conditions would need to be met to potentially raise a person's interest in using a system to facilitate the problem solving it is important to note that none of these conditions should be violated, e.g. the number of interactions might become irrelevant if the operators are much harder to 'use' for the agent.

Depending on the user's preferences or the specific situational setting the user might decide that time constraints (e.g. more interactions necessary) might be acceptable as the interactions that are necessary are significantly easier to perform.

Nevertheless, in conducting a full behavioural analysis for product and system design the analysis of user behaviour and learning preferences can usefully be combined with analysis of blocks to error detection and error utilisation. Failure to do so will serve to increase system redundancy and user dissatisfaction. The intrinsic role of problem solving can also be seen in the context of Lewin's Force Field Theory (Lewin, 1938).

Consequently, when an object serves as a means to an end it is imperative to study the user needs to the fullest extent. This includes all disciplines involved in the design and production process to maximise the understanding of psychological and physical needs as well as technological and design-specific possibilities regarding materials and procedures used. The outcome of such a research will, in the first instance, be a 'fact-oriented' Behavioural Prototype© which contains observable limitations and requirements to objectively quantifiable factors. From Day's (2011) perspective, this is the assessment of the subject in order to define the object.

In a second step these factors should be analysed and assessed within a psychological context – with reference to qualitative user studies involving research into perceptions, attitudes, ambitions, preferences and individual differences among the user group – to be interpreted on a psychological level (Rosenberg, 1965; Rowe & Kahn, 1997; Sveinbjornsdottir & Thorsteinsson, 2008). By identifying how the product can be beneficial to the user on an underlying psychological level (e.g. reduce perceived anxiety) the fact and support oriented 'technological' proto-type can be merged with the actually lived 'psychological' prototype to form the 'full' design-technical prototype. In this way the learning relationship between system and user can be enabled and assessed through the application of discrete psychological categories. These categories combine to provide a design frame-work guided by functional attributes of user access and sub architectural aspects of user learning pathways. This assessment and interpretation could be compared to the quasi-subject, which the user becomes in using a particular piece of equipment in a defined way to reach a goal that, momentarily, governs his or her behaviour or even represents a key factor of who the user is. The issue here is how the potential usability, inherent to a system, is able to shape and encourage the users' pursuit of problem solving beyond that manifest in its initial functional capability.

3.2 Affordances and the Nature of Perceived Use

A key issue for this area of research is to establish viable solutions that allow the designers to match the perceived needs of the user and the mental representation of a possible solution with the affordances offered by a product/ object. As defined by Gibson (Gibson, 1977; Greeno, 1994) an affordance "is a perceived feature of the environment which indicates a possible action in the environment for the per-ceiver" (Almäng, 2008, p.161-162). In this context it is important to note that this opportunity for interaction does not only have to be appealing to the user, i.e. not causing a conflict of interest with different underlying motives and needs, but also be 'performable'. This is based on the fact that the affordance has to be recognisa-ble as a possibility for interaction and that the user must be able to perform the in-teraction (taking into account the multiple physical and mental prerequisites which have to be met in order to accomplish this interaction successfully). An interesting theoretical comparison and discrimination has been made by Day (2011) in article "Death of the user". He differentiates between mechanistic causation and formal causation in human-object interaction, formal causation representing the concept of affordances. These affordances are therefore not mentally stored by the user (comp. Osiurak et al., 2010) but a distinct characteristic of the object itself.

In a more formalized way the characteristics of an object (or even a system) can be illustrated as follows:

$$\sum_{j=0}^{n} s_j = a_j + r_j$$

with

$$r = t + d$$

The factor 'a' represents the affordances of the object. In this equation the variable 'r' is a residual factor. The term residual factor is used here as the factors covered by this factor do not immediately affect the adequacy with which a user need is met, even though the included factors might be seen as needs or affordances in different circumstances (e.g. pieces of art). It consists of factors which have an impact on the product from a technological as well as a design perspective.

From this perspective the issue of how technology is perceived and used is shaped by a 'mental affordance' of the object – an invitation to the user to project certain characteristics onto the object and thereby changing his or her perception of self (as the use of the object defines its 'master'). This framework attempts to offer insight into the notion that users 'intuit' a number of characteristics associated with a system, these intuitions are robust, rather like implicit first impressions, and they can readily anchor initial and subsequent user engagement in both a supportive or inhibitory manner. The primary way in which to mitigate inhibitory impressions is through the quality of the system's support of corrective error feedback combined with a fundamental alignment of its core functionality along a continuum that spans basic needs to exploration/ play behaviours.

With regard to design, the credo 'form follows function' comes to mind. Design, which is not necessary to meet the users need (from an 'assistive devices' point of view) is only an expression of the 'zeitgeist' and the designers perception of technology. This does not apply in cases, when the design ensures the perception of an affordance of the object by the user, as this is clearly a function of the object. Certain psychological factors will be perceived by the user as a need and therefore will have to be met by the object/ system by providing relevant affordances. If the need is of a socio-cultural nature the identification of an appropriate group will be key to design, as establishing the balance of structural accommodations made towards individualistic user requirements and average need can be a key challenge to prompting wider access to the technology.

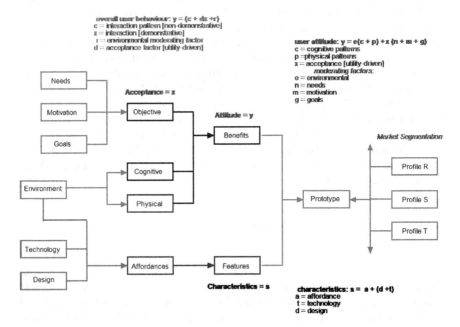

Fig. 2 A method for conceptualising overall user and designer behaviour in a coherent affordance based framework

Matching both, the underlying as well as the verbalized or conceptualized needs of a person to the affordances of an object requires a robust analytical framework that allows for inter- and intra-subject flexibility regarding behaviour and manipulation of the environment. This framework should allow the designers and researchers to identify both demonstrative as well as non-demonstrative concepts that the user holds with regard to needs (Almäng, 2008). In other words the framework should facilitate the categorization of interactions and needs (specified as well as unspecified) into meaningful and demonstrable Behavioural Prototypes©. As shown in Figure 2 the relationship between acceptance and attitudes may be underpinned by the application of affordances and the mediating variable of cognitive and emotive dimensions.

4   Shortcomings of Designs: The Importance of Behavioural Profiles

It has been mentioned earlier in this chapter that research and design processes, without an underlying collaborative and scientific framework, are not covering the full range of possible improvements and use. The essential part not being addressed in these cases is the aspect of 'why' make this design. The issue of 'why' can be divided into multiple subcategories regarding impact factors on design. Physical needs can be regarded as the first and most commonly used category. Nevertheless, interaction with objects always has a psychological impact on the user. This is most obvious when a design does not allow the user to use the object in a 'natural' way, i.e. a way which involves as little mental adjustment of known and intuited procedures as possible (Fisk & Rogers, 2002). Furthermore the psychological impact of a product may be present on different levels and in different facets at the same time. Interacting with the 'value system' of the user positive emotions might be triggered as the object fulfils the need for 'exclusivity' and 'luxury'. In contrast to this a very small sized object combined with a higher price might increase the anxiety induced by the fear of losing this object and might prevent the individual to buy the product or make the end user uncomfortable using it. Objects are connected with the personality of the user on multiple levels which allows the assessment of their impact at several different stages. First of all they should serve a purpose, e.g. provide a solution to a problem or perform a given task. This reduces the general 'need' which they are – hopefully – purpose built for. On a second level the user experiences further needs and related anxieties. These can be connected to other 'tasks' which need to be performed or other, non related factors of everyday life such as personal safety and security.

Referring back to the learning and problem solving aspect as well as the linked mental representations, a user/ agent will have, the number of interactions to solve a problem (while using the assistive device) will determine the success of this technology. If the needs of the user and – based on these – the desired 'end state' of a 'problem space' (Anderson, 1993) is not known to the designer, it is very likely that the benefit of the technology will either not be realised fully or not be realised at all. This lack of 'perceivable potential benefits' in the form of less or easier interactions necessary to solve a problem is likely to cause the user / agent not to accept the technology as it does not seem to facilitate the task at hand (i.e. a reduction of 'distance' between the currently perceived state and the desired end state: "The basic cycle of the problem solver is to look for the biggest difference between the current state and the goal state and try to reduce that difference."(p.37, Anderson, 1993). This phenomenon has been labelled 'Technology Acceptance' (Davis, 1989; Davis, et al., 1989; Venkatesh & Davis, 2000; Venkatesh, et al., 2003), models of which are shown in Figures 3 and 4.

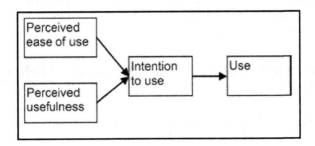

Fig. 3 Basic assumptions of TAM (Technology Acceptance Model)((Heerink, et al., 2010) p. 363)

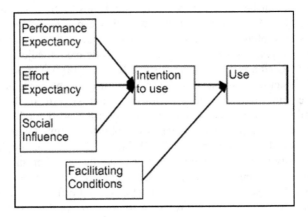

Fig. 4 UTAUT (Unified Theory of Acceptance and Use of Technology) model: Direct influences (Heerink, et al., 2010, p. 363)

5   Prototypes and Archetypes

The underlying personality of the user is likely to determine if a purchase is made or if the product, in general, fits the needs of the user. Perceived needs go beyond the factual requirements of the task the object is meant to perform or the usability of the product. Based on psychological predispositions of the user the needs will be triggered in differing intensities under different situations and be perceived in different ways but might still serve the same purpose. These needs are therefore aligned with the persona that is 'lived' by the user. In psychology the term 'archetype' has been widely used in typology and psychometric assessment procedures which is why the authors suggest for this persona (in a research context) to be referred to as a prototype. This psychological prototype includes relevant; attitudes, preferences, ambitions and – most importantly – needs of a person. An essential ingredient of the prototypes framework is a capacity to represent users' likely learning pathways, as a system will need to balance those hard-wired and open ended attributes into overall capacity.

Modelling a design on an imaginary prototype can promote the 'one size fits' all approach, when in fact, the statistical mean of all user needs and preferences is unlikely to be found in a 'real' person. This results in 'one size fits none' as the median of the derived user needs is unlikely to 'fit' the generic need structure in an efficient manner. A design based on this approach is only ever valid if the object serves no other function than to allow the user/ owner to project a certain 'image' on it and identify him-/ herself with this image as can be found amongst many branded and lifestyle products.

5.1 Inclusive Design

In the recent past a stream of work called 'inclusive design' has developed out of the idea that products (including building structures etc.) should be designed to meet the actual needs of the end users rather than those of a brand strategy. A definition of the concept of inclusive design has been provided by the British Standards Institute in 2005, stating that inclusive design is: "The design of mainstream products and/or services that are accessible to, and usable by, as many people as possible... without the need for special adaptation or specialised design." (www.inclusivedesigntoolkit.com/betterdesign). In the United States the term 'universal design' is more common than 'inclusive design', while both seem to stand for the same concept. Unfortunately, as exhibited in several research papers, this conceptual framework is only followed by one discipline at a time (Heylighen, 2008). When addressing the user needs in urban development the lead (if not sole and dominant) discipline involved is architecture. In a more holistic view many other disciplines are also included in such a development and could possibly provide valuable insights into more detailed aspects of such projects. The research used as a base for such attempts of inclusive design is mostly based on data from a small sample population. These small sample populations are unlikely to give adequate representations of the overall population and the different views on any topic. As a consequence, larger, preferably nationwide or multinational, studies are needed as a base for well balanced and well informed inclusive design.

More recently the research and design community has identified several key factors and benefits which directly relate to inclusive design and that might be food for thought – not only for experts in this field of study. Hosking (2010) stated the benefits of a more general approach to inclusive design and distilled his experiences into the motto: "it is normal to be different" (p. 496). Furthermore he refers to the concept of 'dynamic diversity' (Gregor, et al., 2002), which not only encompassed the facet of individual differences, present in every design and development context, but also the shift in choice preference that this diversity experiences over time. This is especially relevant when the service / product developed is aimed at the market of older customers. Not only do their needs and ambitions differ (in part) from those of younger generation but the defining factors of behaviour also change over time, especially as decline in both mental and physical capability is a significant trend of the older population.

Accordingly a central need of design is to have a capacity to utilise data drawn from behavioural profiles from both the individual and the group level. The combination of this data would provide the basis for constructing a behavioural trajectory (behaviour change over time) such that design can be informed by both the behavioural norm (group means) and the variation to the norm (individual, standard deviation). For example the information derived regarding learning strategies from a 64 year plus population will contain internal variance whilst providing clear differentiation between learning strategies associated with a younger population segment.

We can examine the impact that 'dynamic diversity' could have under such circumstances. Assuming that a static set of personas can sufficiently represent and 'embody' the whole of all users, it can be suggested that different changes within a single persona could affect the research and design outcome. The motivation of a person to interact with and to use technology are bound to the perception and reasoning of the prospective user, which are –in turn – dependent on the mental (goals, ambitions, needs, personality type) and physical (capability to sense and interact in certain ways) predispositions. A change in the ability to interact with the environment in a certain fashion, can – and is very likely to – change the underlying motivations behind an action. Such a change in underlying processes or patterns of reasoning effectively changes the 'non-demonstrative' part of the user interaction.

5.2 Formalized Behavioural Prototyping©

This concept of dynamic change leads to the question of whether the approach of defining personas in order to model user interaction with a product or system is the most appropriate way of including user interaction and behaviour patterns in product design. For further analysis of this topic the term 'user interaction' should be divided into two separate concepts. On the one hand the user has a 'demonstrative' action sequence which can be regarded as the main interaction with a system or product. An example for this is how a person interacts with a computer programme, i.e. which options are selected in which sequence and which errors are made. On the other hand, an underlying factor of the 'demonstrative interaction' is the 'non-demonstrative action', i.e. the underlying motivations and needs that define the goal of the demonstrative interaction. These two dimensions of interaction can be regarded as factors in an equation outlining the key parameters of prototypical behaviour. The equations in the following are included for purposes of illustration, not mathematical robustness.

In the following the basic factors of goal directed interactive behaviour with technology will be expressed as an equation:

$$y = \left| (c + dx + r) \right|$$

With the variable 'y' being the overall individual user behaviour or prototypical behaviour, 'c' being the 'non-demonstrative' interaction pattern, 'x' being the 'demonstrative' interaction and 'r' being a residual factor compensating for the possible environmental factors acting as latent moderating or mediating variables. As indicated by the two vertical lines bracketing the equation, the variable 'y' will always equal the (positive) value of the other part of the equation. This is the case as behaviour cannot be negative under these circumstances. The smallest value the factor 'y' could possible reach is '0', indicating that no interaction whatsoever is taking place. The factor 'd' represents the impact of the general technology acceptance of a person with regard to assumptions of usability and usefulness. In this equation it acts as a slope coefficient, indicating the likelihood and ease of the up-take and interaction with technology. The overall Behavioural Prototyping© operation for a specific product with a clearly specified range of features (index 'j') as well as a potential user group (with modelled need structures [index 'i']) can be illustrated as follows:

A Behavioural Prototype© equals

$$\sum \left(|s| + |y| \right)$$

with $\sum_{i=0}^{n} y_i = \left| (c_i + d_i x_i + r_i) \right|$ and $\sum_{j=0}^{n} s_j = a_j + r_j$

The sum of the individual prototypical behaviours (or Behavioural Prototype©) will – in an ideal scenario – equal two things: First of all it will, by definition, equal the potential for demonstrative action immanent to the defined group of prospective users, taking into account the potential itself, as well as underlying predispositions (moderators) and external influences (mediators). Secondly, the Behavioural Prototype© should equal the affordances of the object / system once it is created. This means that the perceived potential for interaction with an object by the user is identical to the need structure of the user.

When combined, these different facets lead to a stage where the behaviour and interaction of the user, with regard to the developed product, becomes more pre-dictable. Knowledge of the internal processes, when presented with a certain type of product, allows predictions on the use and perception of the product by the user, including how the user learns to maximise the assistive impact of the product. A summary of the potential alignment between user and system behaviour is shown in Figure 5.

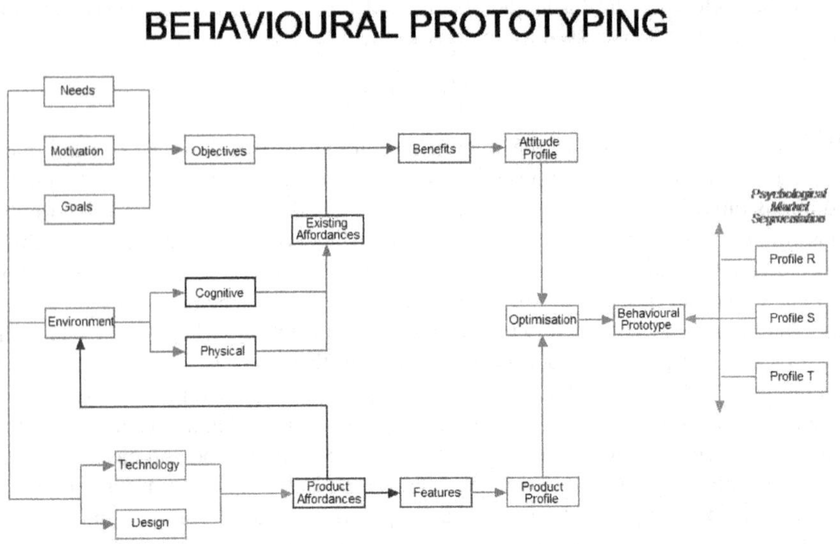

BEHAVIOURAL PROTOTYPING

Fig. 5 A summary of Behavioural Prototyping©. An approach for identifying designer and user expectations and perceived benefits.

6 Conclusion: Strengths of the Framework

The immense diversity of potential response available to the user of a system represents the best and worst of contexts. The ability to be creative and adaptive in learning a system and to personalise the benefits is at the centre of reduced redundancy. However, the greater the capacity of a system to provide flexibility and open ended interactions, the higher the risk of redundancy. As a method for recognising and evaluating fundamental psychological dimensions within such diversity the Behavioural Prototype© should have value. This prototype offers a framework of affordances used for identifying fundamental psychological aspects within both designers and users, thereby increasing the alignment between design, utility and use. Firstly, the psychological impact of solutions derived from a technological application are rarely assessed, in situ, regarding, monitoring and evaluating the psychological and behavioural impact on the end user. Full assessment would require the construction of a behavioural profile that had benefitted from an in situ analysis of both mean and standard deviation of errors made and behaviour observed. Such a consideration would allow the experts, from multiple and particular areas of research, to adjust design as an integrated part of the development process, which from the perspective of a specific field might be a minor or even irrelevant shift but one that could have a large impact on the 'feel-good factor' perceived by the user. Secondly the structured psychological perspective provided by the prototype approach and the associated shared language base allows for more sophisticated information sharing, problem representation and liaising for team building within the research consortium. Increased information sharing, conducted on a basis of equality and not professional primacy, can prompt an increase in different, if not divergent inputs, each of which grounded by a common and overarching framework derived from the Behavioural Prototype©. This 'user-centred' approach provides for a common 'modus operandi', one derived from a range of disciplines yet capable of producing an integrated solution.

References

1. Almäng, J.: Affordances and the Nature of Perceptual Content. International Journal of Philosophical Studies 16(2), 161–177 (2008)
2. Anderson, J.R.: Problem solving and learning. American Psychologist 48(1), 35–44 (1993), doi:10.1037/0003-066x.48.1.35
3. Benton, S., Manning, B.R.M.: Empowering Proactive and Personal Health Care: No Option No Choice. In: Published in Proceedings of Worldcomp 2008 World Congress in Computer Science Computer Engineering and Applied Computing, Las Vegas (2008)
4. Ben-Zur, H.: Coping, affect and aging: the roles of mastery and self-esteem. Personality and Individual Differences 32(2), 357–372 (2002), doi:10.1016/s0191-8869(01)00031-9
5. Bos, L.: Medical and Care Compunetics: The Future of Patient-related ICT. In: Bos, L., Blobel, B. (eds.) Medical and Care Compunetics 4. IOS Press (2007)
6. Carroll, K.E., Kincade, D.H.: Inclusive Design in Apparel Product Development for Working Women with Physical Disabilities. Family and Consumer Sciences Research Journal 35(4), 289–315 (2007)
7. Davis, F.D.: Perceived Usefulness, Perceived Ease of Use, and User Acceptance of Information Technology. MIS Quarterly 13(3), 319–340 (1989)
8. Davis, F.D., Bagozzi, R.P., Warshaw, P.R.: User Acceptance of Computer Technology: A comparison of two theoretical models. Management Science 35(8), 982–1003 (1989)
9. Day, R.E.: Death of the user: Reconceptualizing subjects, objects, and their relations. Journal of the American Society for Information Science and Technology 62(1), 78–88 (2011)
10. Dempsey, P.G., Wogalter, M.S., Hancock, P.A.: What's in a Name? Using Terms from Definitions to Examine the Fundamental Foundation of Human Factors and Ergonomics Science. Theoretical Issues in Ergonomics Science 1(1), 3–10 (2000), doi:10.1080/146392200308426
11. Fisk, A.D., Rogers, W.A.: Psychology and Aging: Enhancing the Lives of an Aging Population. Current Directions in Psychological Science 11(3), 107–110 (2002)
12. Greeno, J.G.: Gibson's Affordances. Psychological Review 101(2), 336–342 (1994)
13. Gregor, P., Newell, A.F., Zajicek, M.: Designing for dynamic diversity: interfaces for older people. In: Proceedings of the Fifth international ACM Conference (2002)
14. Gibson, J.J.: The Theory of Affordances. In: Shaw, R., Bransford, J. (eds.) Perceiving, Acting, and Knowing: Toward an Ecological Psychology, pp. 67–82. Lawrence Erlbaum, Hillsdale (1977)
15. Heerink, M., Kröse, B., Evers, V., Wielinga, B.: Assessing Acceptance of Assistive Social Agent Technology by Older Adults: the Almere Model. International Journal of Social Robotics 2(4), 361–375 (2010), doi:10.1007/s12369-010-0068-5

16. Heylighen, A.: Sustainable and inclusive design: a matter of knowledge? Local Environment: The International Journal of Justice and Sustainability 13(6), 531–540 (2008)

17. Hosking, I., Waller, S., Clarkson, P.J.: It is normal to be different: Applying inclusive design in industry. Interacting with Computers 22(6), 496–501 (2010), doi:10.1016/j.intcom.2010.08.004

18. Lewin, K.: The conceptual representation and the measurement of psychological forces. Duke University Press, Durham (1938)

19. Manning, B., Benton, S.: Setting Core Standards: Privacy, Identity and Interoperability. In: Bos, L., Blobel, B. (eds.) Technology and Informatics: Medical and Care Compunetics 6, pp. 32–39. IOS (2010) ISBN 978 1 60750 564 8

20. Manning, B.R.M., McCann, J., Benton, S., Bougourd, J.: Active Ageing: Independence through Technology Assisted Health Optimisation. In: Bos, L., Blobel, B., Marsh, A., Carroll, D. (eds.) Medical and Care Compunetics 5. Studies in Health Technologies, 137. IOS Press (2008) ISBN 978-1-58603-868-7

21. Nam, J., Hamlin, R., Gam, H.J., Kang, J.H., Kim, J., Kumphai, P., Starr, C., Richards, L.: The fashion-conscious behaviours of mature female consumers. International Journal of Consumer Studies 31(1), 102–108 (2006)

22. Newell, A., Simon, H.A.: Human problem solving. Prentice-Hall, Englewood Cliffs (1972)

23. Osiurak, F., Jarry, C., Le Gall, D.: Grasping the affordances, understanding the reasoning: Toward a dialectical theory of human tool use. Psychological Review 117(2), 517–540 (2010)

24. Palmore, E., Luikart, C.: Health and Social Factors Related to Life Satisfaction. Journal of Health and Social Behavior 13(1), 68–80 (1972)

25. Penninx, B.W.J.H., van Tilburg, T., Kriegsman, D.M.W., Boeke, A.J.P., Deeg, D.J.H., van Eijk, J.T.M.: Social Network, Social Support, and Loneliness in Older Persons with Different Chronic Diseases. Journal of Aging and Health 11(2), 151–168 (1999)

26. Rigby, M.: Holism, health and Data-Managing the person-centred Digital Haystack. In: Bos, L., Blobel, B. (eds.) Technology and Informatics: Medical and Care Compunetics, 6, pp. 181–188. IOS (2010) ISBN 978 1 60750 564 8

27. Rosenberg, M.: Society and the adolescent self-image. Princeton University Press, Princeton (1965)

28. Rowe, J.W., Kahn, R.L.: Successful Ageing. The Gerontologist 37(4), 433–440 (1997)

29. Savikko, N., Routasalo, P., Tilvis, R.S., Strandberg, T.E., Pitkälä, K.H.: Predictors and subjective causes of loneliness in an aged population. Archives of Gerontology and Geriatrics 41(3), 223–233 (2005)

30. Sveinbjornsdottir, S., Thorsteinsson, E.B.: Adolescent coping scales: A critical psychometric review. Scandinavian Journal of Psychology 49(6), 533–548 (2008)

31. Symonds, J., Parry, D., Briggs, J.: An RFID-based system for assisted living: Challenges and Solutions. In: Bos, L., Blobel, B. (eds.) Medical and care Compunetics 4. IOS Press (2007)

32. Thomas, V., Wolfe, D.B.: Why won't television grow up? American Demographics 17(5), 24 (1995)

33. Venkatesh, V., Davis, F.D.: A Theoretical Extension of the Technology Acceptance Model: Four Longitudinal Field Studies. Management Science 46(2), 186 (2000)

34. Venkatesh, V., Morris, M.G., Davis, G.B., Davis, F.D.: User Acceptance of Information Technology: Toward a unified view. MIS Quarterly 27(3), 425–478 (2003)

The Design of a Teacher-Driven Intelligent Agent System for Supervising Lessons in LAMS

Themistoklis Chronopoulos and Ioannis Hatzilygeroudis

Department of Computer Engineering and Informatics, University of Patras, Patras, Greece
{chronop,ihatz}@ceid.upatras.gr

Abstract. This chapter presents the design of an intelligent agent based system that aims to support teachers in supervising and evaluating learners and activities of lessons in the Learning Activity Management System (LAMS). A monitoring agent has been designed to collect and aggregate information from LAMS database, related to the participation of learners in lesson's activities, at time intervals which indicates the teacher through a scheduler agent. A user notification agent diagnoses conflicts in the learning and collaborative processes and issues alert and awareness messages to the teacher, learners and groups of learners via rules that are based on a teacher defined participation model. The system, based on this model, generates also evaluation reports of learners and activities of the lesson, in order to assist the teacher to intervene effectively in the course of a lesson.

Keywords: intelligent agent collaborative learning, intelligent agent based learning supervision, intelligent agent based learning monitoring.

1 Introduction

Learning Management Systems (LMSs) are web-based education systems which support the management and the administration of learning material, activities and tasks, providing a framework for teaching and learning, usually complementary to classroom education. LMSs provide authoring, management, delivering and sharing of learning material, student monitoring, assessment management, and online collaboration. They aim at providing flexibility, accessibility and convenience to their users [1].

A typical LMS, as shown in Fig.1, consists of, a Course Administrator System, which the teacher uses to create and organize the content of a course and manage users and groups, a Run-Time system, for presenting content and making it available to learners, and a Database, in which all data and information relative to the lessons are stored. In small scale systems teachers are usually the authors and also the supervisors or moderators monitoring and coordinating the evolution of lessons [2].

LMSs provide a set of synchronous and asynchronous tools such as, forums, chats, wiki's etc. to generate collaborative activities in courses [3]. The teachers in a LMS can also provide information to the learners, author lesson material, prepare assignments and tests, be engaged in discussions and manage the virtual

T. Daradoumis et al. (Eds.): Intelligent Adaptation & Personalization Techniques, SCI 408, pp. 211–238.
springerlink.com

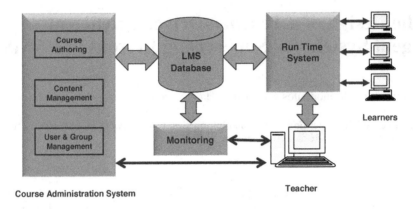

Fig. 1 A general view of a Learning Management System (LMS)

classes [4]. Some examples of commercial systems are: Blackboard [5], WebCT
[6] and Top-Class [7]. Some examples of open source systems are Moodle [8],
Ilias [9], Claroline [10] and Learning Activity Management System (LAMS). The
LAMS, is written in Java and is representative of a new generation of LMSs, that
move e-learning from a content-centric approach to an activity-sequence based
approach [11].

LAMS is an e-learning environment for designing, managing and delivering
online learning activities. Teachers, using its authoring environment, can create
sequences of learning activities that can be individual tasks, group work or class
activities based on content, assessment or collaboration. LAMS can be used either
as a stand -alone system or in combination with other learning management systems, like Moodle, WebCT and BlackBoard.

Teachers, as supervisors of lessons in LAMS, should guide the leaners during
their interaction with the activities and redesign online the learning process according to the needs of individual learners and groups. To succeed in these tasks,
teachers need to become aware of problems experienced by individuals or groups
of students [12, 13]. However, monitoring student activities and deciding which
actions are most appropriate can significantly increase the teachers' overload [14].

Athough LAMS, like most LMSs, accumulates a great deal of log data about
students' activities and interactions and provides management, monitoring and reporting tools for the lessons, it leaves most of the guidance for the collaboration
and the learning process to the teacher, who has to interpret a large volume of information in order to enable and control how the collaborative and other activities
take place. [3]

As a result, there is a high demand for automatic feedback to supervisors of lessons in LAMS, by highlighting important learning situations and pointing at potential collaborative conflicts. This could assist them to effectively intervene in the
learning process or the collaboration, e.g. by assigning new appropriate activities
to learners or by reformulating groups [15].

The autonomy of an agent which means its ability to perform independently a
task assigned to the agent by a person or other software has been exploited in

many educational situations. The autonomous feature of agents reduces users' burdens of learning activities, teaching activities, management activities, and so on [16].

Agents can process a huge amount of data, make direct interventions in the process, and interact with other agents for carrying out tasks. Thus, they can help users concentrate on the contents that they are studying. [17] Also, they can be used for helping teachers.

Formative assessment or continuous assessment is defined as "assessment that is specifically intended to provide feedback on performance to improve and accelerate learning". Feedback is information about how a student has performed in relation to some standard or goal and it is addressed to both teacher and learner. In distance education, formative assessment is an essential part of the learning process since it forms the basis of the main communication between the teacher and the learner and it is also used to support the regulation and the evaluation of the process. [18]

The objective of computer-based interaction analysis (IA), in technology based learning activities (individualised or collaborative), is to provide useful and structured information, via appropriate IA indicators, to assist the teachers and the students, to analyze and to understand the learning process. By that, IA supports them to take part in the control of the activity, contributing to awareness, self-assessment or even regulation and self-regulation. [19]

In this work, we present the application of collaboration and interaction analysis, in the context of formative assessment, in to the learning environment of a LMS, to support different functions, such as awareness, regulation and evaluation, based on the understanding of collaborative and learning processes [20].

More specifically, our proposal consists in the design of an agent-based module that aims to assist teachers in supervising the learning process and the collaboration in a lesson of LAMS, by continuously monitoring and analysing these processes during the lesson and by notifying the teacher of common problems about the participation of students and groups in the activities.

The analysis is based on a teacher defined model for the participation of the learners in the activities, and aims to support different functions, such as awareness, regulation and evaluation, of the collaboration and of the individual learning processes.

The system also provides learners with awareness information about the group work and reminds them of the lesson scheduling and the activities' goals, that teacher has defined. Finally, it aims to support teachers to intervene effectively in the design of the lesson by evaluating the learners and the activities after their conclusion. This work is an extension of [50].

2   Monitoring and Supervising the Learning Process and the Collaboration in LMSs

Collaborative learning is the social process whereby students learn interacting with each other, to accomplish a shared learning goal and maximize their own and their group members' achievements. By these interactions is entailed the social construction of knowledge. There are many benefits from collaboration learning e.g. builds self-esteem in students, stimulates learning, increases motivation, promotes feelings of belonging to a team, promotes a positive attitude toward the subject matter, encourages creativity, eases communication and enhances student satisfaction with the learning experience [21]. Collaborative learning was originally adopted in a classroom-based environment. In Computer-Supported Collaborative Learning (CSCL) the participants are supported in their collaboration using appropriately designed computer-based tools for communication and collaboration. When interactions take place in virtual environments, usually over web or a computer network, then we talk about web or virtual collaborative e-learning.

In the context of CSCL, teachers have to coordinate the collaboration among members by designing activities and assigning roles to participants, by assessing and providing feedback on individual and group learning and performance etc. Collaborators, on the other hand, have to be informed about their relative participation regarding the whole group and the objectives of the learning process, in order to self-regulate their learning behaviour.

The objective of interaction analysis, in the field of CSCL, is to give useful and structured information about the collaboration process to the teachers and to the students, to support and improve collaboration process management [22].

The interaction analysis tools support the users in three major levels: awareness, metacognition and evaluation. The objective is the optimization of the learning activity through: (a) refined participation by the students through reflection, self-assessment and self-regulation, (b) better activity design, regulation, coordination and evaluation by the teacher [19]

Three types of systems that analyze interactions to support collaboration have been identified: *mirroring or awareness tools* that automatically collect and aggregate data about the participants' interaction and reflect this information back to the user, *metacognitive or assesment tools* that display information about what the desired interaction might look like in comparison with the current state of indicators and c) *guiding or advising tools* that perform all the phases in the collaboration management process and issue messages guiding participants (mainly students) during their activity in order to adapt the results of the interactions to the "ideal" model. [23]

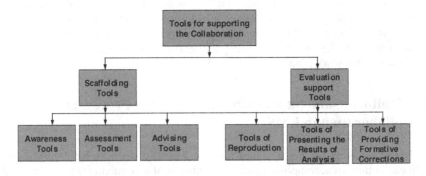

Fig. 2 Tools for supporting the collaboration in CSCL

All these tools are being categorized as scaffolding systems while another category of tools supporting the collaboration in CSCL are the evaluation tools (Fig. 2). Latter tools also support the evaluation of the collaborative learning by performing an analysis of collaboration at the end of each activity without using an ideal model. Their purpose is either to present the results of the analysis to the users, allowing the evaluator to understand the process in an efficient manner and learners to self-regulate their learning actions, or to suggest improvements (i.e. as a formative tool). [24]

Most of the CSCL systems presented in the literature are specific to a domain of knowledge and support problem-solving learning processes or another pedagogical approach. Additionally, many of them rely on data recorded from dialog-based or/and action-based structured environments to perform analysis of interactions or to analyze the solution itself, relating it with the collaboration process [25, 26]. Some of them use the results of the analysis of interactions for the regulation of the collaboration process and/or for the improvement of the collaborative product [27].

CSCL Systems are being differentiated from web-based collaborative learning systems in that their purposes are: a) to promote learning during users activity and not just to perform task b) to enable collaboration and not just to enable communication between partners c) to support collaboration through specific tools and functions and only to allow collaboration [28].

LMSs, in their majority, provide unstructured collaboration spaces that group course participants and offer an open domain-independent interface for communication and sharing of knowledge and experiences. Students interact with LMSs in a stand alone or a collaborative mode. LMSs gather a lot of log data about the interaction of students with the activities, such as reading lesson material, taking tests, communicating or collaborating with other learners. All this information is usually stored in a database and it is related to the user's profile, his/her scores in assessment activities, his/her interaction data, etc [29].

However, LMSs pose some limitations for the efficient monitoring and the regulation of the collaboration learning process. First of all, LMSs gather data of interactions performed at a very low and primitive level ('user x added a page in wiki y'). Therefore a quantitative analysis is needed to be performed to obtain aggregated results (e.g. 'number of pages added by user x in wiki y') in order to build indicators with higher semantic value [3]. Another limitation is that the reporting and management tools of the LMSs provide the basic functionalities to support participation monitoring, but without relation to planned criteria. As a result, they cannot help efficiently a teacher to extract useful information for the learning process, especially when there are a large number of students, since teacher needs to take extra steps to monitor, to analyse and to evaluate the learning process. [30]

Some problems that have been observed in online distance learning environments, such as students' feeling of isolation, lack of motivation and high drop-out rates could be reduced by the continuous monitoring and assessment of students' participations in a lesson [31].

Awareness is also very important for learners in a LMS. Through awareness information, participants are informed about the performance and the progress of the work of their team mates in the activity, detect changes in plans and understand

how the group work is getting along [32, 33]. Without this information, participants are not able to measure the quality of their own work compared to the objectives and progress of the group.

So teachers who use LMSs need more specific tools, which could assist them to
manage the large amount of data those systems can produce, in the directions of
the assessment of the activities performed by the learners and/or the evaluation of
the structure and the contents of the course. This knowledge can also be useful to
learners, for reminding them of the goals of the lesson or assisting them in reflecting upon their activity, as well as the overall activity, allowing them to selfregulate their actions and/or behavior.

Such tools have been implemented in several e-learning environments by developing intelligent agents, which help teachers by automatically detecting problems,
instead of leaving this task to them to be done by inspection. In addition, they can
automatically warn teachers and learners about conflictive learning cases [34].

The common roles of agents that support collaborative learning are: (1) monitoring the collaborative learning process; (2) giving feedback and guidance to activate interaction and collaboration among participants; (3) giving information
about the current state of a learner's interaction in the collaborative learning
process; and (4) giving advice about the learning process according to the followed strategy by comparing the current and ideal states [23, 35]

Our proposal consists in the design of an intelligent system that aims to assist
teachers in supervising the learning process and the collaboration in a virtual class
in LAMS, by continuously monitoring all the activities of a lesson according to
planned criteria for the attendance and the performance of learners that teacher defines in the participation model of the lesson. The proposed system informs by
messages, the teacher and the learners, about several conflictive learning cases and
supports also the teacher by the automatic evaluation of both learners and activities according to the same planned criteria.

Based on this support, the teacher is able to redesign on line the lesson or re-
formulate the groups in order to achieve the goals that he/she has set for the lesson. Our system, finally, supports awareness between the learners and notifies or
warns them about their participation regarding the participation of the group or the
whole class and regarding the objectives the teacher has set, in order to regulate
their learning behavior.

3 Related Work

Several types of agents have been developed in e-learning environments. Such
types are: (1) Collaborative agent which has to negotiate in order to reach mutually acceptable agreements on some matters (2) Interface agent which is a personal
assistant who is collaborating with the user in the same work environment; (3)
Mobile agent who has the ability to move around some network; (4) Information
and internet agent that performs the role of managing, manipulating or collating
information from many distributed sources; (5) Reactive agent that monitors and
reacts to the state of its environment; (6) Hybrid agent that refers to the one whose

constitution is a combination of two or more agent philosophies within a singular agent; (7) Heterogeneous agent that contains one or more hybrid agents which belong to two or more different agent classes. [36]

Rules have been proved suitable to define abstract and real agent architectures and they have been used for realizing the so called "rule-based agents", that is, agents whose behavior and/or knowledge is expressed by means of rules. "Rules-enhanced agents", on the other hand, are agents whose behavior is not normally expressed by means of rules, but they use a rule engine as additional component to perform specific reasoning, learning or knowledge acquisition tasks.

All the previous types of agents have been developed and integrated in several e-learning collaborative environments as facilitators, monitoring agents, teacher or student assistants, tutor agents or assessment agents to improve the teaching, the management and the performance of the collaboration learning and/or to advise the participants according to an ideal model of interaction. Agents have been supported in synchronous and asynchronous collaborative environments. [37]

A number of researchers have proposed the development of software agents to support the teachers in the management and the supervising of web courses and/or the learners facing conflictive learning situations in several web learning environments. Jafari [38] conceptualized three types of software agents to assist teachers and students: a Digital Teaching Assistant to assist the human teacher in various teaching functions, a Digital Tutor to help students with specific learning needs and a Digital Secretary to act as a secretary to assist students and teachers with various logistical and administrative needs. Choy et al. [39] proposed an interface agent for a distance learning environment to assist the course coordinator in performing supervising a web course. The agent can access information regarding students' learning progress and study behavior, aggregate the information, and report status to the coordinator through e-mails. The course coordinator can also update agent duties, through a web interface, to send e-mails to tutors or students to remind or inform them for important dates or events of the web course. The agents that have been developed for those systems don't deal all with the collaboration or the communication between learners.

There have been developed some intelligent systems for collaborative learning web environments, whose objective is to infer information through structured interfaces on the learning process and/or the collaboration in domain-independent learning systems in order to inform teachers and/or to advise students. Casamayor, Amandi and Campo in [34] have implemented an interface agent that observes the learning behaviour of students and assists teachers to perform supervising of collaborative activities in SAVER, a web-based collaborative distance learning environment. The agent tracks in user profiles their learning behaviour and provides teachers with statistical information about the individual progress of each group member and the types of his/her participation in structured discussions. It also detects collaboration conflicts among students and notifies teachers through on-screen alerts, reasoning about them, by using a set of rules based on the action plan defined for a work, the learning styles of students and the tasks being held in a group. The rules that relate the students' learning styles and their profiles are being generated using association rule mining, but new rules can be generated

also, either automatically or semi-automatically or even provided by an expert in
the area. The DEGREE system [40] is an asynchronous news-group style system
that provides a structured interface for conversation in which the students have to
select the type of their contribution each time they add something to the discussion. By doing so, it is trying to capture the quality of the conversation. The system manages and stores all user interventions, which uses to make quantitative
and qualitative analysis and to rate the individual performance and the collaboration between pairs of students along different dimensions. DEGREE advises students using a software agent and informs teachers by sending messages, depending on the calculated collaborative indicators. A limitation of the previous systems
is their dependence on the ability and the willingness of the learners to select the
right contribution on their structured interfaces.

Some other intelligent systems have been developed to support the collaborative activities that have been designed using asynchronous discussion tools like forums, by performing quantitative and/or qualitative analysis and by informing
and/or advising the teachers and the learners during the development of the activities or at the end of them. Dolonen et al [41] integrate a pedagogical agent system
in FLE, a distributed open source CSCL environment, in which students can post
messages. The agent system consists of two components: a Student Assistant
agent and an Instructional Assistant agent. Both agents observe and detect problems in the collaboration and the knowledge-building process among students.
Student Assistant agent identifies possible problems and gives advice to each student based on principles of collaboration and knowledge building. Instructor Assistant agent computes statistics for viewing and presents possible problems and
advice to the instructor who can be engaged in a dialog with the students. The
suggested agent system for FLE3 is limited to the support of a structured environment using messages for progressive-inquiry learning. Scutelnicua et al [42]
integrated software agents into Moodle. The agent system has been developed to
monitor and measure the interaction of students in forum activity by performing
data aggregation on the Moodle database. The implemented agent system provides
teachers and students with the evaluation of the collaboration work they have done
for a course. The agent system that has been developed focuses only in forum activity and has not been addressed for other collaborative activities of Moodle.
Joice Lee Otsuka et al. in [30] proposed a multiagent system (MAS) to support
formative assessment in LMSs, by allowing the continuous monitoring and orientation of the learning process. The system mainly consists of agents that monitor
and analyze the participation of learners in each activity that has been designed
and implemented by the Discussion Forums or Portfolio tools on two different
LMSs. For the analysis of the interactions of learners with the activities, the agents
use knowledge bases and rules that reflect the criteria that teachers select for the
assessment of each activity. The MAS presents the results of the assessment by
means of statistical reports that are being update after the conclusion of each activity, but it doesn't issue any kind of messages to the learners or to the teacher.
Moreover, the intelligent system focuses only in activities developed in the LMS'
communication tools.

Kosba et al. [15] developed the Teacher ADViser framework (TADV) which describes the behaviour of an intelligent assistant that: (a) analyses tracking data collected in a Web Course Management System to derive models of individual students, groups and the whole class; (b) recognises situations that have to be brought to the teacher's attention; (c) generates feedback to the teacher pointing at possible problems and actions that can address these problems. The kernel of TADV is the feedback generation mechanism which relies on advice generation algorithms based on user models extracted from student tracking data. The proposed framework doesn't use any planned criteria for the evaluation of the performance of the users neither evaluates the activities of the course.

There have also been developed intelligent systems that use data mining methods to identify and characterize the interactions or the collaboration that has occurred or is occurring in open learning systems [3, 43, 44, 45]. Their objective is to explore and analyze data with limited semantic information in order to identify and discover useful patterns that assist the teacher in interaction and collaboration assesment. However, those approaches rely on the existence of large data sets.

Our proposal consists in the design of an agent-based system that aims to assist teachers to supervise the learning process and the collaboration in a lesson of LAMS, by continuously monitoring those processes during the lesson. The system using rules that are based on a participation model that the teacher has to update, issues suitable messages to notify all the users of the lesson for common problems about the participation of students and groups in the activities.

The supervising system, additionally, processes and analyzes the interactions' data to provide learners with awareness information relating to the group work and also to remind them of the lesson schedule and the activities' goals, that teacher has defined. Finally, it aims to support teachers to intervene effectively in the design of the lesson by evaluating the learners and the same the activities after the conclusion of each activity.

4 Supporting the Supervisor of a Lesson in LAMS

LAMS is an open source e-learning tool developed by Macquarie University, which provides teachers with an open infrastructure to create sequences of learning activities that students complete online. LAMS includes environments for user administration, student run-time delivery of sequences, teacher run-time monitoring of student sequences and, most importantly, teacher authoring/adaptation of sequences.

LAMS provides the appropriate support to develop a collaboration approach, especially through shared workspaces (MindMap tool), Wiki edition and other communication tools, such as forums and chats, where group members can exchange ideas and make progress in developing activities collaboratively [46].

The main categories of the activities – tools of LAMS are: Informative Tools (Noticeboard Tutorial, Create a Resource Sharing Activity, Create a Task List), Evaluative Tools (MCQ - Multiple Choice Quiz, Submit Files Activity), Collaborative Tools (Chat and Scribe Tutorial, Forum Tutorial, Dimdim Conference

Tutorial, Pixlr Image Suite Tutorial, MindMap), Reflective Tools (Question and Answer, Survey Tutorial, Voting Tutorial) [11].

LAMS records in a database the data concerning learners' interactions with the activities that are related to the attendance of the learners (e.g. the time a learner is starting and ending an activity and the time he is stopping and restarting an activity). LAMS also stores some indicators of the performance of the learners in evaluative activities (e.g. the number of right answers in a MCQ activity) or the contribution of the learners in collaborative activities (e.g. number of postings on the discussion forums).

The objective of those indicators is to characterize learner's behaviour. The supervisor can be use them to obtain information about the way the students take part in a collaboration task in order to detect some problems in the collaboration process such as poor collaboration of specific students, the failure of a collaborative task or conflicts within the groups.

The teachers, as coordinators of the learning process in LAMS, have to ensure that the activities are being executed according to the lesson schedule, the restrictions and the objectives that they have defined for each lesson. The teachers, also, as moderators of the collaborative activities, have to deal with conflictive learning cases and so they need to be aware of the contribution of learners in the activities and also of the quality of the activities.

In LAMS, especially when many students are participating in a lesson with many activities, it is difficult for the supervisors to overview and to regulate the learning process through the monitoring interface. It is also difficult for the students to be aware of others' activities.

Some of the limitations of using LAMS for designing and implementing teaching experiences for students, as pre-service teachers have reported in [47], are: "Students may be left behind or group may be slowed down", "Difficult to collaborate", "Poor feedback to students" and "Poor access to teacher assistance".

Some problematic situations in collaboration in open-collaborative environments as have been reported in [44] are: team members do not communicate well (lack of communication), work at different rates (lack of coordination) or some members dropped out of the collaboration experience (abandonment).

Formative or continuous Assessment is a process used by teachers and students during instruction in distance e-learning systems. Its purpose is to provide regular feedback to the teachers and to the learners to adjust ongoing teaching and learning in order to stimulate learning and improve students' achievement of intended instructional outcomes [48].

The formative assessment process is a repeated process consisting of four phases. The teacher, in the first phase, sets the instructional goals, based on learning targets, objectives, or standards that could be shared with students. In the next phase instruction takes part, based on the pre-set learning goals and objectives. Measuring is the third phase of the assessment process and refers to the collection of information about student learning, to find out if the students are meeting the instructional goals. The last phase of the assessment process is giving feedback to the teachers and to the students. The goal of providing feedback to students and teachers is to promote action to set new goals or to re-teach or re-instruct students to make sure they meet those goals.

Feedback is a key element in formative assessment, and is usually defined in terms of information about how a student's present state relates to goals and standards. Feedback is addressed to the teachers and to the students. Teachers use feedback to make programmatic decisions with respect to readiness, diagnosis and remediation. Students use it to monitor the strengths and weaknesses of their performances, so that aspects associated with success or high quality can be recognized and reinforced, and unsatisfactory aspects modified or improved. [49]

In order to overcome the problems and the limitations that the environment of LAMS poses to the supervisor and the learners of the lessons, we propose the application of interaction analysis, based on a teacher-defined participation model, in the context of a formative or continuous assessment of the collaboration and individual learning processes of a lesson in LAMS.

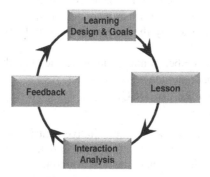

Fig. 3 The phases of formative assessment based on the Participation Model of a lesson in LAMS

The proposed formative assessment process which aims to support the monitoring and supervising process of a lesson in LAMS, consists of a continuous recycling of four phases, as it is shown in figure 3.

In the first phase the teacher designs - authors the lesson and sets his learning standards for each activity in the *Participation Model* and in the *Lesson Calendar*. In the second phase an activity of the lesson is executed and the system collects data from the interactions of the users with the activity. In the third phase the system performs Interaction Analysis on the data that have been collected. The IA process consists of recording, filtering and processing data regarding system usage and user activity variables, in order to produce the analysis indicators. The interaction analysis results are presented to the participants of the learning activities' so as to be aware of and regulate their behavior, either as individuals or as groups. [19]

In the fourth phase, the feedback of the system is generated from rules that are based on the *Participation Model* and on the *Lesson Calendar*, in the form of alert and evaluation messages. The teacher using this feedback can redesign the lesson e.g. by assigning new activities (through Live Edit), by reformulating the groups etc. in order to meet the objectives he has set for the lesson. The cycle of the assessment is repeated every time an activity of the lesson is executed.

In the proposed monitoring application the supervisor can define in the Participation Model of the lesson, his objectives for the attendance and the performance of the learners for each activity, through the *Activity Parameters* form, as follows:

- Minimum/maximum time required to spend on each activity
- Minimum/maximum contribution in each collaborative activity
- Minimum/maximum score in each evaluative activity

The *Participation Model* consists of some quantitative indicators which are retrieved from LAMS database, as students participate in collaborative or evaluative activities. Some of them are:

- *Forum activity:* Number of the user's messages, number of new topics started by the user, number of replies to messages sent by the user
- *Wiki activity:* Number of pages that were added by the user, number of pages that were edited/modified by the user, number of links that were added by the user
- *MindMap activity:* Number of nodes that were added by the user
- *Chat activity:* Number and length of messages
- *For evaluative activities,* like MCQ, a learning progress indicator is the number of right answers (score).

Although the evaluation of students' quantitative contribution is not sufficient for effective and fair marking, it could be helpful to detect any individual's reduced contribution early on and try to help them to overcome it.

Additionally, the Supervisor defines, through the *Lesson Calendar* form, the 'time start' and the 'time duration' for each activity. These values will be used by the system to locate those learners that have stretched the time limits of the activity.

After the completion of an activity the system processes and analyzes the data concerning the interactions of the learners with the activity and computes, also, some higher semantic level indicators e.g. the mean level of the performance of a group or of the whole class in the activity. These indicators will be necessary for the generation of the awareness information.

Additionally to the awareness information, the system is expected to generate, as feedback, alert messages to the users of the lesson and evaluation reports to the teacher, based on the results of the analysis and on several rules that have been extracted from the learning experience of previous lessons or have been edited by the teacher.

By these types of feedback the supervisor could be capable to intervene effectively in the learning process by e.g. reformulating groups or by assigning more adequate activities (with the Live Edit of LAMS) or by advising a student through Instant Messaging. The student that receives feedback from the system is expected either to modify an unsatisfactory learning behaviour or to recognize and reinforce a successful performance.

5 An Agent-Based Architecture for Intelligent Monitoring and Supervising of Lessons in LAMS

As already mentioned, the main objective of this work is the design of an intelligent system to assist teachers in supervising LAMS collaborative learning environment and providing learners with awareness information. From an intelligent agent's system point of view, this assistance can be seen as the combination of different tasks. More specifically the monitoring and supervising system is expected to: (a) analyze tracking data collected in LAMS for monitoring the participation of individual students; (b) recognize conflictive learning situations that have to be brought to the teacher's attention (c) issue warning messages to notify the teacher and the students for possible learning conflicts (d) issue awareness messages to the students regarding their relative participation in the group or in the whole class and (e) generate evaluation reports for learners and activities.

As a result, the intelligent system has to collect data from LAMS' database, to collate that data with the standards and the objectives of the lesson, to process the data in order to extract the awareness and the evaluation information and finally to issue alert messages using rules about deficient attendance of learners in the lesson based on the *Lesson Calendar*, about learners' insufficient performance and/or meagre contribution in the collaborative activities based on the values of *Activity Parameters* that teacher has defined.

The proposed architecture consists basically of the Monitoring Agent, which is an information agent for managing, manipulating and collating information from LAMS' database and of the User Notification Agent, a reactive rule-based agent communicating with the users of the lesson with messages through the Noticeboard window in the lesson of LAMS.

According to Nwana's [36] typology, the User Notification Agent can also characterized as an 'interface agent' since it can be described as autonomous agent reacting to changes in its environment, communicating with other agents and communicating directly with the users.

The *Monitoring Agent* has been designed to retrieve and collate data from LAMS database at time intervals that the supervisor of the lesson sets through a *Scheduler Agent*. It has also been designed to compute statistical information, after the completion of each activity.

The *User Notification Agent*, by using that data, the Participation Model and a rule base, detects conflictive learning situations and notifies the users of the lesson (the supervisor and the learners) by issuing and delivering appropriate messages, depending on the situation.

The architecture of the information processing by the agents and the interaction with the users is shown in Fig. 4.

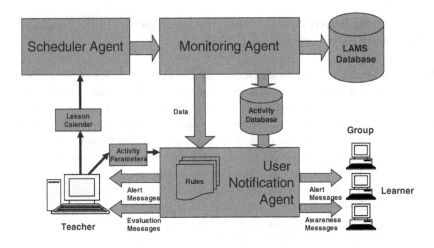

Fig. 4 LAMS' Lessons Monitoring & Supervising System

5.1 The Scheduler Agent

The *Scheduler Agent* is a non-intelligent agent, which is used to set one-shot alarms that go-off at a specified time that supervisor can define, or it can be used for recurrent alarms at specified intervals, based on the *Lesson Calendar*.

5.2 The Monitoring Agent

The Monitoring Agent checks LAMS's data base about learner's attendance of the activities of a lesson, at time interval instances that have being defined by the teacher in the *Lesson Calendar* and they are being "announced" by the *Scheduler Agent*.

After the completion of an activity, the *Monitoring Agent* also collects and aggregates, from LAMS database, the learners' interaction data with an activity and stores them in a local database (*Activity Database*).

More specifically, after the completion of an activity by all learners, the *Monitoring Agent* updates the Activity Database with the following information: user_id, group_id, activity_id, start_time, end_time, total_time, num_entrances, final_contribution, final_score.

Finally, the agent processes all the relevant information calculating statistical information regarding the participation of users and groups in each activity. More specifically, the *Monitoring Agent* calculates the mean levels for the participation time and performance of all learners of a group for each activity.

All this information will be used by the User Notification Agent to issue the suitable messages.

5.3 The User Notification Agent

The *User Notification Agent* is an interface agent that uses a rule base and forward chaining to determine if, when, what and to whom a message has to be issued depending on a problematic learning situation. It provides also a centralized user interface for the display of notification messages that can be either alert messages that have been issued from the rules or awareness messages with statistical info that has been calculated by the *Monitoring Agent*. This prevents the duplication of functions and code, because all messages are displayed in a single window for the users of the lesson.

The *User Notification Agent*, using the *Participation Model*, the *Lesson Calendar* and the information about learners' participation and contribution in each activity, detects problems and conflicts in the learning and collaborative processes issuing, according to the corresponding rules, alert messages to the users of the lesson (teacher and learners). Additionally the agent, after the conclusion of each activity, generates messages for each learner with awareness information about his relative participation in the activities, regarding the mean level of the corresponding indicators of the whole group. Finally, after the conclusion of an activity, it evaluates and classifies the learners and the activity, and informs the teacher via evaluation reports and messages.

5.4 Examples of Rules Producing Feedback Messages

Some of the rules that *User Notification Agent* uses refer to learners with insufficient contribution in collaboration activities or low score in assessment activities and/or deficient time_participation in introducing activities. In all the above cases, the suitable messages are issued based on the corresponding rules which, in their turn, rely on the teacher-defined *Lesson Calendar* and *Participation Model*.

Lets consider two learning cases in which a learner hasn't started yet an activity, as it is defined in *Lesson Calendar* and another in which a learner has ended an activity without significant contribution, as it was defined in the *Participation Model*.

In those cases the agent is able to determine and notify teacher and learner accordingly, by issuing alert messages, based on rules like: *if (learner x has not started activity y **and** time>start_time_for_activity_y) **then** ((issue_message_learner_x("You have been delayed for activity y") and (issue_message_teacher ("Learner_x has been delayed for activity y")) or **if** (learner x has ended activity y **and** contribution(x,y) <min_contribution(y) **then** (issue_message_learner_x("User x your contribution in activity y was too low, try harder")*). Similar rules can be edited by the teacher for several conflicting learning cases that have been recorded in the literature or that could be deducted from his experience.

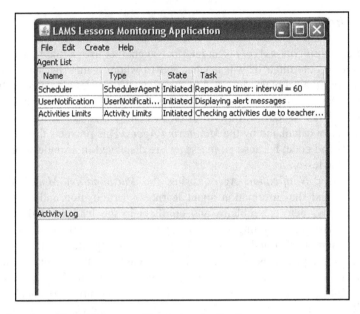

Fig. 5 The Interface of LAMS' Lessons Monitoring Application

A prototype of the intelligent module described above has been partially developed, using the CIAgent agent development framework [51]. The agents can be created and started up using the interface of LAMS Lessons Monitoring Application (Fig. 5). The application listens to events from all the running agents and presents them in the Activity Log. We are going to develop the application in an open-source agent environment (like e.g. JADE).

6 Feedback Messages Taxonomy

The main goal of our system is aiding the teacher during the difficult and time consuming task of manually tracking students' activities and notifying him/her via messages about the existence of a variety of problematic learning situations.

The proposed messages taxonomy was based on our analysis of the problems with Web-based distance learning courses as discussed in the literature and on interviews with a few teachers. In the design of the proposed messaging system we have also taken into account the TADV framework [15] and the structure of the advising systems of COLER and DEGREE [27, 40]

Messages are issued by the system to the teacher, to individual students and to the groups. Messages are related to the attendance, to the participation or to the performance of students and groups in the learning process.

The system produces alert messages to the users of the current lesson that are inferred by rules based on data that Monitoring agent is querying from LAMS data base, on calculated information and on the *Activity Parameters* (for user participation and performance in each activity) and *Lesson Calendar* (for learners' attendance) that are being defined by the teacher.

Table 1 The taxonomy of messages of LAMS' Lessons Monitoring Application

Activity Parameters / Type of Message	Attendance	Participation	Performance or Contribution
Alert message	Learner, Teacher	Learner, Teacher	Learner, Teacher
Awareness message	Learner, Group, Teacher	Learner, Group, Teacher	Learner, Group, Teacher
Evaluation message	Teacher	Teacher	Teacher

The feedback messages provided by the system, as it is shown in Table 1, can be distinguished into three main types: *alert messages, awareness messages* and *evaluation messages*. Each type of message may refer to a different activity parameter (attendance, participation, performance or contribution) and to a different recipient (learner, teacher, and group). So, from each main type several sub-types come out, to indicate different situations that have to be recognized by LAMS Monitoring and Supervising system. More specifically:

- *Alert messages* refer to students that do not participate at all or do not regularly participate (passive students) and to students with low contribution in collaborative activities or low score in evaluative activities.
- *Awareness messages* inform a student (and/or the teacher) for his/her participation and his/her contribution in an activity regarding the whole group or the whole class.
- *Evaluation reports and messages* inform teacher about the quality of an activity and for the classification of learners and groups based on their participation.

Thus, the teacher avoids having to look over every particular case, obtaining the related information quickly.

6.1 Alert Messages

The system issues alert messages to the teacher (T), to an individual student (S) and to a group (G) according to the participation model of the lesson. Some cases that the system issues Alert messages are:

- Alert1,2 (T, S): Student hasn't started an activity as it was defined by the Lesson Calendar.

By the notation *Alert1,2 (T, S)* we mean that two types of messages are issued: A message of type *Alert1* to the teacher and a message of type *Alert2* to the student.

- Alert3,4 (T, S): Student hasn't ended an activity as it was defined by the Lesson Calendar.
- Alert5,6 (T, S): Student has bypassed an activity
- Alert7,8 (T, S): Student had a low contribution in a collaboration activity
- Alert9,10 (T, S): Student had a low time participation in an activity
- Alert11,12 (T, S): Student had a low score in an evaluative activity
- Alert13,14 (T, G): Group had a low mean_contribution in a collaboration activity
- Alert15,16 (T, G): Group had a low mean_score in an evaluative activity

6.2 Awareness Messages

Awareness messages are issued after the execution of an activity to inform the learners (and also the teacher) about their relative performance regarding the mean performance of the group (or the whole class), the mean performance of groups in the learning process etc. Some types of awareness messages are:

- Aware1 (G): Informs about the start of a synchronous collaboration activity
- Aware2 (G): Informs about a change in the workspace of collaboration tools (Forum, Wiki, MindMap).
- Aware3,4 (T,S): Informs about the mean_performance (mean_contribution or the mean_score) of the whole class in an activity
- Aware5,6 (T,G): Informs about the mean_performance of a group in an activity.
- Aware7 (T): Informs about the mean_time_participation of a group in an activity.
- Aware8 (T): Informs about the mean_time_participation of the whole class in an activity.
- Aware9,10 (T,S): Informs about the relative performance of a student in an activity regarding the whole group.
- Aware11,12 (T,S): Informs about the relative performance of a student in an activity regarding the whole class.

6.3 Evaluation Reports and Messages for Learners and Activities

As opposed to other types of numerical information related to concepts such as grades that is easy to interpret, data reflecting quantity or averages are difficult to understand. A solution that has been implemented in several e-learning environments, like TADV framework [15] and DEGREE [40], is to discretize those numerical features into ranges that provide a much more comprehensible view of the data for an average person.

This process is performed automatically by the Monitoring Agent, which, by a discretization process, creates ranges of values (low, medium, high) for the parameters that are defined in the participation model.

The system, after the conclusion of the activity, classifies the learners and the activity, regarding the participation data of a learner in this activity and the statistical indicators that have been calculated for the activity, correspondingly.

Table 2 The classification of learners of LAMS' Lessons Monitoring Application

Classes Indicators	Min	Medium	Max
Contribution	Non Collaborative	Collaborative	Highly Collaborative
Performance	Weak	Good	Excellent
Time_Participation	Slow	Normal	Precipitant
Attendance	Punctual	Late	Absent

By this process, the system classifies the learners (see Table 2). Most of the classes we are going to use are based on the TADV framework. [15]:

- Based on its contribution in a collaborative activity (related to min_contribution and max_contribution which have been defined by the teacher) a learner is characterized as *Non Collaborative, Collaborative* or *Highly Collaborative.*
- Based on its score in an evaluative activity (related to min_score and max_score which have been defined by the teacher) a learner is characterized as *Weak, Good* or *Excellent.*
- Based on its time_participation in an activity (related to min_time and max_time which have been defined by the teacher) a learner is characterized as *Slow, Normal* or *Precipitant.*
- All those classifications are extracted by the system, for the groups also.
- Based on learners' attendance in an activity (related to start_time and max_time which also have been defined by the teacher) a learner is characterized as *Absent, Late* or *Punctual.*

All the above classes of users are presented, at the end of each activity to the teacher, in the form of a report like Table 7.

By the same process, activities, too, are classified as follows:

- Based on learners' mean_score in an evaluative activity and the mean_time_participation of learners, the activity is characterized as *Too Easy, Easy, Normal, Difficult* or *Too Difficult.*
- Based on learners' mean_time_participation in an evaluative activity, the activity is characterized as *Uninspired, Interesting,* or *Exciting.*

Table 3 shows the classification of an activity, based on the levels of mean_performance and mean participation time the users have achieved in this activity.

Table 3 The Classification of Activities of a lesson in LAMS

Mean Contribution or Performance / Mean Time Participation	Low	Medium	High
Low	Unispired	Easy	Too Easy
Medium	Difficult	Interesting	Normal
High	Too Difficult	Normal	Exciting

Additionally, the system issues evaluation messages to inform the teacher and the learners about high or low participation of learners in each activity:

- Evaluation1,2 (T,S): Informs about Excellent students and Weak students of the whole class.
- Evaluation3,4 (T,S): Informs about the Slow or Precipitant students of the whole class. Supervisor can use use this information in the formulation of groups.
- Evaluation5,6 (T,S): Informs about the High Collaborative students and Non Collaborative students of the whole class.
- Evaluation7 (T): Informs about the quality of an activity.
- These evaluation messages are being delivered to the teacher so that he/she can redesign the learning process through LiveEdit, e.g. by assigning roles of coordinators to the most communicative users or by inserting more suitable collaborative activities for the least collaborative users or by directing Excellent and High Collaborative students to assist some Weak students.

6.4 Response of the Messaging System to a Learning Scenario

We are going to depict the operation and the functionality of the underlying messaging system using a virtual learning scenario which is based on a part of a lesson in LAMS.

As it is shown in fig. 6, the first part of the lesson consists of the following activities: An introductive activity designed with the Notebook tool of LAMS, two informative activities that will be implemented with the Q&A (Questions & Answers) and with the Noticeboard tools, two collaborative activities which have been designed with the MindMap and with the Wiki tools respectively, and finally an evaluative activity which will be implemented in the MCQ (Multiple Choice Questions) tool.

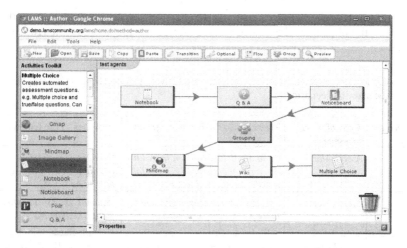

Fig. 6 The lesson in LAMS for the learning scenario

Four students (S1, S2, S3, S4) are participating in the lesson and they are assigned by the teacher into two groups (G1, G2).

The learning scenario includes the following learning events which we assume that violate the *Lesson Calendar* and the *Participation Model*, as they have been defined by the teacher:

1. Student *S1* hasn't started the second activity as it was defined by the Lesson Calendar.
2. Student *S3* had a low time participation in the *NoticeBoard* activity.
3. Student *S1* had a low contribution in MindMap activity.
4. The group G2 had a low mean_score in MCQ activity.

As a result, the monitoring system will issue the alert messages that are briefly presented in table 4. The types of alert messages refer to the codification of 6.1.

The awareness messages that the Supervising Module will issue are presented below in table 5. The types of awareness messages refer to the codification of 6.2.

After each activity an evaluation report of the performance and the participation for each learner will be generated by the system in order to provide the supervisor, with a picture of the learning behavior of each learner. A possible evaluation report of the learning scenario is shown in Table 6.

7   An Evaluation Scenario

In this work, we propose an agent-based system for the collaboration and interaction analysis within the context of formative assessment in the learning environment of a LMS, to support different functions, such as awareness, regulation and evaluation, through messages and reports that are being generated from rules reflecting the objectives that teacher has set for the participation of the learners in the lesson.

Table 4 The alert messages that will be issued by LAMS' Lessons Monitoring Application due to the learning scenario

Learning case	Type of alert messages	Messages to the learner / group	Messages to the teacher
1	Alert1,Alert2	"You are delayed in dealing with the starting activity A1. Try to follow the Lesson Calendar"	"Student S1 is delayed in dealing with the starting activity A1"
2	Alert9,Alert10	"You haven't studied enough activity A2. Maybe, you need to come back"	"Student S3 hasn't studied enough activity A2"
3	Alert7,Alert8	"You haven't contributed enough in MindMap activity. You need to add more concepts"	"Student S1 hasn't contributed enough in MindMap activity"
4	Alert15,Alert16	"Your group had a low mean_score in MCQ activity. You should try harder".	"Group G2 had a low mean_score in MCQ activity They might need to be supported".

Table 5 The awareness messages that will be issued by LAMS' Lessons Monitoring Application due to the learning scenario

Learning activity	Type of awareness messages	Messages to the learner or to the group	Messages to the teacher
After the activities Q & A, Mindmap, Wiki	Aware3,Aware4	"The mean_performance of the whole class in the activity x was y"	"The mean_ performance of the whole class in the activity x was y"
	Aware5,Aware6	"The mean_ performance of your group in the activity x was z"	"The mean_ performance of group [G1] / [G2] in the activity x was z"
	Aware7,Aware8	-	"The mean_time_participation of the[whole class]/[G1]/[G2] in the activity x was w"
	Aware9,Aware10	"Your relative performance in the activity x, regarding [your group] /[the whole class] was v%".	-

Table 6 An Evaluation Report of Learners that would be issued by LAMS' Lessons Supervising Application due to the learning scenario

ACTIVITY	LEARNER / GROUP	PERFORMANCE	CONTRIBUTION	TIME	ATTEND-ANCE
-	-	-	-	-	-
MindMap	S1	-	Non-Collaborative	Normal	Punctual
MindMap	S2	-	Non-Collaborative	Percipitant	Punctual
MindMap	**G1**	-	**Non-Collaborative**	**Normal**	**Punctual**
-	-	-	-	-	-
MCQ1	S3	Weak	-	Slow	Late
MCQ1	S4	Weak	-	Slow	Late
MCQ1	**G2**	**Weak**	-	**Slow**	**Late**
-	-	-	-	-	-

We claim, that the proposed agent-based system will not only help the learners to self-regulate their attendance and participation in the lesson, based on the objectives of the teacher for this lesson, but also it will help teachers to maintain an overview of the learning process and the collaboration, so that they could intervene when they consider it necessary.

To validate our hypothesis, we plan to perform an evaluation of a prototype of the agent-based module, that we are developing, in a lesson that will be implemented in the environment of LAMS. The lesson will be refer to teaching "informatics" in two classes of students in two different schools of Secondary Education in Greece.

The evaluation will be implemented using two teams of students: the control team which will use LAMS without the intelligent module and the experimental team in which the agent-based system will generate feedback messages to the supervisors and to the students. The supervisors for the control and experimental team will be the same.

During the study, various types of data will be collected, including log files, pre-and post-test, teacher interviews and observations and student questionnaires.

The evaluation is going to be contacted in two phases. In the first phase we intend to assess the system's behaviour and to discover potential problems. The prototype will be thoroughly tested during all development phases. Comments and suggestions will be elicited from supervisors and students to ensure appropriate system usability. The prototype will be then modified to satisfy the users' requirements.

In the second phase of the evaluation we are going to use an experimental study methodology to assess the usefulness and benefits of the overall approach. We are going to use quantitative and qualitative methodologies to enable the examination of the collected data from different perspectives, in order to investigate into the suitability of generated messages and benefits to supervisors and students.

The evaluation will focus on the effectiveness and efficiency of the support that the module will provide to the participants of the lesson.

More specifically we plan to investigate whether our system can help educators and learners on behavior perception and problems identification in time to provide a useful remediation.

It will also be tested whether our model provides autonomous, appropriate, periodic and relevant feedbacks for the educator and the learners throughout the lesson according to the planned goals and criteria.

8 Conclusions and Future Work

The approach discussed in this work explores the idea that supervisors in LAMS environments should be automatically provided with information about their virtual classes which focuses on performance, behavioral, and social aspects of students, groups and the whole class. Moreover, supervisors should be supported in the task of sending feedback to the students in an easy and quick manner.

In order to facilitate the monitoring and the supervising of the individual learning process and the collaborative learning in LAMS, we propose in this work, an intelligent module for the continuous assessment of those processes in order to support self-regulation and evaluation.

The system has been designed in order to enable teachers to supervise both synchronous and asynchronous collaborative activities of a virtual class in LAMS and to allow them to dynamically intervene in the evolution of the learning process by redesigning a lesson based on the evaluation results of users and activities.

The proposed intelligent system has been designed to issue alert messages, awareness messages and evaluation reports for learners and activities and, therefore, enables the teacher to automatically inspect a particular learning situation.

Considering this information, teachers may decide to intervene, the same way they would do in a regular presence course when students are not making any progress with their activities.

The system aims to support also the learners of a virtual class in LAMS, since it provides them with awareness information and reminds them of the schedule and the standards of the lesson.

The completeness of such messages taxonomy has not been assured and the likelihood of new needs for other types of advice is possible. We intend to use interviewees with supervisors and learners in LAMS for requirements capture which may result in a more comprehensive taxonomy.

There is also an obvious need for reduction in the amount of generated messages, maybe by using some filtering and aggregation mechanisms to reduce or to merge the messages displayed to the supervisors.

By the same system the teachers can evaluate the structure of course content and its effectiveness in the learning process and also classify learners into groups based on their participation and attendance.

So, the teacher could redesign, through Live Edit of LAMS, the learning sequence and also can reassign roles to learners, to achieve better learning results.

A future direction of our work is to parameterize the Participation Model to enable a supervisor to include his own criteria for the evaluation of the learners and the activities.

Another research direction that is in our plans is to support the interoperability between the proposed intelligent supervising system and more LMS. This would make our tool more autonomous and valuable. Some problems that will be faced in such direction are the data exchange and the synchronization between the two parts.

Finally we plan to modify the design of the system to be capable to determine regular and irregular patterns of learners' interactions, allowing the most frequently conflictive learning cases to be identified and by that allowing more effective feedback to be generated.

References

1. Yacef, K.: Some thoughts on the synergetic effects of combining ITS and LMS technologies for the service of Education. In: 11th International Conference on Artificial Intelligence in Education, AIED Society, Sydney, Australia (July 2003)
2. Mencke, S., Reiner, D.: A Framework for Agent-Supported E-learning. In: Conference ICL 2007, Villach, Austria (2007)
3. Talavera, L., Gaudioso, E.: Mining student data to characterize similar behavior groups in unstructured collaboration spaces. In: Workshop on Artificial Intelligence in CSCL. 16th European Conference on Artificial Intelligence (ECAI 2004), pp. 17–23 (2004)
4. Romero, C., Ventura, S., García, E.: Data mining in course management systems: Moodle case study and tutorial. Computers & Education 51, 368–384 (2008)
5. BlackBoard (2010), http://www.blackboard.com/ (accessed May 26, 2011)
6. WebCT (2010), http://www.webct.com/ (accessed May 27, 2011)
7. Topclass (2010), http://www.topclass.nl/ (accessed May 28, 2011)
8. Moodle (2010), http://moodle.com/ (accessed May 29, 2011)
9. Ilias (2010), http://www.ilias.de/ (accessed May 30, 2011)
10. Claroline (2010), http://www.claroline.net/ (accessed May 26, 2011)
11. LAMS (2010), http://www.lamsinternational.com/ (accessed May 26, 2011)
12. Galusha, J.: Barriers to learning in distance education. Interpers. Comput. Technol. Electron. J. 21st Century 5(3/4), 6–14 (1997)
13. Rivera, J., Rice, M.: A comparison of student outcomes and satisfaction between traditional and web-based course offerings. Online Journal of Distance Learning Administration 5(3) (2002)
14. Helic, D., Maurer, H., Scherbakov, N.: Web-based training: what do we expect from the system. In: Eighth International Conference on Computers in Education, Taiwan, pp. 1689–1694 (2000)
15. Kosba, E., Dimitrova, V., Boyle, R.: Adaptive feedback generation to support teachers in web-based distance education. User Modelling and User-Adapted Interaction 17(4), 379–413 (2007)

16. Lin, F., Esmahi, L., Poon, L.: Integrating Agents and Web Services into Adaptive Distributed Learning Environments. In: Lin, F.O. (ed.) Designing Distributed Learning Environments with Intelligent Software Agents, pp. 184–217. ISP, Hersey (2005)

17. Suh, H.J., Lee, S.W.: Collaborative learning agent for promoting group interaction. ETRI Journal 28(4), 461–474 (2006)

18. Royce Sadler, D.: Formative assessment and the design of instructional systems. Instructional Science 18, 119–144 (1989)

19. Dimitracopoulou, A.: Computer based Interaction Analysis supporting Self-regulation: Achievements and Prospects of an Emerging Research Direction. In: Kinshuk, Spector, M., Sampson, D., Isaias, P., (eds.) Technology, Instruction, Cognition and Learning (TICL), vol. 6(3) (2008)

20. Dillenbourg, P., Baker, M., Blaye, A., O'Malley, C.: The evolution of research on collaborative learning. In: Espada, E., Reiman, P. (eds.) Learning in Humans and Machine: Towards an Interdisciplinary Learning Science, pp. 189–211. Elsevier, Oxford (1996)

21. Plantamura, P., Roselli, T., Rossano, V.: Can a CSCL environment promote effective interaction? In: Proceedings of the IEEE International Conference on Advanced Learning Technologies (ICALT 2004), pp. 675–677. IEEE Computer Society, Washington, DC (2004)

22. Dimitracopoulou, A., Hoppe, U., Dillenbourg, P.: Workshop on Interaction analysis supporting participants during technology-based collaborative activities. In: CSCL Symposium, Kaleidoscope NOE, Lausanne, October 7-9 (2004)

23. Jermann, P., Soller, A., Muhlenbrock, M.: From mirroring to guiding: A review of state of the art technology for supporting collaborative learning. In: Proceedings of European Conference on Computer-Supported Collaborative Learning 2001 (Euro-CSCL 2001), pp. 324–331. Maastricht McLuhan Institutt, Maastricht (2001)

24. Martínez, A., Dimitriadis, Y., de la Fuente, P.: Interaction Analysis for Formative Evaluation in CSCL. In: Llamas, M., Fernandez, M.J., Anido, L.E. (eds.) Computers and Education. Toward a Lifelong Learning Society, pp. 227–238. Kluwer Academic (2003)

25. Dimitrakopoulou, A., et al.: State of the Art on Interaction Analysis: Interaction Analysis Indicators. Kaleidoscope Network of Excelence. Interaction & Collaboration Analysis' Supporting Teachers and Students' Self-Regulation. Jointly Executed Integrated Research Project Deliverable D.26.1 (2004)

26. Duque, R., Bravo, C.: A Method to Classify Collaboration in CSCL Systems. In: Beliczynski, B., Dzielinski, A., Iwanowski, M., Ribeiro, B. (eds.) ICANNGA 2007 Part I. LNCS, vol. 4431, pp. 649–656. Springer, Heidelberg (2007)

27. Constantino Gonzalez, M., Suthers, D., Escamila de los Santos, G.: Coaching web-based collaborative learning based on problem solution differences and participation. International Journal of Artificial Intelligence in Education 13 (2003)

28. Dimitracopoulou, A., Petrou, A.: Advanced Collaborative Distance Learning Systems for young students: Design issues and current trends on new cognitive and meta-cognitive tools. Themes Journal, Special Issue, Issues and Trends regarding the Application of Information and Communication Technologies to Distance Learning (2004)

29. Mostow, J., Beck, J., Cen, H., Cuneo, A., Gouvea, E., Heiner, C.: An educational data mining tool to browse tutor–student interactions: Time will tell! In: Proceedings of the Workshop on Educational Data Mining, Pittsburgh, USA, pp. 15–22 (2005)

30. Otsuka, J.L., da Rocha, H.V., Beder, D.M.: A Multi-Agent Formative Assessment Support Model for Learning Management Systems. In: ICALT (2007)
31. Galusha, M.: Barriers to learning in distance education. University of Southern Mississippi (1997), http://www.infrastruction.com/barriers.html (accessed July 26, 2011)
32. Gutwin, C., Stark, G., Greenberg, S.: Support for workspace awareness in educational groupware. In: Proceedings of the ACM Conference on Computer Supported Collaborative Learning, pp. 147–156. ACM Press, Indiana (1995)
33. Dourish, P., Belloti, V.: Awareness and coordination in shared workspaces. In: Proceedings of Computer Supported Collaborative Work (CSCW 1992), Chapel Hill, NC (1992)
34. Casamayor, A., Amandi, A., Campo, M.: Intelligent assistance for teachers in collaborative e-learning environments. Computers & Education 53, 1147–1154 (2009)
35. Hmelo, C.: Collaborative ways of knowing: Issues in facilitation. In: Proc. Computer Supported Collaborative Learning (CSCL) Conf., pp. 199–208 (2002)
36. Nwana, H.S.: Software Agents: An Overview. The Knowledge Engineering Review 11(3), 1–40 (1996)
37. Erlin, N.Y., Azizah, A.R.: Overview on agent application to support collaborative learning interaction. US-China Education Review 5(1) (Serial No.38) (January 2008) ISSN1548-6613
38. Jafari, A.: Conceptualizing Intelligent Agents for Teaching and Learning. EDUCAUSE Quarterly 25(3), 28–34 (2002)
39. Choy, S.O., Ng, S.C., Tsang, Y.C.: Software Agents to Assist in Distance Learning Environments. EDUCAUSE Quarterly (2) (2005)
40. Barros, M., Verdejo, M.: Analysing student interaction processes in order to improve collaboration. The DEGREE approach. International Journal of Artificial Intelligence in Education 11, 221–241 (2000)
41. Dolonen, J., Chen, W., Anders, M.: Integrating software agents with fle3. In: Proceedings of the International Conference on Computer Support for Collaborative Learning, pp. 157–161. Kluwer Academic Publishers, Dordrecht (2003)
42. Scutelnicua, A., Linb, F., Kinshukb, Liuc, T.C., Graf, S., McGreal, R.: Integrating JADE Agents into Moodle. In: Supplementary Proceedings (Workshops/Doctoral Student Consortium) of the International Conference on Computers in Education, APSCE, pp. S.215–S.220 (2007) ISBN: 978-4-924861-19-0
43. Zorrilla, E., Menasalvas, E., Marin, D., Mora, E., Segovia, J.: Web Usage Mining Project for Improving Web-Based Learning Sites. In: Moreno Díaz, R., Pichler, F., Quesada Arencibia, A. (eds.) EUROCAST 2005. LNCS, vol. 3643, pp. 205–210. Springer, Heidelberg (2005)
44. Anaya, A.R., Boticario, J.G.: Reveal the Collaboration in an Open Learning Environment. In: Mira, J., Ferrández, J.M., Álvarez, J.R., de la Paz, F., Toledo, F.J. (eds.) IWINAC 2009 Part I. LNCS, vol. 5601, pp. 464–475. Springer, Heidelberg (2009)
45. Anaya, A.R., Boticario, J.G.: A Data Mining Approach to Reveal Representative Collaboration Indicators in Open Collaboration Frameworks. Educational Data Mining (2009)
46. Dalziel, J.: Implementing learning design: The Learning Activity Management System (LAMS). In: 20th ASCILITE, Adelaide, Australia (2003)
47. Cameron, L.: Picture this: My Lesson. How LAMS is being used with pre-service teachers to develop effective classroom activities. In: Proceedings First International LAMS Conference 2006: Designing the Future of Learning, Sydney, pp. 25–34 (2006)

48. Popham, W.J.: Classroom assessment: What teachers need to know, 5th edn. Allyn & Bacon, Boston (2008)
49. Nicol, D.J., Macfarlane-Dick, D.: Formative assessment and self-regulated learning: A model and seven principles of good feedback practice. Studies in Higher Education 31(2), 199–216 (2006)
50. Chronopoulos, T., Hatzilygeroudis, I.: An intelligent system for monitoring and supervising lessons in LAMS. 2nd International Conference on Intelligent Networking and Collaborative Systems, Thessalonica, Greece, November 24-26 (2010)
51. Bigus, J.P., Bigus, J.: Constructing Intelligent Agents using Java, 2nd edn. Wiley Computer Publishing (2001)

The Text-Based Computer-Mediated Communication in Distance Education Fora: A Modelling Approach Based on Formal Languages

Kiriakos Patriarcheas, Spyridon Papaloukas, and Michalis Xenos

Hellenic Open University, School of Sciences & Technology, Computer Science,
13-15 Tsamadou Street, GR 26222, Patras, Greece
{k.patriac,s.papaluk,xenos}@eap.gr

Abstract. Electronic asynchronous discussion fora are increasingly becoming part of the distance education process and are a dynamically evolving field which needs to be constantly updated and redefined. This chapter presents a system development approach for automated interpretation of discussion threads' messages in asynchronous distance education fora by using the content category as unit of analysis. This system inputs discussion threads from distance education fora and outputs specific strings representing the messages of these threads according to concrete modelling based in a formal language.

Keywords: Distance education, Electronic forum, Discussion thread, Formal languages, Content category, Modelling.

1 Introduction

Electronic asynchronous discussion fora (hereinafter "fora") are a key tool, supporting a large part of the distance education process. An important issue which, in recent years, has concerned field researchers, designers, coordinators and tutors is how they might gain, at any given moment, an overall view of the situation from a number of discussion threads in a distance education forum, not only on the quantitative level of participation (something which has been dealt with by maintenance of statistical data from the relevant environments) but also on the level of what is discussed and where the discussion is focused. This chapter presents a system development approach for automated interpretation of discussions in distance education fora.

According to the above, it is important to provide feedback to the designers to help them understand the parameters of fora (as a component part of distance education) in a concise manner at a strategic level. For example, knowledge of the focus of interest of the discussion may constitute a source of substantive knowledge during the phase of the redesign of an existing distance education course, or the original design of another (similar course) and affect the (re)design of the educational material, both from the aspect of the curriculum and from the aspect of its

T. Daradoumis et al. (Eds.): Intelligent Adaptation & Personalization Techniques, SCI 408, pp. 239–266.
springerlink.com

structure-presentation (e.g. in a subsequent version of the education material, more emphasis may be placed on exercises or case studies with indicative comments, etc), or the desired profile of the corresponding teaching staff, the proposed teaching techniques, etc., always bearing in mind that design in distance education is a constant process of redefinition. Fittingly, Posner [1] adopts the term "reflective eclecticism". This feedback is a crucial factor in guiding the correlation of decisions in the selection and development of each component part with regard to the targets set by the e-learning strategy.

Moreover, coordinators who are responsible for the pedagogical soundness of the courses need to have a comprehensive understanding during the review process. It is important to be able to know, with an automated method, certain aspects such as: what the focus of interest of the discussion in the fora is, whether the tutor is consistent (replies to students, and in the amount of time is allocated to each content category), whether there is same intensity and the same view throughout the academic year, if there are typical behaviors in fora so they can be interpreted and addressed (both by Tutors and by students).

Also, the Tutor (who in distance education has a complex role where he must have an in depth knowledge of the teaching process and of ways to support his students, solve their problems, guide them, or facilitate, manage and evaluate their learning process) needs, among other things a tool that can represent discussions in fora in an automated way thus enabling the detection of any "alarm-raising" discussions so he can intervene and either encourage or discourage them.

In technological level, the problem that must be therefore tackled is that the text developing in fora is in a free (i.e. unstructured) form. It is therefore a real challenge to attempt to "codify" and subsequently analyze this quantity of unstructured data that comprise the texts in the fora of distance education.

A vast amount of the research presented in international literature concerning distant education's fora, refers to content analysis technique. Despite the fact that this research technique is frequently used, no standards have been established as of yet. There are a variety of approaches, as to the level of detail and the type of categories of analysis used (e.g. thematic unit, message, sentence, paragraph etc.).

The Hellenic Open University (hereinafter HOU) has turned to the modeling of messages in recent years in order to classify the interventions of participants in its fora into large categories in order to detect where the subject of interest of the discussion is focused. This is indeed the educational purpose of the system developed by HOU and presented in this chapter.

With this goal in view HOU decided to employ, the message content category as a unit of analysis because in the observation of the discussion threads, it was noticed that there are cases of messages which may comprise two (or/and more) content categories, e.g. in the same message may be a question on study of educational material and as well as assignment.

This chapter addresses the development of a system which inputs discussion threads from distance education fora and automatically outputs strings representing the messages according to a model developed in HOU by simultaneously recording the date and the time difference for each message content category in each discussion thread.

The aim of this system is to operate as a complementary tool to assist designers, coordinators and tutors with the issues referred to above, in an effort to understand what happens in the fora. For example, continuous questions and replies where the same symbol is repeated show that the discussion focuses on one aspect or that an interpersonal conflict may have taken place. When we have many replies to the same message content category (e.g. study of educational material) then it seems that students simply reply to a question without a substantive discussion developing. Therefore the Tutor can intervene when the interaction level decreases. When there are threads that do not end with a reply, this means that the discussion did not reach any conclusions, which is possibly the Tutor's fault. If this happens repeatedly, then the Coordinator must take specific measures. Given that the system records the corresponding dates, periods of low and high participation (in quantity and quality) per message content category may also be diagnosed. Also, given that time differences (in days) that are recorded, the tutor's consistency with regard to time may be determined, and comparisons can be made between groups. This does not mean of course that the tutor must reply immediately - only after a certain time elapses and if another student has not replied.

Another use of the system is its utilization in the comparison of educational techniques (e.g. Brainstorming vs Snowballing) with regard to their participation and effectiveness [2,3] or in finding the appropriate group size when the working group technique is used [4], by comparison of the number of appearances per content category, and outcomes in the relevant papers submitted by students.

From the above indicative presentation, the contribution of this chapter is obvious. In other words, this system was not created as an addition, but it was created to operate as a complementary tool for extracting educational conclusions within the framework of upgrading distance education.

2 Theoretical Framework

2.1 Generally

Numerous surveys have been conducted to study the behaviour of students in distance education fora. Researchers such as Newman [5] advocate giving students ample time to think in order to cultivate classroom thoughtfulness. Similarly Cohen and Ellis [6] and Hara et al. ([7]) support that the students can provide feedback and mentoring within electronic discussions. Therefore, the development of a learning group into a community is a major accomplishment that requires special processes and practices [8], is a complex process that relies on the willingness of the students to adopt a collaborative learning style [9].

Consequently, there is a long discussion on the role of a tutor in supporting interaction [10,11]. A tutor may want to have an overview of students' discussions, such as the top ten hot topics, without reading every message in the learning forum. Hence, it is necessary to model the relationships between discussion messages during a specific time interval for showing the major discussion topics [12]. The written protocols that are available provide an exact record of the instructional transactions at a given time in the online discussion [13].

Relative with the above is, also, the approach of Bratitsis and Dimitracopoulou [52] that is reported in the DIAS (an Asynchronous Discussion Forum Software, mainly developed in order to offer extended monitoring and interaction analysis support, by providing a wide range of indicators jointly used in various situations, to all discussion forae users).

According to the above, a main issue in e-learning is student modelling. It is difficult to monitor the students' learning behaviours. A proposed solution is the exploitation of automatic systems for the generation and discovery of user profiles, to obtain a simple student model based on his/her learning performance and communication preferences, that in turn allows to create a personalized education environment [14].

Given that tutors can use and manage online discussions at the message level to promote critical thinking [15,16], the question arises as to how data mining may offer promise as a strategy for discovering and building alternative representations for the data underlying asynchronous discussion fora. This could be dealt with by intersecting the information (i.e., participation indicators) an instructor may wish to extract from the forum with viewable and useful information that the system could produce from the tutor's query [17].

2.2 The Complexity of Text Based Computer-Mediated Communication (CMC)

According to Harasim [18] communication via text-based messages provides a high level of interactivity, which encourages collaboration and influences the learning process. The asynchronous capabilities of text-based CMC allows for more thought, reflection and processing of information [7]. These two factors indicate that electronic messages are potentially a rich source of data for researchers. Text-based messages commonly used in CMC have unique characteristics [19]. A large part of the complexity related to the analysis of human communication through the exchange of electronic messages in asynchronous discussion fora is due to the fact that while they are written texts they do not share the same features as traditional written communication and contain more characteristics of spoken communication [20]. Similarly, Kern [21] argues that the CMC is somewhere on the continuum between paper-based writing and speech. According to McCreary [22] the written word demands an exactness and coherence of thought, indicating that text-based communication results in more well planned and structured interactions. In more detail, Kol and Schcolnik [23] argue that the language complexity focuses on lexical or syntactic factors. Specifically, lexical complexity is reflected in two dimensions: *range* (lexical variation) and *size* (lexical sophistication) [24]. An analysis of lexical complexity looks at how many different words are used or how sophisticated the words are. Syntactic complexity reflects elements such as sentence length, amount of embedding, and range and sophistication of structures [25]. This complexity leads to the question of Kol and Schcolnik [23] "how can forum discussions be analyzed?" who note that many studies focus on the nature of student messages or their length, depth, or purpose.

2.3   Theoretical Approaches

An important issue in this framework is the content analysis, a technique frequently used in the approach of issues concerning asynchronous computer mediated discussion groups in distance education. De Wever et al. [53] presents an overview of different content analysis instruments, building on a sample of models commonly used in the Computer-supported collaborative learning (CSCL) literature. During the last years, numerous efforts to approach this issue were made, stemming from different theoretical backgrounds. Indicatively, Barrett & Lally [26] have used content analysis to investigate learning and socio-emotional behaviour in learning community from the Gender differences view. Henri [20]) uses the point of Cognitive and metacognitive knowledge, while Newman et al. [27] and Bullen [28] the point of Critical thinking.

Many, though, start from social constructivism using different variations. Indicatively, some [29-32] utilize the approach of social constructivism in combination with knowledge construction; Jarvela and Hakkinen [33] in combination with perspective taking, while Lockhorst *et al.* (2003) in combination with learning strategies. Rourke et al. [34] utilize the approach of community of inquiry from the point of social presence, rather than [35] from the point of cognitive presence or [36] the teaching presence.

In agreement with the above, an important issue which arose is that of which unit of analysis is to be used for content analysis. Researchers as Fahy et al. [37] consider each single sentence as one unit of analysis, while Pena-Shaff and Nicholls [31] uses the sentence as unit of analysis, trying to approach it at a paragraph level. Others choose the definition thematic unit (or otherwise of a "theme" or an "idea") as a unit of analysis [5,20, 34, 38]. Another approach [28-30, 34-36, 39] is to consider the whole message that a student enters at a specific moment in the conversation as the unit of analysis. Jarvela and Hakkinen [33] choose a complete discussion. Further down, a comprehensive review is presented in a table form (Table 1), referring to the unit of analysis used by this field researchers.

Given that the choice of a unit of analysis is dependent on the context and should be well-considered, because changes in the size of this unit will affect coding decisions and comparability of the outcome between different models [40], given the fact that Schrire [13] refers to a dynamic approach in which data is coded more than once and the grain size of the unit of analysis is set, depending on the purpose and the research question, it was decided not to take into consideration the discussion thread, not even the message as unit of analysis, nor the paragraph or the single sentence.

From the extensive observation of the discussion threads, it was noticed that there are cases of messages which may comprise two (or/and more) content categories, e.g. a question about the next advisory meeting and a reply to a question concerning the study of the educational material.

Table 1 Overview of the content analysis scheme

Unit of analysis	Theoretical background	Researcher
Thematic unit	Cognitive and metacognitive knowledge	Henri (1992)
Thematic unit	Critical thinking	Newman *et al.* (1995)
Message	Theories of cognitive and constructive learning – knowledge construction	Zhu (1996)
Message	Social constructivism – knowledge construction	Gunawardena *et al.* (1997)
Message	Critical thinking	Bullen (1997)
Thematic unit	Community of inquiry – social presence	Rourke *et al.* (1999)
Sentence	Social network theory – Interactional exchange patterns	Fahy *et al.* (2001)
Message	Social constructivism – knowledge construction	Veerman & Veldhuis-Diermanse (2001)
Message	Community of inquiry – cognitive presence	Garrison *et al.* (2001)
Message	Community of inquiry – teaching presence	Anderson *et al.* (2001)
Complete discussion	Social constructivism – perspective taking	Jarvela & Hakkinen (2002)
Thematic unit	Social constructivism – knowledge construction	Veldhuis-Diermanse (2002)
Thematic unit	Social constructivism – learning strategies	Lockhorst *et al.* (2003)
Paragraph	Social constructivism – knowledge construction	Pena-Shaff & Nicholls (2004)
Micro and macro-level	Social constructivism –knowledge construction	Weinberger & Fischer (2006)

3 Methodological Framework

3.1 HOU Fora and Choices Justification

HOU is the distance education institution par excellence in Greece. The HOU has currently 25,418 students (16,066 graduate, 9,301 post-graduate, and 51 PhD students) and 1,485 professors (27 of which are tenured and the rest are external tutors-consultants).

An educational unit at HOU is a course module; today, 184 course modules are offered at HOU. The tutor and all students of a module may participate in the discussion threads of each module.

As far as Computer Science students are concerned, at the time of this survey, 6,067 discussion threads with 26,246 messages had been created in the 16 Computer Science modules (at graduate level) offered by HOU.

With reference to the evolution of the use of HOU for a (Figure 1), by way of illustration, in the last academic years there has been a large increase in the number of messages in the module Introduction to Information Technology (INF10), with a (relatively) invariable number of discussion threads, as shown in Table 2.

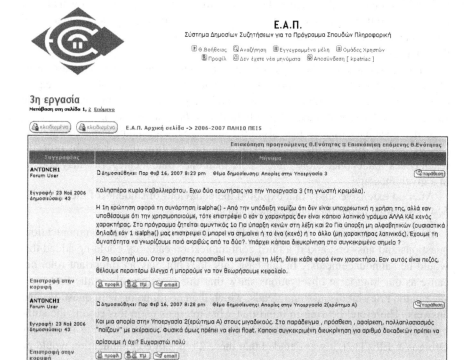

Fig. 1 Messages in the Computer Science forum of the HOU.

Table 2 The evolution of the number of messages for the academic years 2005-6, 2006-7 and 2007-8 in the module INF10

Year	Threads	Messages	Messages/Thread
2005-6	237	982	4.14
2006-7	236	1205	5.11
2007-8	219	1942	8.87

Given that the choice of a unit of analysis is dependent on the context and should be well-considered, because changes to the size of this unit will affect coding decisions and comparability of outcome between different models [40], as well as given the fact that Schrire [13] refers to a dynamic approach in which data is coded more than once and the grain size of the unit of analysis is set, depending on the purpose and the research question, it was decided to use, as unit of analysis, the category of message content.

On the observation of the discussion threads, it was realized that there are cases of messages which may comprise two (or/and more) content categories, e.g. a question about the next advisory meeting and a reply to a question concerning the study of the educational material (see below *modelling* subsection).

Thus, in such a case, the analysis at the message level used by some researchers [28-30, 34-36, 39] is deficient in exploiting information that arises in order to reach educational conclusions, since more than one content categories may coexist in a given message.

After an extensive study that explored the behaviour of students of the HOU [2-3, 41] we came to the following conclusions:

a) The vast majority of messages in fora relate to two types of communication: questions and answers; and if a third communication category could be added this would be "announcements". In our case this does not play an important role, because as our long-term observations show that this pertains only to the first message in the relevant discussion threads, which then continue on with questions and answers. Therefore, bearing in mind that what interests us in this case is the summary view of the discussions within the framework of "practicality", as defined by Dringus & Ellis [17], it was decided that these 2 types of communication would be used.

β) HOU fora make a significant contribution to the learning process as they help with organising the study of a module and the processing of and elaboration on what learners have already studied as follows:

i) For organization of studies during the course module:

- To the communication between the teacher and the students (regularity of contacts, subject, resolution of "technical" problems etc.).
- To the organization of homework (method of use of the teaching material and the preparation of the activities, exploitation of the literature and the other sources, timetables, encountering problems related to it et. al.)
- To the supply of information about the advisory meetings (their number, their duration, the timetables, the goals, their content and methodology applied, problems encountered in the ability to attend them et. al.).
- To supply clarification about the procedure of preparation and evaluation of the written assignments (form, method of preparation, evaluation criteria, means of support by the teacher et. al.).
- To inform about the procedure of final exams (student preparation, support by the teacher, marking criteria, means and time of examination et. al.).

ii) For the elaboration and development of what the students have already studied, the HOU's fora may be exploited for:

- The presentation of consolidation exercises, short suggestions, presentation of examples, methodologies, literature et. al.,
- The resolution of questions and the supply of clarifications about the teaching material.
- The interconnection between what is already studied, subsequent chapters and the written assignment to follow.

Given the heavy flow of information carried through HOU fora, the designing and development of the system presented in this chapter simulated the development of a formal language to interpret messages in the fora of HOU.

3.2 Modelling of HOU Distance Education Forum

Based on observations at HOU fora the following became evident:

a) There are two categories of communication actors: Tutors and Students. For brevity, tutors will be symbolised with a T and students with an E.
b) As regards message types, these are distinguished into questions and answers. Hereinafter, symbolised with q and a respectively.
c) As to their content, messages are distinguished into those relating to (the respective symbols are given in brackets):

- The study of educational material (M)
- Questions/answers for exercises – assignments (X)
- Presentation of sample assignments by tutors (P)
- Instructions (I)
- Assignment comments, corrections (F)
- Student comments on assignments (D)
- Sending – receiving assignments (J)
- Sending - receiving grade marks (G)
- Notification of advisory meeting (V)
- Pointless message (L)

Finally, the order in which above symbols will be written is: a) message carrier b) message type and c) the content of the category to which the message belongs.
  Thus, we have a language which contains:

a)*Terminal symbols alphabet* V_T, where $V_T = \{T, E, q, a, M, X, P, I, F, D, J, G, V, L\}$
b) *Non terminals alphabet* V_N, where $V_N = \{u, r, y, c\}$, more specifically :
    r: represents the message carrier (T for tutors and E for students)
    u: represents a pair yc i.e. a message type y (whether it is a question q or an answer a) followed by its content category.

c) The grammar P

A set of rules of the form α → β, where α and β sequences containing terminal and non-terminal symbols and α is not an empty sequence, as follows:

1.S → *ruS*	*7.r* → *ε*	*13.c* → *P*	*19.c* → *V*
2.S → *ε*	*8.y* → *q*	*14.c* → *I*	*20.c* → *L*
3.u → *uyc*	*9.y* → *a*	*15.c* → *F*	*21.c* → *ε*
4.u → *ε*	*10.y* → *ε*	*16.c* → *D*	
5.r → *T*	*11.c* → *M*	*17.c* → *J*	
6.r → *E*	*12.c* → *X*	*18.c* → *G*	

Where   stands for an empty symbol

d) Symbol S where every sentence generated starts with this symbol.

An indicative example is presented that contains a series of messages represented by the sequence *EqVMTaVMEqMXEaXTaM,* which, according to the above, represent a discussion thread as follows: in the beginning is a message whose sender is student *E* who is asking a question *q* referring to forthcoming advisory meeting *V* and also concerning the study of educational material *M.* This message is replied to by tutor *T* who is answering *a* referring to forthcoming advisory meeting *V* and also about the study of educational material *M.* This message is replied to by student *E* who is asking a question *q* concerning the study of educational material *M* and also about the forthcoming assignment *X.* This message is replied to by other student *E* who is answering *a* about the forthcoming assignment *X.* In the end of discussion is found a reply from tutor *T* on the question concerning the study of educational material M.

According to the above, the sequence *EqVMTaVMEqMXEaXTaM* constitutes a sentence of the *Language* because:

Rule: (1)   (1)   (1)   (1)
   S—>ruS—>ruruS—>rururuS—>rurururuS

(1)   (3)(5)(6)
—>rururururuS————>EuycTuycEuycEuycTuyc

Rules: *(3)(8)(9)*
    ————>EquycuycTauycEquycuycEauycTauyc

Rules: *(4)(10)(11)(19)*
    —————————> EqVMTaMEqMyEayTaM

Rules: *(4)(10)(12)*
    —————————>EqVMTaVMEqMXEaXTaM

As it is obvious from the example, while one content category corresponds to the 4th, and 5th messages (*M* and *X* respectively), in the 1st, 2nd and 3rd messages there are two content categories *VM* and *MX* (Figure 2).

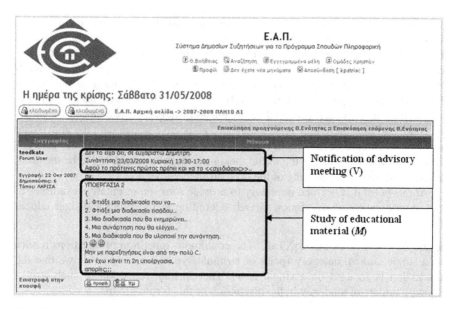

Fig. 2 In the fist message there are two content categories V and M.

The above procedure can be represented schematically with the figure 3.

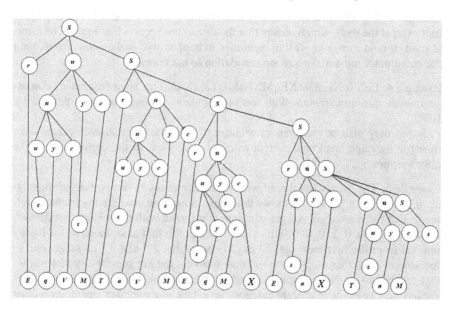

Fig. 3 A graphical representation for the sequence EqVMTaVMEqMXEaXTaM.

It is worthy to note here that this system incorporates the sense of time along with its association with each of the 10 categories of message content chosen as unit of analysis. More specifically and given that within a message (as it is deduced both from literary review and from the observation of the fora of HOU) more than one contents may exist, the dates are recorded for each such case and not simply in each message. Consequently, time differences may automatically exist (in days, if from each current date, by content category, it is deduced the previous one) and thus there may arise another 10 respective stacks with the above date references. Of course, the length of these stacks is equal to the length of dates minus one (-1), i.e. apart from the initial message, which is considered to be the point zero (0), where the numbering of the time differences begins.

At this point it is deemed necessary to provide some representative examples of the representation of discussion threads according to modeling and their educational importance.

Example 1: EqXEqXEaXEqXTaX. The continuous repetition of questions/replies of a single content category (pairs of terminal symbols q or aX) shows that the discussion is focused on a single issue. The constant repetition of such strings shows a unilateral tendency in the discussion of this forum group.

Example 2: EqMEaMEaMTaM. The constant appearance of the aM pairs shows that there are many replies in the same content category, essentially that no substantial dialogue is developed.

Example 3: EqMXEqXDEqDEqX. The terminal symbol a is not found anywhere (not even at the end), which means that the thread has received no reply. The tutor, at least, should intervene. If this continues to happen within the same group, then the coordinator must make a recommendation to the tutor.

Example 4: EqXTaXEqDMXEqMXEaMDTaMX D. This is an "expected" discussion, with questions/replies with various contents among the ones desired by HOU.

Issues may also be detected in relation to how the discussion develops over time (on message level and content category level) and to the consistency in the tutor's replies.

Example 5: The discussion thread with time differences in the posting of messages: 0, 10, 12, 9, 8, 2, 15, 19 shows that there is no time consistency in the development of the discussion compared to another thread with the following time differences: 0, 1, 0, 0, 2, 0, 0, 0. This means that, in the first case, there was a 10 day interval from the first message, 12 days between the 2nd and 3rd message, etc. In the second case, there a 2 day interval at most (at the 5th message) which shows that the discussion is developing smoothly (from a time aspect). The same observations may be made at content category level (see in *exemplary operation section in Fig. 11*).

Example 6: With regard to the tutor, the discussion thread with time differences in the posting of messages by the tutor: 0, 7, 8, 11 shows that the tutor does not intervene in the discussion frequently. This does not mean of course that he should always intervene on the same day, because he may often wait for another student to reply to a question. This means that a thread with time interventions by the tutor of the form: 0, 2, 2, 1, 3, is an expected behavior (from a time aspect, see in *exemplary operation section* in Fig.12).

4   System Development

In accordance to this approach, it was deemed necessary to develop a system that would automatically carry out the above process and output the corresponding strings. It was important to determine firstly the message carrier, then the type (i.e. whether it was a question or answer), and finally the content category(ies) of the message (i.e. study of educational material, practice-assignment, instructions, etc.).

A significant problem which was faced was how to "filter" the plentiful information (contained in distance education fora) and then how to classify the free-form (not structured) Greek text and correlate it to the symbols of the specific modelling. In other words, we would have to deal with the problem of text management in distance education fora. A first approach was to determine the keywords (or better yet, word roots) or symbols according to which the message text under examination will enter a category of message content. The following procedure was followed for this purpose.

4.1   Data Filtering

Web pages hosting discussion threads in a distance education forum contain several data that doesn't contain essential information relating to the educational process (e.g. captions, icons, etc.), therefore, there was a need to weed out needless information, and maintain only the relevant part of discussion thread, which would be used as a source of information to draw educational conclusions.

Therefore, an algorithm was created (procedure "data_filtering") that would input a file containing one or more discussion threads in their original form and output a file of documents containing the following information: a) User name; b) date; c) message content.

To this end, used commands that will allow discussion threads to be read and stored per character in order to be processed were used. In addition, a variable was used as an index to determine the position of the character table, whose value increases until a condition is met to find the mark word that defines the forum user name. Each time the condition is met, the user name is stored with a temporary variable. In the same way, the index is used to find the starting points where the day literal, month literal, day number and year number are entered, with respective allocations to temporary variables.

As far as the message content is concerned, the index was used to detect both the opening and the ending mark word of the message content. During this process, the message content is temporarily stored and at the end of the process all contents of temporary variables are used to enter "useful" information in the records file.

4.2 Roots Files Storage

Given that, by knowing the User name it is easy to determine the message carrier (i.e. whether it is an instructor or a student), the issue of storing information that is necessary had to be dealt with to decide whether a message is a question or answer.

The experimental results of the algorithm execution proved wrong the initial estimate that when a message contains the question mark (English "?" or Greek ";") then it is a question, seeing that there were messages comprising a question, although they did not include a question mark, yet other phrases such as "I would like to ask", etc.

Therefore, should be designed a dynamic method to store the information required to determine the type of message (whether it is a question or answer). This effected the decision to create an algorithm (procedure "message_type") that takes pairs of information: a) word or phrase root or symbol, and b) terminal symbol q or a if it is a question or answer, and to create a records file of two respective fields containing the above pairs.

The same reasoning (procedure "content_category") was used to store information needed to determine the content category of a message, namely if it refers to a study, assignment, comment, etc., or a combination thereof (e.g. when a message refers to both a study and assignment). Therefore, was created an algorithm that would input pairs of information as follows: a) word or phrase root, and b) terminal symbol of content category ($M, X, P, I, F, D, J, G, V, L$). Thus, it is possible to add more content categories if needed.

As for the roots file creation on the message content category was followed the basic syntax rule in Greek Language that the endings are created by combinations smaller and simpler endings.

Constituted by 6 steps is each one by which is executed sequentially (Figure 4). The rules that are used are in form: A1 → A2 (conditions) with the significance of replacement of ending A1 from the A2 if the letters that remain from the A1 satisfy the condition.

At the first step of algorithm becomes handling of plurals and aorists (pasts). This step is separated in three sub-steps. The first handles plurals (e.g. in English language: caresses → caress).

The second removes and/or changes the endings, if this is required (e.g. in English: ed and ing). The process is continued and, if the ending has been removed, the root that remains is converted (e.g. in English language: conflated → conflate, motoring → motor, agreed → agree).

The third (sub-step) converts the final letter (e.g. in English language: happy → happi). The steps deal mainly with the different sequence in the ending groups. For this reason they convert the double endings in one while they remove also endings that satisfy certain criteria as they appear also in the following figure (4).

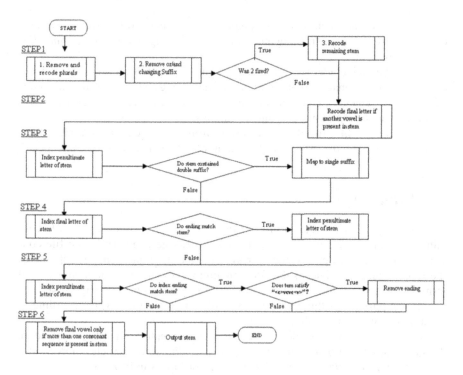

Fig. 4 First stage: the followed steps for the creation of roots of words as for the message content category.

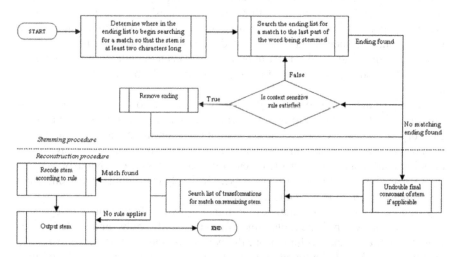

Fig. 5 Second stage: the followed steps for the creation of roots of words as for the message content category

The second stage materialises an algorithm of roots export of words that produces the result with one parsing and removes the endings based on the *Quick Fitting (QF)* principle.

The algorithm includes two sub stages (Figure 5). First, is the sub stage of roots export (stemming phase) where is removed the endings and is checked the application by any chance exceptions between the steps. The second stage uses rules for the reconstruction of words from the endings.

In the end of the procedure, the results from the execution of 2 algorithms (stages) are compared and created the common roots on the message content category.

4.3 String Output

The next stage was to create a process that would input: a) the records file containing useful information (User name, date, message content); b) the file containing pairs of word/phrase roots or symbols and terminal symbols relating to the type of message; and c) the file containing the pairs of words/phrases and terminal symbols referring to the content category of the message and, evidently, the purpose of this process was to obtain the desired string according to the aforementioned modelling.

Accordingly, it was decided to design a corresponding algorithm (procedure "symbol_sequence") that would not only display the string but also store it in a relevant extensible file in order to keep results for further utilization.

On these grounds, and aside from the string, it was worthwhile to find a way to include information such as user name, in addition to symbols T (Tutor) and E (Student), as well as the publication dates of messages in each discussion thread (procedure "symbol_sequence_ with_users_and _dates") so that at a next stage (see future goals) these data could be used to draw educational conclusions along with information such as grades obtained by students in examinations, or how many times they had to take a test in order to pass each module, or if they gave up the specific module, as well as information relating to the profile of students (sex, age, family status).

During the algorithm control stage, there was the issue of "unnecessary" repetition of symbols in relation to both the type of message (question/answer) and the category of content. Namely, a message could contain more than once the Greek (";") or English ("?") question mark; by way of illustration, there was a case of the question mark appearing four (4) times in the same message. Obviously, according to both the model and common sense, the symbol q should appear once, not four times (i.e. $qqqq$) in the same message.

A similar issue arose in relation to terminal symbols of the content category of messages. For example, the roots "stud" from the word "study" and its derivatives (e.g. studied, etc.) and "volum" from the word "volume" and its derivatives (e.g. "volumes", etc.) could be found in the same message more than once. Therefore, a message relating to study could repeatedly include the symbol M, when in fact it concerned only one message. Thus, we decided, for each message of the thread, to temporarily allocate the respective terminal symbols to a (temporarily used) intermediate file and to follow the sorting method during the output. Namely, only one of the (unnecessary) repeated identical symbols will appear, i.e. instead of *TaXXXMMM*, *TaXM* will appear, which is the desired result according to the model.

4.4 System Integration

Based on the above presentation, the overall graphic display of the system operation would be as follows (Figure 6):

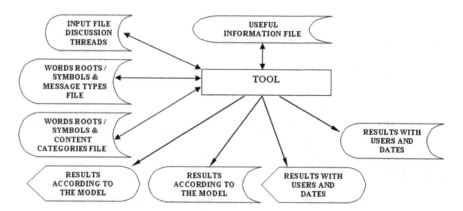

Fig. 6 Overview of the system.

The initial input is one or more discussion threads. Then, once "data_filtering" is completed, a file is created that contains user names, dates and contents of messages. Moreover, following the execution of "message_type" and "content_category", the respective files are created which contain pairs of root words or symbols and type of question or content category of message, respectively. It is worth noting that "message_type" and "content_category" may be executed independently of "data_filtering". In the final stage, the execution of "symbol_sequence" results in the string representing the thread messages according to the modelling. Furthermore, the execution of "symbol_sequence_with_users_dates" results in the replacement of message carrier symbols (*T* and *E*) by user names and in the viewing of message publication dates.

5 Exemplary Operation

This point follows the illustration of an exemplary operation of system (Figure 7 in indicative HOU discussion threads.

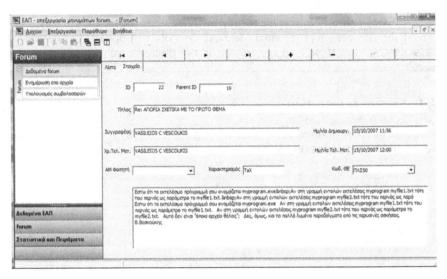

Fig. 7 The HOU's environment for automated interpretation of discussion threads' messages in asynchronous distance education fora (commands' menu is in Greek).

5.1 Input

At first, the addresses of discussion threads are entered (Figure 8); the user may enter as many threads as wanted to have their messages interpreted as follows:

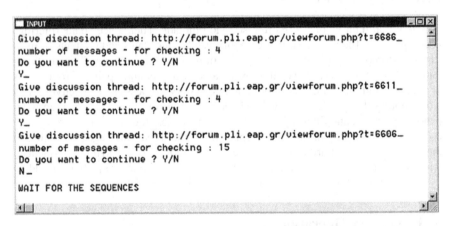

Fig. 8 Entering of three discussion threads of the HOU Computer Science forum for message interpretation.

5.2 Output

The execution of "symbol_sequence" results in the automated output of strings (Figure 9) where each one represents the messages of the respective discussion thread and finishes with the word END, as follows:

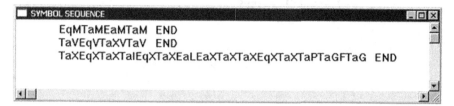

Fig. 9 Strings output according to the model.

Figure 10 shows the same results after the execution of "symbol_sequence_ with_users_and_ dates" where the symbols of the message carrier have been re-placed by User names and date per message is also displayed.

```
SYMBOL SEQUENCE WITH USERS AND DATES
       ANTONCHI Δευ 02 Απρ 2007 :qM
        ergina Τρι 03 Απρ 2007 :aM
       ANTONCHI Τετ 04 Απρ 2007 :aM
        ergina Τετ 04 Απρ 2007 :aM
       END

        ergina Τετ 21 Φεβ 2007 :aV
       ANTONCHI Τετ 21 Φεβ 2007 :qV
        ergina Πεμ 22 Φεβ 2007 :aXV
        ergina Κυρ 25 Φεβ 2007 :aV

       END

       vagelat Τρι 20 Φεβ 2007 :aX
     gskalidis Πεμ 01 Μάρ 2007 :qX
       vagelat Παρ 02 Μάρ 2007 :aX
       vagelat Δευ 05 Μάρ 2007 :al
     kseimenis Δευ 05 Μάρ 2007 :qX
       vagelat Τρι 06 Μάρ 2007 :aX
     gskalidis Τετ 07 Μάρ 2007 :aL
lambrosskantzis Τετ 07 Μάρ 2007 :aX
       vagelat Τετ 07 Μάρ 2007 :aX
       vagelat Τετ 07 Μάρ 2007 :aX
     kseimenis Παρ 09 Μάρ 2007 :qX
       vagelat Παρ 09 Μάρ 2007 :aX
       vagelat Πεμ 15 Μάρ 2007 :aP
       vagelat Κυρ 25 Μάρ 2007 :aGF
       vagelat Τετ 28 Μάρ 2007 :aG
       END
```

Fig. 10 Results after the replacement of symbols T and E by the respective user names and display, at the same time, of message publication dates (Days: Κυρ=Sun, Δευ=Mon, Τρι=Tue, Τετ=Wed, Πεμ=Thu, Παρ=Fri, Σαβ=Sat, Months:Φεβ=Feb, Μάρ=Mar, Απρ=Apr).

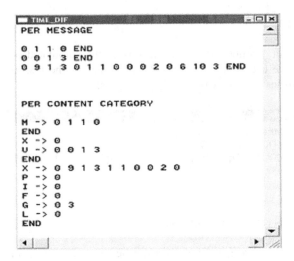

Fig. 11 Time differences per message and per content category.

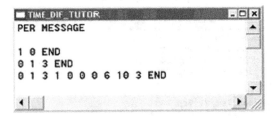

Fig. 12 Time differences in tutor's replies.

6 Experiments

During the development of the system, we followed the experimental control process.

6.1 Phase A

At first, experiments were carried out by using 80 discussion threads of the INF10 module of the academic year 2007-8. Given that 219 threads with 1,942 messages had been created throughout the year, there was the ratio of approximately 9 messages (in particular 8.87) per discussion thread. Therefore, out of the 80 selected threads, we tried to use those containing 8 or 9 messages for the purpose of experimental control. Thus, we finally chose 80 discussion threads with 712 messages in total (average 8.90 messages/thread).

At the first experimental operation, the word root files in relation to both the type (question/answer) and (mainly) the content category of message contained 18 and 92 entries respectively. Under these conditions (Table 2), we ended up having 58 discussion threads with no errors and 16 threads with only one wrong symbol (compared to what was expected). Namely, out of (approximately) 9 messages (of each of the 16 threads), 8 of them were correct and one message was wrong because it did not contain not even one of the 92 provided word roots. Respectively, there were 5 threads with two errors and 1 thread with more errors (this thread was created before Christmas holidays and its messages contained mainly wishes). We should note here that there has been no error regarding the type of messages (question/answer), only in terms of determining the content category.

Following the observation/study of messages in the 21 threads that contained 1 or 2 errors, we recorded 49 additional word roots (concerning the content category) and we decided to enter them in the root file. The experimental operation performed in the same 80 threads had, clearly, better results, with total success in 70 threads, one wrong symbol in 8 threads, two errors in 1 thread, and 1 thread that did not actually refer to educational content (Table 3).

At this point it should be clarified that the control of the results produced by the system in this phase (A), was conducted with manual comparison of all the messages in the discussion threads that were used in order to control system reliability at the first degree.

Subsequently, terms and concepts were extracted as features (word roots) from the messages in the training and test corpus. The feature extraction process consisted of the stages described in system development section (Roots files storage subsection) using the stemming algorithms (stage 1 and 2), resulting in a total number of 92 (in 1st experimental operation) and 141 (in 2nd) distinct term features (word roots).

Table 3 Experimental operation - Phase A'

2007-8	1st Exp. operation	2nd Exp. operation
Threads	80	80
Messages	712	712
Messages/Thread	8.9	8.9
Success (threads with no errors)	58	70
Threads with one error	16	8
Threads with two errors	5	1
Threads with more errors	1	1
Correct messages interpretation	677	693
Wrong messages interpretation	35	19

6.2 Phase B

Given that the 8 discussion threads with one error were found not to have any common word root feature that would adequately correspond, we decided to initiate the second experimental phase (B'). Classification was performed according to international literature [42-38], using the algorithms indicated for this purpose:

Naive Bayes (NB), 1-Nearest Neighnor (1-NN), WINNOW and discrete AdaBoost (in the form generalized by Nock & Nielsen [49] based on Freund & Schapire [50].

During this phase every algorithm was formed using the data collected from the academic year 2007-8. Subsequently, a group of data for two other academic years (2005-6 and 2006-7) was also collected. The results show that the discrete Ada-Boost algorithm produced the greatest accuracy. This result agrees with Bloehdorn & Hotho [48] who used the discrete AdaBoost algorithm in a similar experiment. The accuracy is denoted in the Table 4.

Table 4 Accuracy of algorithms for the academic years 2005-6, 2006-7 and 2007-8

	2005-6	2006-7	2007-8	Average accuracy	Average accuracy
In thread level					
AdaBoost	75.11	80.08	87.21	80.64	80.80
Naive Bayes	72.47 *	77.83	86.18	78.66	78.82
1-Nearest	73.45	76.66	83.65*	77.77	77.92
WINNOW	70.13 *	73.24*	83.10*	75.34	75.49
In message level					
AdaBoost	92.36	95.19	97.89	94.96	95.15
Naive Bayes	89.11*	92.51	96.73	92.59	92.78
1-Nearest	90.31	91.13	93.89*	91.60	91.78
WINNOW	86.23*	87.06*	93.27*	88.67	88.85

The star (*) indicates that used algorithm performed statistically better than the specific classifier according to t-test with $p<0.05$. In all the other cases, there is no significant statistical difference between the results (Draws).

The average accuracy (1) corresponds to the total number of threads and messages, while in (2) the years have an equal participation (1/3) in the total average.

It is worth noting that in all cases, the type of message has been correctly identified (i.e. whether it is a question or answer), and therefore any errors concerned the content category (as noticed in the initial experimental operation). Results are shown in Table 5.

Table 5 Results for the academic years 2005-6, 2006-7 and 2007-8

Year	2005-6	2006-7	2007-8	Total
Threads	237	236	219	692
Messages	982	1205	1942	4129
Messages/Thread	4.14	5.11	8.87	18.12
Threads with no errors	178	189	191	558
Percentage	75.11 %	80.08 %	87.21 %	80.64 %
Correct messages interpretation	907	1147	1901	3955
Percentage	92.36 %	95.19 %	97.89 %	95.79 %

The above results, which followed a calibration process of repeated readjustment, were deemed satisfactory (97.89% correct message interpretation for 2007-8) and in the end, the development of this system gives a clear affirmative answer to the question "is there an automated method to interpret messages in a distance education forum?" Therefore, by using this system, it is now possible to read study and classify, within a few minutes, a large number of messages (4,129 messages) which took 12 months to be completed within the framework of this chapter.

7 Discussion

In the last few years, distance education practice has taken on new features, in terms of methodology and systems employed [51]. Fora are a communication key tool, supporting a large part of the distance education process, and contributing to both the organizing of studies and the processing and development of the material already studied by students. Consequently, distance education fora are a dynamically evolving field which needs to be constantly updated and redefined. A big part of the research presented in the international literature concerning distant education's fora, refer to the content analysis, which principally aims to reveal information invisible at first sight. There are a variety of approaches, varying both in the level of detail's and in the type of categories of analysis used stemming from different theoretical backgrounds.

The development of this system was stimulated by the heavy flow of information in HOU's distance education fora, and it aspires to cover a gap in the interpretation of messages in an asynchronous discussion forum for distance education, by creating a system that automatically classifies messages according to a modelling built to this effect. This system uses the content category as unit of analysis for the messages' interpretation.

The creation of this system makes an important contribution to the decoding of discussions in fora, and aims at summary identification of discussions which do not develop in the desired way. Therefore this approach can be used as a tool which may assist in "intelligent" coordination, in order to limit potential malfunctions, and it could ultimately be interpreted as a step towards a procedure for formulating quality indicators for the educational value of a forum in distance education.

As described in the methodological framework, the modeling on which the system was based refers to 10 message content categories (see Grammar P- terminal symbols that stem from terminal symbol c) based on the long-term observations of the discussions in the HOU fora. It is clear however from the definition of the language rules, which follows the form a b, and specifically from the total of non-terminal symbols V_N, as well as the description of the roots file storage – on the implementation level – that both more types of messages and more types of categories may be added (theoretically an infinite number) thus satisfying the logical requirement for generalization of our methodology. Consequently more categories could be added to this system, however, always bearing in mind the concept of "practicality" according to Dringus & Ellis [17], the use of a large number of categories would lead to disorientation from the basic idea of the summary review of a large total of discussion threads "at a glance", which is what led us to design and develop this system.

Consequently, despite the fact that the present system is context-sensitive in terms of the underlying formal language and algorithms, is nevertheless forecasted the generalisation of present approach in different contexts, such as general purpose forums, web 2.0 tools such as wikis and blogs, etc.

Even though the approach presented may apply to other distance education institutes that use fora, there are limitations.

It is obvious that satisfactory results in the operation of the system are based on the fact that they concern specific subjects with a defined field of knowledge, and therefore more standardized dialogues compared to similar systems of text classification that refer to more open forms of discussion. Furthermore, the system that is presented was designed for students who are attending courses in the Greek language; therefore the results may be different in other languages. Another parameter is that in the HOU Forum environment, after an initial agreement between tutors and students, (Greek) words are unabbreviated, therefore the satisfactory results of the system may have been different if abbreviations or greeklish (Greek words in the Latin alphabet) were used, as used profusely in other forms of communication (e.g. SMS, mobile learning). An important parameter which could also be a future goal is the use of the system in the case where postings are not signed, as they are now, but anonymous, something which would not be possible now under HOU's legislative framework of operation, but only after obtaining the relevant permission from the Greek authorities.

In our effort to obtain the greatest possible reliability we executed 4 algorithms, which are indicated for this purpose, and compared their results which is appropriate procedure in the case of discussions on a specific subject, and therefore a more limited range [44, 46-48].

Among others, the prediction for future research actions are long-term studies in the central question, what reinforces the participation at Fora and how this contributes to the educational process effectiveness by investigating side questions, such as how much it affects the person who starts the thread (tutor or student), how it starts, the period when the thread starts, how important is the time of response in threads, the groups' size etc. The results from using this system will serve as data to create a database aiming to explore the effects of fora on the educational process in terms of causal explanation. Given that the HOU is not a conventional university (having the characteristics of a homogeneous student population), yet addresses to adults with special educational needs and diversity (in terms of age, professional and family obligations etc.), it is very important to do research on such issues in the future.

Finally, as future research actions are forecasted the incorporation of system in environments LMS after test process as well as the creation of more intelligent version of system that will be able to analyse messages "just in time" and produce triggering and alert messages according to the results of the interpretation of the messages.

References

1. Posner, G.J.: Analyzing the curriculum. McGraw-Hill, Inc., New York (1995)
2. Patriarcheas, K., Xenos, M.: Asynchronous Distance Education Forum - Brainstorming vs. Snowballing: A Case Study for Teaching in Programming Didactics. In: Spaniol, M., Li, Q., Klamma, R., Lau, R.W.H. (eds.) ICWL 2009. LNCS, vol. 5686, pp. 322–331. Springer, Heidelberg (2009)
3. Patriarcheasm, K., Xenos, M.: Educational techniques comparative study by using combined environment via computer and mobile devices in asynchronous discussion forum. In: Proceedings of the 9th Global Mobility Roundtable (ICMB/GMR 2010), pp. 297–304. IEEE Computer Society Press (2010)
4. Patriarcheas, K., Papaloukas, S., Xenos, M.: Suitable asynchronous distance education fora size for working groups in informatics teachers training. In: Proceedings of the 4th IEEE Balkan Conference in Informatics (BCI 2009), pp. 157–162. IEEE Computer Society Press (2009)
5. Newman, F.W.: The prospects for classroom thoughtfulness in high school social studies. In: Collins, C., Mangieri, J.N. (eds.) Teaching Thinking: An Agenda for the 21st Century, pp. 105–132 (1992)
6. Cohen, M.S., Ellis, T.J.: Predictors of success: a longitudinal study of threaded discussion forums. In: 33rd Annual Frontiers in Education IEEE Cat. No. 03CH37847C.T3F-14-18 vol. 1 (2003)
7. Hara, N., Bonk, C., Angeli, C.: Content analyses of on-line discussion in an applied educational psychology course. Instructional Science 28, 115–152 (2000)
8. Kling, R., Courtright, C.: Group behavior and learning in electronic forums: a sociotechnical approach. Information Society 19, 221–235 (2003)

9. Sheard, J., Ramakrishnan, S., Miller, J.: Modelling learner and educator interactions in an electronic learning community. Australian Journal of Educational Technology 19, 211–226 (2003)

10. Kneser, C., Pilkington, R., Treasure-Jones, T.: The tutor's role: an investigation of the power of exchange structure analysis to identify roles in CMC seminars. International Journal of Artificial Intelligence in Education 12, 63–84 (2001)

11. Mazzolini, M., Maddison, S.: Sage, guide or ghost? The effect of instructor intervention on student participation in online discussion forums. Computers & Education 40, 237–253 (2003)

12. Chang, C.K.: Scaffold learners' qualitative browsing of major topics in learning forum. In: Proceedings 3rd IEEE International Conference on Advanced Technologies, pp. 370–371. IEEE Computer Society, Los Alamitos (2003)

13. Schrire, S.: Knowledge building in asynchronous discussion groups: going beyond quantitative analysis. Computers & Education 49, 49–70 (2006)

14. Licchelli, O., Basile, T.M.A., Di Mauro, N., Esposito, F., Semeraro, G., Ferilli, S.: Machine Learning Approaches for Inducing Student Models. In: Orchard, B., Yang, C., Ali, M. (eds.) IEA/AIE 2004. LNCS (LNAI), vol. 3029, pp. 935–944. Springer, Heidelberg (2004)

15. Chen, G., Chiu, M.M.: Online discussion processes: Effects of earlier messages' evaluations, knowledge content, social cues and personal information on later messages. Computers & Education 50, 678–692 (2008)

16. Collison, G., Erlbaum, B., Haavind, S., Tinker, R.: Facilitating On-line Learning: Effective Strategies for Moderators. Atwood Publishing, Madison (2000)

17. Dringus, L.P., Ellis, T.J.: Building the SCAFFOLD for evaluating threaded discussion forum activity: Describing and categorizing contributions. In: 34th ASEE/IEEE Frontiers in Education Conference, Savannah, GA (2004)

18. Harasim, L.M.: Online Education: An environment for collaboration and intellectual amplification. In: Harasim, L.M. (ed.) Online Education: Perspectives on a New Environment, pp. 39–64. Praeger, New York (1990)

19. Duncan-Howell, J.A.: eCAF: A new tool for the conversational analysis of electronic communication. In: Proceedings British Educational Research Association (BERA) Annual Conference 2008. Herriot-Watt University, Edinburgh (2008)

20. Henri, F.: Computer conferencing and content analysis. In: Kaye, A.R. (ed.) Collaborative Learning Through Computer Conferencing. The Najadan Papers, pp. 117–136. Springer, London (1992)

21. Kern, R.: Perspectives on technology in learning and teaching languages. Teachers of English to Speakers of Other Languages (TESOL) Quarterly 40, 183–210 (2006)

22. McCreary, E.K.: Three behavioral models for computer-mediated communication. In: Harasim, L.M. (ed.) Online Education: Perspectives on a New Environment, pp. 117–130. Praeger, New York (1990)

23. Kol, S., Schcolnik, M.: Asynchronous forums in EAP: assessment issues. Language Learning & Technology 12, 49–70 (2008)

24. Wolf-Quintero, K., Inagaki, S., Kim, H.-Y.: Second Language Development in Writing: measures of fluency, accuracy and complexity. In: Second Language Teaching and Curriculum Center. University of Hawaii, Honolulu (1998)

25. Ortega, L.: Syntactic complexity measures and their relationship to L2 proficiency: A research synthesis of college-level L2 writing. Applied Linguistics 24, 492–518 (2003)

26. Barrett, E., Lally, V.: Gender differences in an on-line learning Environment. Journal of Computer Assisted Learning 15, 48–60 (1999)

27. Newman, D.R., Webb, B., Cochrane, C.: A content analysis method to measure critical thinking in face-to face and computer supported group learning. Interpersonal Computing and Technology 3, 56–77 (1995)

28. Bullen, M.: A case study of participation and critical thinking in a university-level course delivered by computer conferencing. University of British Columbia, Vancouver (1997)

29. Gunawardena, C.N., Lowe, C.A., Anderson, T.: Analysis of a global online debate and the development of an interaction analysis model for examining social construction of knowledge in computer conferencing. Journal of Educational Computing Research 17, 397–431 (1997)

30. Veerman, A., Veldhuis-Diermanse, E.: Collaborative learning through computer-mediated communication in academic education. In: Euro CSCL, pp. 625–632. University of Maastricht, McLuhan institute, Maastricht (2001)

31. Pena-Shaff, J.B., Nicholls, C.: Analyzing student interactions and meaning construction in computer bulletin board discussions. Computers & Education 42, 243–265 (2004)

32. Weinberger, A., Fischer, F.: A framework to analyze argumentative knowledge construction in computer-supported collaborative learning. Computers & Education 46, 71–95 (2006)

33. Jarvela, S., Hakkinen, P.: Web-based cases in teaching and learning: The quality of discussions and a stage of perspective taking in asynchronous communication. Interactive Learning Environments 10, 1–22 (2002)

34. Rourke, L., Anderson, T., Garrison, D.R., Archer, W.: Assessing social presence in asynchronous text-based computer conferencing. Journal of Distance Education 14, 51–70 (1999)

35. Garrison, D.R., Anderson, T., Archer, W.: Critical thinking, cognitive presence, and computer conferencing in distance education. American Journal of Distance Education 15, 7–23 (2001)

36. Anderson, T., Rourke, L., Garrison, D.R., Archer, W.: Assessing teaching presence in a computer conference context. Journal of Asynchronous Learning Networks 5, 1–17 (2001)

37. Fahy, P., Crawford, G., Ally, M.: Patterns of interaction in a computer conference transcript. International Review of Research in Open and Distance Learning 2, 1–24 (2001)

38. Lockhorst, D., Admiraal, W., Pilot, A., Veen, W.: Analysis of electronic communication using 5 different perspectives. In: Paper Presented in a Symposium Conducted at the 30th Onderwijs Research Dagen (ORD), Kerkrade, The Netherlands (2003)

39. Zhu, E.: Meaning negotiation, knowledge construction, and mentoring in a distance learning course. Proceedings of Selected Research and Development Presentations at the 1996 National Convention of the Association for Educational Communications and Technology. Available from ERIC documents: ED 397 849, Indianapolis (1996)

40. Cook, D., Ralston, J.: Sharpening the focus: methodological issues in analysing online conferences. Technology, Pedagogy and Education 12, 361–376 (2003)

41. Kotsiantis, S., Patriarcheas, K., Xenos, M.: A combinational incremental ensemble of classifiers as a technique for predicting students' performance in distance education. Knowledge Based Systems 23, 529–535 (2010)

42. Aggarwal, C., Gates, S., Yu, P.: On the merits of building categorization systems by supervised clustering. In: Proceedings of the Fifth ACM SIGKDD International Conference on Knowledge Discovery and Data Mining, pp. 352–356. ACM Press (1999)
43. Yang, Y., Liu, X.: A re-examination of text categorization methods. In: Proceedings of the 22nd Annual International ACM SIGIR Conference on Research and Development in Information Retrieval, pp. 42–49. ACM Press (1999)
44. Dumais, S., Chen, H.: Hierarchical Classification of Web Content. In: Annual ACM Conference on Research and Development in Information Retrieval, pp. 256–263. ACM Press (2000)
45. Kongovi, M., Guzman, J.C., Dasigi, V.: Text Categorization: An Experiment Using Phrases. In: Crestani, F., Girolami, M., van Rijsbergen, C.J.K. (eds.) ECIR 2002. LNCS, vol. 2291, pp. 213–228. Springer, Heidelberg (2002)
46. Sebastiani, F.: Machine Learning in Automated Text Categorization. ACM Computing Surveys 34, 1–47 (2002)
47. Sebastiani, F.: Classification of Text, Automatic. In: Encyclopedia of Language & Linguistics, Section: Applications of Natural Language Processing, 2nd edn., vol. 14. Elsevier Science Publishers (2006)
48. Bloehdorn, S., Hotho, A.: Boosting for Text Classification with Semantic Features. In: Mobasher, B., Nasraoui, O., Liu, B., Masand, B. (eds.) WebKDD 2004. LNCS (LNAI), vol. 3932, pp. 149–166. Springer, Heidelberg (2006)
49. Nock, R., Nielsen, F.: A Real Generalization of discrete AdaBoost. Artificial Intelligence 171, 25–41 (2006) (2007)
50. Freund, Y., Schapire, R.E.: A Decision-Theoretic generalization of on-line learning and an application to Boosting. Journal of Computerand System Sciences 55, 119–139 (1997)
51. Lewinson, J.: Asynchronous discussion forums in the changing landscape of the online learning environment. Campus Wide Information Systems 22, 162–167 (2005)
52. Bratitsis, T., Dimitracopoulou, A.: Monitoring and Analyzing Group Interactions in Asynchronous Discussions with the DIAS System. In: Dimitriadis, Y.A., Zigurs, I., Gómez-Sánchez, E. (eds.) CRIWG 2006. LNCS, vol. 4154, pp. 54–61. Springer, Heidelberg (2006)
53. De Wever, B., Schellens, T., Valcke, M., Van Keer, H.: Content analysis schemes to analyze transcripts of online asynchronous discussion groups: A review. Computers & Education 46, 6–28 (2006)

Part III
Collaborative Learning Systems

Antecedents of Collaborative Learning in Massively Multiplayer Online Games

Iro Voulgari and Vassilis Komis

Department of Educational Sciences and Early Childhood Education,
University of Patras, Patras, Greece
{avoulgari,komis}@upatras.gr

Abstract. Massively Multiplayer Online Games (MMOGs) are rich in goal-oriented activities and collaborative and social interactions, both essential for learning the game and progressing. In this chapter we employ a theoretical framework for linking learning and collaborative learning principles with MMOGs and investigate, through an exploratory and qualitative approach, features of the tasks, groups, and player interactions that may support the emergence of collaborative interactions and learning. The critical role of both the design of the environment and of the community of players is highlighted and it is concluded that their balanced inter-connection is critical for the emergence of effective collaborative interactions. Implications on further research are also discussed.

1 Introduction

Over the past few years Massively Multiplayer Online Games (MMOGs) have triggered the interest of research in education on their potential for learning (Dickey 2007; Steinkuehler 2004, 2006). MMOGs are not merely environments where the players accomplish tasks and solve puzzles, but rather they are highly social, interactive and dynamic, with players forming groups, socializing, co-operating and competing with each other (Ducheneaut and Yee 2008; Kolo and Baur 2004; Nardi and Harris 2006; Williams et al. 2006). The gaming experience and the progress of the player in the game are linked to both the design of the environment and the interactions with other players (Ang and Zaphiris 2008). Players progress, learn the rules of the game and build up their skills through an incremental process of accomplishing tasks and attaining goals; they not only have to learn the game, but also interact with others, compete, co-operate and exhibit problem solving, planning, organisational, social, and communication skills. The complexity and depth of the virtual environment and of the community of players, and the interconnection of different factors, particularly in relation to collaborative learning, require novel research frameworks and tools (De Freitas and Griffiths 2009; Schrader and McCreery 2008).

This study is situated with this context. We employ an exploratory, qualitative approach and focus on two main components which seem to be decisive for the quality and outcomes of a collaborative learning situation: the tasks and the players' interactions (Jonassen 2000). The tasks convey the cognitive content and

T. Daradoumis et al. (Eds.): Intelligent Adaptation & Personalization Techniques, SCI 408, pp. 269–293.
springerlink.com

scaffold the interactions of the participant with the environment and with others, while the quality of interactions defines the construction of shared knowledge and the attainment of the learning outcomes (Barron 2003; Van Den Bossche et al. 2006). The investigation of these issues may provide valuable insights on the way people interact, build up their knowledge, progress, and learn within such a computer-mediated setting, and on the ways the design of the environment can affect the community of the participants, their interactional processes, behaviours and the knowledge acquired. Such insights may provide the foundations for the development of effective collaborative environments for learning which will benefit from the potential of both the design of MMOGs and of the emerging and vibrant community of participants. Our goal in this chapter is to combine learning principles and game design, identify aspects of the tasks and group practices in MMOGs and, based on our theoretical framework, examine them from the perspective of their potential for supporting collaborative interactions for learning.

2 Theoretical Background: MMOGs and Collaborative Learning

The rich potential of MMOGs for learning seems to emerge from the effective and functional "marriage of emotion to cognition" (Gee 2008). MMOGs integrate a quest system – problems to be solved and tasks to be accomplished – through which the players construct their knowledge of the environment and develop their skills. From this perspective, quests can be perceived as the learning tasks of an interactive learning environment (Dickey 2007). Acquisition of skills and expertise in MMOGs is also closely related to the interactions with other players. Players can only progress in the game if they cooperate with others, form teams, if they acquire a social capital – friends willing to help them, give them advice and support them (Schrader and McCreery 2008; Huffaker et al. 2009). Players have to exhibit both content-specific knowledge –knowledge of the environment content, the game mechanics, and the affordances (Huffaker et al. 2009; Wang et al. 2009), as well as social, interpersonal, and communication skills (Ducheneaut and Moore 2004; Nardi and Harris 2006; Reeves et al. 2009). Furthermore, motivation and engagement seem to be two of the most prominent features of MMOGs. Through mechanisms such as the narrative context, difficulty and challenges appropriate for the level of the player building up skills and self-confidence, the sense of control, the graphic environment, interactions and relationships with other players, a variety of tasks for different playing styles and preferences, motivation is stimulated and the players remain committed to the game (Dickey 2007; Yee 2006).

Research in the area of collaborative learning has identified the interconnection of multiple factors that support or inhibit effective learning processes and it indicates the complementarity of motivational, social and cognitive perspectives (Slavin 1996). Learning is not only viewed as an internal cognitive process for the acquisition of knowledge, but also as interaction with others and participation in community practices (Wenger 1998). Social and cognitive approaches are equally critical and complementary for understanding learning (Sfard 1998). Motivation and positive emotions have a positive impact on engagement, achievement and learning (Boekaerts 2001; Järvelä and Volet 2004). Motivation may derive from the personal interest of the individual, from the features of the activity and the tasks, or from the social context and the interactions with others (Dillenbourg et al. 2009; Zimmerman 1989). The social context, the quality and properties of the communication and interactions among collaborating individuals seem to affect learning outcomes (Barron 2003). Features of the team such as the psychological safety of the members, the common goals, their interdependence for attaining the goals, and the social bonds, affect team cohesion, team effectiveness and individual learning (Garrison et al. 1999; Jonassen and Kwon 2001; Van Den Bossche et al. 2006).

Problem solving, individual or collaborative is situated at the heart of learning (Jonassen 2000). Acquisition of problem solving skills constitutes one of the most critical objectives for education and learning (Gagné 1980). Features of the tasks and the problems to be solved such as representation, complexity, structure, goal, rules, rewards and penalties, the flexibility and the possibility to solve the problem through multiple approaches, engaging and motivational elements, and the correspondence with the skills and knowledge of the individuals, seem to impact the cognitive processes involved and the learning outcomes (Csikszentmihalyi 1992; Dillenbourg et al. 2009; Jonassen 2004; Mayer 1998).

Based on these three axes –motivational, cognitive and social- our main focus in this chapter is to look at aspects of the in-game tasks and player interactions in relation to their potential to support learning. For situating these two components within a meaningful framework we were based on McGrath's general conceptual framework for the study of group interactions and performance (McGrath 1984, p.12) and adapted it to our theoretical framework and our goal. McGrath's framework describes the characteristics of the group members, the properties and structure of the group, the properties of the task and of the surrounding environment, as input sources shaping the emerging group interaction processes, which in turn can impact these input sources. Qualities of the interactions among participants such as elements indicating social, cognitive and teaching presence can enhance the quality of the "educational experience and learning outcomes" (Garrison et al. 1999).

The tasks and the group are situated at the centre of our conceptual framework (Fig1). Features and properties of the tasks, and in-group behaviours, structure, and practices affect the quality of the interactions, and these interactions affect the learning outcomes, as described earlier. In MMOGs, it is not only the design of the environment that defines the player's experience but also the social environment emerging, the community of the players, their attitudes and behaviours (Ducheneaut and Moore 2004; Voulgari and Komis 2010). The community of the players establish their own rules, norms, and codes of behaviour in parallel to those defined by the design of the game (Myers 2008; Voulgari and Komis 2011). The main constructs of our conceptual framework are, therefore, a) the tasks, their format, rules, structure, rewards, penalties, motivational elements, and cooperation opportunities, b) the group, its structure, hierarchy, roles, and goals, c) the interactions among the players, the communicative practices, behaviours and attitudes, d) game design elements relevant to the three previous constructs, e) the social environment, the behaviours and practices of the players' community, and f) learning outcomes, which in our study involve mainly knowledge and skills which seem to be essential for playing the game.

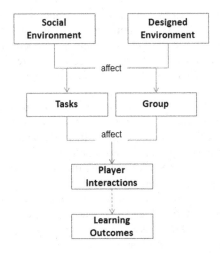

Fig. 1 Conceptual framework for the study of factors involved in learning in MMOGs

3 Research Methodology

For this study, we employed a qualitative and interpretive research approach, which will allow us to explore the area and identify emerging issues, trends, and concepts, and situate them within a meaningful framework (Charmaz 2006, p.125). As a research method we employed virtual ethnography which is an appropriate method for the investigation of behaviours and practices of individuals and groups (Creswell 2002), for observing and interpreting little-known phenomena within complex and dynamic environments, and for situating them within context-specific settings (Hoepfl 1997). It also seems to be an appropriate method for exploring and understanding the practices in online games (Hjorth 2011, p. 62).

3.1 Data Collection

We collected data through participant observation in two MMOGs and through interviews with MMOG players. We also referred to relevant information from external resources such as websites, blogs, and fora, for enhancing our understanding of the environments and of the player practices and interactions.

For the participant observation one of the authors created accounts in the MMORPG (Massively Multiplayer Online Role-Playing Game) *Lineage II* and the MMORTS (Massively Multiplayer Online Real Time Strategy) game *Tribal Wars*

-one account in each game- and was involved in the game for 18 and 7 months respectively. These environments were selected both because of their popularity and the number of players they attracted, as well as for the rich task-directed interactions they supported. Both games require that the players interact with each other for accomplishing goals of the game, such as defeating a common enemy or completing a task. *Lineage II* is set within a fantasy 3D world, where the players interact with others and with the environment through their virtual representations, the avatars, they complete quests, they participate in the in-game economy, and they collaborate or compete with other players or groups of players. *Tribal Wars* is a browser-based game, where the players have to manage their resources and their army, form groups, collaborate with other players, expand their territories by attacking others or defend them against any attacks. Through the participant observation, we collected notes, screenshots, and video recordings which included actions and discussions of players. No usernames or other identifying information of the players involved were disclosed, for protecting their anonymity.

Table 1 Characteristics of interview and focus groups participants

Characteristic	Frequency	%
Age		
Younger than 18	6	23,1
18-24	2	7,7
25-34	14	53,8
35-44	3	11,5
45-54	1	3,8
Education		
High school	7	26,9
Vocational training	4	15,4
Higher education	12	46,2
Postgraduate or more	3	11,5
Game experience level		
Novice	1	3,8
Average	4	15,4
Expert	21	80,8
Residence		
Urban	25	96,2
Rural	1	3,8
Gender		
Male	20	76,9
Female	6	23,1

N=26

We also conducted 20 semi-structured interviews and 2 focus groups. The interviewees were self-selected, following a participation call through emails and posts to relevant fora. We addressed the call to players of any MMOG so as to obtain a wider overview of features and practices in different environments. One of the two focus groups was conducted with the cooperation of a secondary education teacher and a group of her students (3 male and 3 female, aged 16) who were playing various MMOGs. The second focus group consisted of 3 participants -2 male and 1 female- members of the same in-game group. The characteristics of the participants in the interviews and focus groups are summarised in Table 1. The focus groups and most of the interviews were conducted face-to-face, while 7 of the interviews were conducted via instant messaging or VoIP. All the interviews and focus groups were audio recorded and transcribed for later analysis.

The interviews and focus groups were semi-structured as we wanted to focus on specific aspects relevant to our theoretical framework, but also provide a degree of freedom to the players to elaborate on emerging issues of interest. The main themes around which our interviews and observation were constructed were based on the main constructs of our conceptual framework and they mainly involved the gaming habits, the perceptions and attitudes on the available tasks and activities, the communication and collaboration processes, perceptions on the group processes, learning the game, acquisition of knowledge, problem-solving, personal preferences, motivations, demographic details, and general information on the game the respondent played.

For further understanding the area of MMOGs and also for further investigation of issues reported by the player during the interviews and focus groups, we looked at websites and fora relevant to the games of the participant observation and the games reported by the players, such as the official and unofficial user-developed sites and fora.

3.2 Data Analysis

At the first level of coding, we employed an in-vivo approach, identifying emerging themes and topics and summarizing our data. We then compared these themes and topics with the constructs of our theoretical model, and further re-coded for identifying emerging patterns. A second independent coder also participated in the coding process. The codes were revised after discussions between the two coders, in cases of any inconsistencies or clarifications required. During coding, trends, patterns and relationships between codes were identified for the interpretation of our data (Creswell 2002, p.191). For the coding and analysis the qualitative software *QSR Nvivo 8* was used. The codes analysed for this study are summarised in Table 2. The interpretation of the findings is discussed in the next section.

Table 2 Codes, description and examples for the main themes "Interactions", "Tasks", "Learning" and "Group"

	Code	Description	Examples (from interview transcripts)
Interactions	**In-group interactions**	Interactions among group members	Note: Any interaction type coded in one of the following codes regarding interactions is coded in this code too, if the interaction is taking place among group members.
	Help	Interactions involving help among players (e.g. information exchange, support, guidance)	*"A player with less experience points may win, because he may have strong friends or a strong tribe supporting him"*. *"[a new player] will need guidance in the beginning, otherwise he will get bored"*
	Competition	Interactions involving competition among players.	*"In Tribal Wars it's all about space. If you want to expand your territory, you have to attack players situated around your area"*
	Negative behaviour	Behaviours which are considered inappropriate by the other players	*"When you play in a team for a common goal, it is considered bad behaviour to leave the team in the middle of the battle"*
	Social interaction	Social interactions among players, or lack of. Exchange of social comments and personal information	*"[we exchange] personal information as well. You may talk about a personal problem or you may make a joke or talk about a bar you went the previous night"*
	Cooperation	Collaboration among players for a common goal	*"If you make a game too simple, just go out and shout that, it will be incredibly boring and it will not promote collaboration in any way"*
	Punishment for negative behaviour	Punishment of player behaviours and actions.	*"If someone makes offensive comments, you can report him to CCP, and there is punishment –a warning or even banning of the virtual character"*
Tasks	**Rewards**	Rewards by the game or by other players.	*"The game rewards you by opening, for example, new areas where you can go and face stronger opponents. From the players, you gain respect, or even fear if they are your competitors"*
	Individual tasks	Activities that can be completed by one person, individually.	*"If someone is a better miner, he will focus on mining and provide the market with minerals"*
	Disadvantages-negative aspects	Negative aspects or difficulties of tasks and activities.	*"[for a quest] you have to go to areas you have never been before. And for a new player... you have to look for a long time, and this may be tiring"*

Table 2 (*continued*)

	Code	Description	Examples (from interview transcripts)
	Group tasks	Activities involving the team	*"It gives you problems that cannot be copped by one person or even two. You have to cooperate with players with different abilities"*
	Planning	Planning and deciding on strategies and tactics for a task or activity.	*"You see a new player, but he may win, because he may have set up his ship in a way you did not expect"*
	Punishment for failure	Any punishment or penalty for failure in a task.	*"If you lose, you die. I died four times over there, because I was the leader. You can't imagine the [experience points] I lost"*
Learning	**Skills-knowledge**	Skills required or practiced during the game e.g. social skills, features of good player, content knowledge. Skills exhibited by expert players.	*"I learned how to use the pc and the internet better"* *"A good player is a player who communicates with others and knows how to manage critical situations"* *"[you need to have] communications skills. Imagine if you have to elaborate a complex strategy plan to a non-native English speaker".*
	Learning the game	How do the players gather information and learn how to play the game.	*"When I started playing, I was reading the help documents of the game"* *"For four days, I couldn't find the solution. Eventually I searched in the internet, and eventually I found it"* *"I was asking other people all the time. I was asking about everything"*
	Knowledge/skills transfer to real life	In-game skills or knowledge that may be applied to real life (perceptions of players in interviews)	*"No [I don't learn anything], I just pass the time"* *"Maybe you learn about other people's behaviours"*
	Knowledge/skills transfer from game to game	Indications of player' skills or knowledge from previous games, applicable to the current game (from player reports in interviews)	*"Most MMOs are more or less the same. Similar class system and roles"*
	Self-assessment/evaluation	Instances of self-assessment, assessment of success or failure, individually or in a team.	*"Except of the 'killboard', there is also the aftermath, where you say that 'I have killed that many opponents and I have lost that many times"*

Table 2 (*continued*)

Group	**Members leaving the group**	Instances of members banned from the team, or leaving the team, and possible reasons	*"I have been dismissed from teams quite a few times, because I was inactive"*
	Negative aspects	References or examples of negative aspects or features of teams.	*"In teams, very often you get arguments, disagreements, and quarrels. That's why I avoid getting into large teams with people I don't know, if I can help it"*
	Structure	Information relevant to the structure of the group, the leader, the hierarchy, ranks, officers, and member roles	*"The leader has a great responsibility for raising the level of the clan"* *"An Academy Manager has to have a lot of in-game currency, for buying staff for the new players"*
	Member selection	Information on criteria for the admission of new members in a team, or to criteria for selecting players to cooperate with. Group recruitment messages	*"Mainly where on the map is the player situated. If he is more than 30 hours away, it's impossible for him to support others in the team, or for the team to support him"* *"Mainly with people we have cooperated well in the past"*
	Decision making	Instances of decision-making processes in the team (e.g. for selecting a leader, for distributing rewards, for deciding on a strategy for a task)	*"If the leader is not there, then there are the 'sub-leaders' experienced players, who make the decisions and manage the team"* *"For distributing the loot, we have a council that makes this decisions"*
	Benefits	References or observations on the benefits of a team for the members – what are the advantages of grouping	*"We defeated the 3rd best player in the game. We took all his villages. Because we acted as a team. While his team remained inactive"*
	Obligations of members	The obligations and responsibilities of the members to the team	*"We have this statute as a corporation, describing the main rules members have to abide to, if they want to take advantage of the privileges of the corp"* *"Raid time is generally considered sacred. You have to be focused"*

4 Tasks, Groups and the Player Community

The players' success in the game is related, at a large extent, to their interactions with other players. Even for expert players it is impossible to reach the higher levels and explore the entire content of the environment unless they cooperate or compete against others. Features such as the economic and trading system of the game, the tasks that can only be accomplished collaboratively, competition with other players, and the complexity of the environment affordances lead the players to rely directly or indirectly on their peers, make friends and allies, and form groups for the attainment of common goals. Perceived learning of the players, as emerged from our interviews, also seems to be linked both to the knowledge of the interface, the mechanics and rules of the game, and also to communication, collaboration, coordination and team working skills. Indicatively, an interviewee reported:

> "You need to be fair, to communicate, to talk to others, to make friends, and to win of course" [Female, 20]

Interaction and collaboration with others is critical not only for progressing in the game but also for learning the game. Design functionalities, such as brief tutorials and Non-Player Characters (NPCs), guide novice players in their initial steps in the environment, through simple tasks in safe, protected, new players' areas. For every new virtual character we created in *Lineage II* we had the option to follow a short and simple interactive tutorial guiding us through the main features of the environment. Players familiarise themselves with the interface, the controls, the affordances, the rules of the game and build up their skills and their avatar's skills. But reaching higher levels of expertise requires interactions with other real players. As it emerged from our participant observation and interviews, due to the complexity of the environment and the lack of detailed manuals, peer-mentoring, apprenticeship, the support of others, and resorting to external player-developed resources, such as web sites and fora, constitute common practices for learning the game in-depth, for solving problems, and acquiring expertise. Quite a few times during our observation in the two games, we had to ask other players, mainly teammates, or resort to external, usually player-developed, websites for more information on quests, on the properties of different equipment and attire, on guidance on where to find resources and how to craft a new material.

It seems, therefore, that the tasks of the environment and the interactions of the players, within and beyond the group, situated within the designed and the social environment, are at the core of the development and practices of game-specific and social skills. In the next section we discuss their features and trends, as emerged from our study, which are relevant to principles for the support of collaboration and learning, as described in the theoretical background.

4.1 Tasks and Activities

Task Types and Structure

Based on problem-solving literature regarding the typology of problems, we distinguish the tasks in an MMOG into two broader types: a) well-structured and b) ill- structured. Well-structured tasks are activities with well-defined initial and goal states and usually predefined solution tactics. Such tasks are, for example, quests or missions where the players have to collect specific items, find an NPC, or kill computer generated monsters -a particularly repetitive and standardised task, often criticized by players and referred to as "grinding". Such tasks, as observed during the participant observation and reported by players, provide little opportunities for planning and the search of the most appropriate approach. They simply require that the players have the appropriate gear and equipment and know how to manipulate their virtual characters.

On the other hand, ill-structured tasks involving the collaboration or competition among players seem to be more dynamic; large numbers of players have to coordinate, groups of players have to be managed, pacts among groups have to be negotiated, terms of war or peace have to be discussed, and the enemy's next move has to be anticipated. Such tasks require not only extensive knowledge of the game and the environment affordances, but also communication and interpersonal skills, and the emergence of processes such as persuasion, diplomacy, negotiation, argumentation, conflict resolution, and decision-making. In our group forum in *Tribal Wars* there were extensive discussions in threads dedicated to the planning of strategies and tactics for expanding the group territories, attacking to opponents, and organising the defence against common enemies. Indicatively, from our interviews:

> *"I consider the players that are good in PvP as smart players, because [in PvP] you don't have to cope with a static thing which you can practice on. [...] In PvP you need many tricks in order to win - even use unorthodox and imaginative methods"* [Male, 29].

> *"In Tribal Wars there are many different strategies you can use; make alliances with other players, with other tribes, different tactics for warfare [...]. There is not a predefined path you can follow"* [Male, 26].

Incremental Complexity and Progress

Novice players, as mentioned, begin the game in a "protective" mode. The game guides the player through tutorials with simple tasks and clear goals, similar, though, to the format of later tasks (see Fig2). At this stage, interaction with other players is not always essential, allowing the player to familiarise with the environment and practice skills and tactics. As the players progress and master the basic skills, they gain access to more elaborate quests and activities, relevant to their level.

Fig. 2 Tutorial for novice players in the World of Warcraft: tips on using the interface.

For attaining higher levels the players are required to have game related knowledge and skill, as well as to be able to work as a team. In higher level tasks, grouping with others is essential: tasks, activities and quests are mainly multi-player, collaborative or competitive (Fig3). Their difficulty, in combination with the high penalties for failing, discourages players from undertaking these tasks alone. In addition, multi-player tasks also provide higher rewards to the players, such as valuable items or experience points. Even specific areas in the environment are difficult to access without a group.

> *"[...] but you don't have either the skills, or the experience, or the ship, to cope. [...] So you gather 3-4 more [people] and you all go in together, so as to be able to cope" [Male, 37]*

> *"Up to a point, you can do it by yourself. But the further you go, [the more] you need a team. At the end of the game, nobody can do anything alone" [Male, 37]*

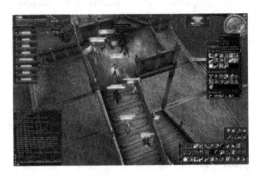

Fig. 3 Siege of a castle by a group of players in Lineage II

Variety and Motivation

For attracting and sustaining the interest and engagement of a wider range of players of different preferences and requirements, for as long as possible, a variety of different tasks is integrated. The design of the environment directs at a large extent the features and the format of the tasks. The option to select or not a task and in which order to accomplish them was one of the main motivators reported by players in the interviews, as a feature providing a sense of freedom of choice.

Although MMOGs are highly social and interactive, integration of individual tasks, as well as collaborative and competitive ones, provides the opportunity for a more individualised gaming experience, as it addresses the requirements of a wide range of players' preferences and playing styles. A balance seems to be kept between solo and group play since the progress of the virtual character is defined by the participation to both individual tasks, such as the practice of a skill or a "profession", and collaborative tasks, such as "raids" or sieges of enemy castles. In our interviews and participant observation, we identified players who prefer to socialize and communicate with others, players who prefer to co-operate with others (e.g. for a quest, or killing a computer generated monster), players who prefer to compete against others (e.g. in PvP, the siege of a castle, and battles against other groups of players), players who do not enjoy continuous interaction with others and prefer to complete tasks individually (such as individual quests, exploration of the virtual world, participation in the in-game economy system, exchange of virtual items, trading, manufacturing, gathering virtual materials), or players who enjoy all different types of tasks, depending on their available time or mood.

> *"I prefer tasks that I can complete by myself, because usually I don't have the time to play with other [players]" [Male, 37]*

> *"One of my most memorable experiences [in the game] was this particular siege. We were 24-25 people and resisted against a horde of barbarians. [...]We were out-numbered. The odds were against us. But we had a good team. We managed to win [...] Epic moments!" [Male, 32]*

These findings are relevant to Bartle's typology of players who also identified four main motivators for play: achievement, competition against others, socialisation, and exploration of the environment (Bartle 1996).

Rewards and Reinforcement

In their study of the MMOG *World of Warcraft*, Ducheneaut et al. (2006) characterized MMOGs as "virtual Skinner boxes" drawing the players into the game and committing them through instant rewards, and gradual increase of difficulty and rewards value. Even though sociability and interaction with other players were ranked high among the players' motivations in our interviews, the rewards were also reported as a definitive task selection criterion. Most of our players reported that they would only take up a task if the rewards were worth the time invested.

> *"When [the quest] does not offer any in-game currency, I won't be bothered at all" [Male, 29]*

Tasks and activities reward the players with virtual items, in-game currency, experience points and levelling up. Players are rewarded for almost any activity in the game. Participation and completion of a task by a group, also rewards the members with experience points distributed and added to the progress bar of each player, respectively to the level of each player, and to the size of the group. In *Lineage II*, for example, at the time of the participant observation, higher level members gain less experience points than lower level, and the more the members of the group, the less the experience points gained by each member. Although such allocation of experience points and rewards may discourage higher level players from grouping, it seems necessary for the matching of players of similar levels, and the matching of tasks with the relevant team level and size.

> *"[...] even from a financial perspective, it's not to my interest to join a party for quests in that area [of the game], because rewards are distributed to more people" [Male, 29]*

Distribution of rewards, though, among team members seems to be a critical issue for the reinforcement of participation to group tasks. Although, for individual tasks, rewarding is a straightforward process by the design of the environment, distribution of rewards among members of a team in collaborative tasks seems to be more complex raising issues relevant both to the affordances of the game and the group processes. The environment may incorporate different reward distribution mechanisms, such as random distribution among group members, collection by the leader, finder's-keeper's or rolling a dice, but these mechanisms do not always suffice for ensuring the fair rewarding of all members. We observed a number of conflicts among players and even teams' disbandment, over the distribution of the rewards. And from this point on, the community of the players takes over. "EverQuest" players developed the Dragon Kill Point system (DKP), currently employed by WoW players as well, a point system which also considers the participation of the members in group activities (see also Malone 2009; Silverman and Simon, 2009; WOWWiki). Through this system, players who are more active in group tasks and committed to the group acquire better rewards. In task-oriented or achievement-oriented groups players resort to such systems of distribution for encouraging participation of players and rewarding commitment to the group. In smaller, more social groups, groups of 5-10 players - usually real life friends as well - distribution is often based on the negotiation and trust among members, and prizes are granted to the player who needs them more. Achievement-oriented and social group types are discussed in the next section.

4.2 Team Practices

Team Types and Structure

In MMOGs it seems that although the design of the environment defines team formation and structure processes, it allows players the freedom to shape, to an extent, the type of the team, the orientation, goals, composition, and member roles. From the perspective of game design, the main types of teams that can usually be identified are a) the clans, or guilds, or tribes, more persistent and long term teams, and b) the groups or parties, temporary and task-oriented. In some cases, the environment divides the players into opposing sides (e.g. the Horde and the Alliance in the *World of Warcraft*, the Elyos and the Asmodians in *Aion*), fostering camaraderie among members of the same side and competition against the opposing side.

> *"It has to do mainly with the sense of team. [...] people that don't even know each other, with nothing in common except of the race, consider it their duty to defend the new players" [Male, 29]*

In many MMOGs, players may create their own personal networks of players: the friend lists - players create a list of people they enjoy playing or socially interacting with. The temporary, task-oriented teams, the parties, usually have a limited maximum number of members, depending on the difficulty of the task, and the leader is the only distinctive role as defined by the design. Clans and guilds present a more elaborate structure and ranking system, allowing for even more than 150 members, with a leader and a hierarchy of officers.

On the other hand, players take advantage of the environment affordances and manipulate them to their preferences, goals, and requirements. As observed and reported by players, there are many cases in the game *Lineage II*, for example, where the players take advantage of the party-formation functionality of the game and form "constant parties", a hybrid long-term team between a clan and a party, of players who enjoy to play and complete tasks together, without having to go through the process of joining or forming an official clan. The criteria for the assignment of a role to a team member in MMOG groups go beyond the skills and level of the virtual characters, and even beyond the player's experience and knowledge of the game; interpersonal characteristics are also considered, such as communication skills, trust, commitment to the group, relations with the group leader. Beyond the ranking system of the environment, the players may define additional roles in their groups, depending on their needs and objectives: an officer for maintaining the group's website, or for technical support, or for following the prices in the in-game market.

Beyond the typology of teams based on the game design, other types of groups can also be identified, based on players' decisions and team dynamics. We identified, consistently with Williams et al. (2006) findings, two main different types of groups based on their orientation, goals, and interactions: a) achievement-oriented groups, with success in the game tasks as a main goal, and b) social-groups, where social interactions and friendship seem to be more prominent. Although most teams tend to one of the two directions, the achievement component and the social component co-exist in all teams and are perceived as complementary. Achievement-oriented groups seem to value the social component as essential for their success and progress in the game, and social groups do not disregard the importance of succeeding in game tasks, as essential for motivating the members and strengthening team cohesion. Indicatively, a member of a large, achievement-oriented group, reported in an interview:

> "When the group has strong links and bonding, the members will not start blaming [the leader, for a failure]. [...] That is the difference between a good team and an inexperienced team" [Male, 37]

In-Group Behaviours

It seems that the different types of groups described in the previous section entail the emergence of different behaviours, attitudes, and team processes. In the long-term clans, guilds, or tribes the commitment of the members is stronger than in the temporary parties. Parties are formed by players that may not even know each other, and may never see each other again in the future, with the main purpose to accomplish a task. Players joining such a party feel less committed to the group and do not always employ their full potential. A high rate of party members leaving the party during the task with no prior notice was observed during the participant observation; such behaviour would be punishable by marginalisation or dismissal in a long-term clan.

> "In a pug [pick up group, random groups], I wouldn't really care at all [about] how much damage I did, or anything. Things are more irresponsible there. I don't know you and you don't know me. We don't care about anything [in pugs]" [Male, 29]

Discourse, sociability and decision-making also seem to differ in long-term and temporary teams. While in temporary parties more task-oriented and less social discourse was observed, in long-term clans, social interactions among members were not uncommon. Members of clans often communicate, via text or voice, about issues irrelevant to the game, exchange information about their personal lives, and engage in social talk. They may become friends and in many cases they even meet in real life. Long-term groups, and especially achievement-oriented groups, seem to also have more robust and rigorous hierarchy, structure, and decision-making system than the temporary teams. In temporary teams, although there is a clearly defined leader, by the design of the game, it was usually the most experienced member who gave directions to the members and everyone in the group was free to disagree and argue. In long-term teams, although in most of the cases members can propose ideas, alternative solutions and approaches, discuss, agree or disagree, it was usually the leader or the elite that had the final word on the decisions.

Group cohesion is also demonstrated differently in long-term and in temporary groups. While in temporary groups the main motivator keeping the members in the group is the effectiveness of the team and the success in the task, in long-term groups affective and social factors seem to emerge, in addition to the effectiveness of the team. In long-term groups players choose to remain in the team not only because their goals and progress depend on the participation to group tasks and the effectiveness of the group, but also because they feel that they communicate and socialize with other members, they find common interests and develop a sense of camaraderie and even real life friendships.

> "This team offers me nothing, but I am staying there, because my friends are there. I don't want to give them up" [Male, 29]

Members that do not comply with the rules or code of behaviour are marginalised or dismissed. At this point it seems that in long-term groups, team cohesion and homogeneity rely both on game oriented aspects as well as on the social environment and the players.

Peer-Mentoring

Peer-mentoring and player support were very common practices among MMOG players, and in many cases the most critical practice for learning how to play the game. The complexity of the environment leads the new players to often resort to the help of other players. Players supported each other with advice, information, resources such as in-game currency and equipment, or help with difficult tasks. Peer-mentoring as a learning practice is more intense in long-term groups. Members of the same group are helped, advised and supported by other group members, and then feel obliged to help other novice players, members of their group. One of our interviewees reported that that such practices actually benefit the group, since the stronger each individual member is, the stronger the group.

"There were certain things that I had, initially, misunderstood. Then, I joined a group. The group [helped me], I learned things, I became more expert, I understood things, and I got to the point where I can also help other, novice players, like a chain - you have been trained and you know how to train the new players" [Male, 37]

"[the novice player] needs guidance, otherwise s/he will get bored. It's tiring in the beginning. If you don't have someone to guide you through your first steps, you will get discouraged by the complexity of the game; you will get stuck" [Male, 46]

There were cases where the design of the environment promoted this mentor-student relationship, through the integration of quests, for instance, that the player can only accomplish under the supervision of a higher level player (Fig4). Cooperation of novices with higher level players was also controlled by the environment so as to ensure that the novice players do not rely on other players' skills but rather assume an active role in their progress. Grouping for accomplishing a task was in many cases restricted among players of the equivalent levels.

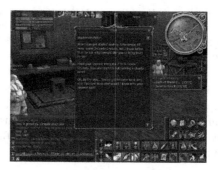

Fig. 4 An NPC quest in Lineage II requiring the participation of another player – a sponsor

Players would also seek support beyond the game environment, in external fora or websites where they could exchange experiences, ask questions, and give answers. The active and vibrant in-game community transcended the boundaries of the game. Players form communities of practice and communities of learning, specifically oriented towards the game. There are hundreds of user-developed sites and fora for the exchange of game-specific information, tactics, and solutions. Many of our respondents admitted that it would be impossible to progress in the game without referring to such external sites.

Multimodal Communication

MMOGs often integrate a variety of available communication modes and channels. Players may interact non-verbally - through actions, and verbally - through discussions (Manninen 2001). Discussions are taking place through text, in synchronous chat or asynchronous fora or through voice, while actions involve the animations, movement, and placement of the virtual characters. When the available channels do not suffice or present limitations, players find ways to overcome the obstacles.

The most common mode of communication is through text, via the integrated functionalities of the environment: synchronous chat or asynchronous fora. Chat channels and fora are further separated into different channels: dedicated to discussions among clan or guild members, among members of a party, among the members of an alliance of groups, channels for trading items, channels for players within proximity. Instant chat channels were usually the easiest, more direct, and with a faster response rate way of asking a question or making a comment. Typing text, though, can be time consuming when trying to keep up with the pace of the game and of the other players (e.g. during a battle). Players have developed a jargon for faster text communication; a jargon which might be difficult to follow by an outsider or a new player. Textual discourse in MMOGs and relevant fora and websites is filled with bizarre acronyms and abbreviations for faster communication among players: LFP instead of "Looking for party", WTB and WTS instead of "Want to buy" and "Want to sell".

> *"[In the beginning it was] Very difficult! Mainly [because] I couldn't understand the language. For every word I needed a dictionary. [...] It was not the English. I am good at English. But I didn't have a clue of the English the players there were speaking. [...]. They would say 'gn' instead of 'goodnight', 'soz' instead of 'sorry'"* [Female, 20]

Since synchronous chat does not seem to suffice for the cases of elaborate discussions and fast coordination and communication, players in many cases resorted to voice communication, through integrated functions of the environment, or through third party VoIP software when they did not conflict with the interface and functionalities of the game.

> *"I prefer to use Ventrilo when possible. Because [...] I am probably involved in an activity that prevents me from chatting [typing]. But I also consider Ventrilo a better mode of communication. It's more direct. You don't have to type. Very often you can pass a message through the tone of your voice, and it's definitely a more direct mode of communication"* [Male, 32]

> *"[In Diablo 2] [VoIP] is not very convenient, because you have to be able to hear the drops. Every item dropping has a specific sound. You know if the dropped item is an amulet, a ring, a shield, a helmet. It has a distinct sound. And you have to know, to search fast, and to see what has been dropped. You can't use Skype at the same time"* [Female, 25]

Fora on the other hand, although not appropriate for fast communication, supported more elaborate discussions. In *Tribal Wars* we observed extensive discussions in the dedicated tribe forum, detailed planning of strategies, advice, tips and tricks of the game, exchange of opinions, negotiations, argumentations, introductions of new members, announcements, publication of the tribe rules and even social discussions, threads about music, jokes, films, and general subject discussions (in Fig5 a screenshot of a thread dedicated to social talk, from our *Tribal Wars* tribe forum. Identifying information has been blurred).

Fig. 5 Screenshot from a Tribal Wars tribe forum (in Greek). Thread dedicated to social talk.

Awareness tools so as to know when your friends are online in the game and interact directly with them through personal messages are usually available, motivating the players and facilitating communication.

> "Good cooperation leads to the addition of new people in the friend list. The players know which of their friends are online and talk through the chat channel, so as to cooperate again in the future" [Male, 37]

> "As soon as I log in [the game] I will first see who is online, so as to chat a bit, before starting to play" [Female, 25]

In some games, players can only communicate if they have mutually accepted to be added to each other's friend list, or players of opposing sides cannot communicate at all.

Animations of the virtual character and gestures seem to increase players' motivation and engagement in the game. Gestures of the virtual character such as bowing, waving, cheering, dancing, laughing, constitute a further means for "breaking the ice", socially approaching other players, and non-verbally interacting even with strangers. As one of our interviewees also reported:

> "I have this friend, for the past 1.5 years. In the beginning, he would just dance in front of me, in game. He was doing it just for the fun of it and eventually we became friends" [Female, 27]

4.3 Community Practices and the Social Environment

The design of the environment seems to be in constant negotiation with the social environment - the community of the players - shaping structures and practices with a direct impact on the interactions, the relations, the acquisition of knowledge and expertise, and the overall gaming experience. In 5 of our interviews, there were 25 direct references to the community of the players and the positive or negative implications. A persistent community constitutes the distinctive feature of MMOGs differentiating them from other online multiplayer or social games.

Players form relationships with others over time and not merely complete tasks together. This community of players seems to present the main elements of Wenger's "communities of practice": a shared domain of interest defining the identity of the members, commitment to the domain, joint activities, helpful relationships, shared resources and information, shared practice and sustained interactions (Wenger 2006). A good community was a game environment where the players help, support and respect each other.

> *"The game is extremely difficult, but luckily there is the community. The guy who helped me in the beginning, I had asked him 'How can I repay you?' and he replied 'Don't give me anything. Just, when you'll be ready, help others as well'"* [Male, 28]

The gaming experience, the practices and behaviours of the players, the way the players learn the game and progress is defined by both the design and the social environment. The design of the environment guides, up to a point, the interactions among the players through, for example the Terms of Use, the End User Licence Agreement, and other relevant official documentation prohibiting offensive or abuse behaviour, and also through the affordances for social interactions of the environment that Yee described as "social architectures" (Yee 2008). The community of the players though, also plays an active role on the interactions among players, through the emergence of informal rules of conduct, and the ethical codes of the game (Bainbridge 2010, p. 92).

> *"In EVEOnline there are no standardised tactics, so the community has to discuss so as to find the best way to progress in the game. This creates a better and more bonded community"* [Male, 28]

Our team in *Tribal Wars* was disbanded when members of the leadership considered the behaviour of an opposing tribe offensive and violating the ethos of the game. They deleted their accounts and left the game. Players also reported that they stopped the game, when their in-game friends decided to leave the game.

> *"[...] conversations, discussions and actions of other players, in other tribes, have made me feel really stupid for even being in the same game with them. [...] what I am really trying to say is that I intend to hit 'delete' very soon"* [Tribe Leader in TW tribe forum]

Offensive behaviour from other players, combined with the lack of appropriate penalties from the designed environment and the administrators of the game, can discourage, disappoint, or lead to the rejection of the game by players.

> *"The first thing I avoid is 'flaming' - the use a bad language for offending others. [...] And if I see a game where the administrators don't take care of [such behaviour], I leave it immediately"* [Male, 26]

Table 3 Summary of MMOG task and group related features for supporting learning

MMOG Feature	Features
Variety of tasks	Well-structured and ill-structured problems for different strategies and cognitive processes Individual and collaborative tasks Different format and content to address preferences and play styles of players
Rewards	Rewards as mechanisms for managing group synthesis and size and for indicating progress Players may manage reward distribution based on social practices Rewards motivate players. Players are continuously rewarded
Long Term Groups & Task-Oriented Groups	Long term groups seem to have increased social bonds and group cohesion, in relation to task-oriented groups. Discussions and decision-making among members In long term groups increased peer-mentoring and support in relation to task-oriented groups The social aspect in long term groups motivates players. Task-oriented groups support achievement and progress of players, but decreased sociability
Complexity & Minimal Game Guidance	May trigger collaboration, interactions, peer-mentoring and the development of communities for learning Collaboration is necessary for learning the game and progressing
Multimodal Communication	Multiple interaction channels for supporting multiple modes of communication (verbal in chat, fora, voice and non-verbal through actions, animations) Richer and more effective interactions among players Players may build their own networks of friends Awareness tools for supporting communication in networks of friends

5 Discussion and Conclusions

Building upon previous research describing the potential of MMOGs for learning and education, we looked at good practices of MMOGs integrating elements that have been described in the literature as conducive to effective conditions for learning. Certainly, not all MMOGs are successful. In fact, quite the opposite is true. Of the many MMOGs produced, only a few attract and sustain the interest of large numbers of players and succeed at developing and fostering a community of players.

We argued that for investigating the potential of MMOGs for learning we have to consider a framework including cognitive, social, and motivational dimensions. We employed an exploratory and qualitative approach and focused on features of the tasks, the groups, and interactions among players that seem to promote collaborative interactions for learning and the acquisition of expertise. We mainly addressed issues as they emerged from our interviews, participant observation and information from relevant websites and fora.

MMOGs motivate and engage the players through a combination of techniques and design decisions as well as through their flexibility that favours the development of viable and active communities of players. A number of interconnected elements were identified which corresponded to learning principles for the design of educational environments, effectively integrated within an engaging and functional system (for a summary of issues described see Table 3). Such elements could be applied to the design of interactive networked learning environments.

MMOGs integrate a specific structure guiding the players through the environment, with tasks, activities, and goals that have to be attained, and progress levels that have to be reached. It seems, though, that they also provide opportunities to the players to personalise their gaming experience, and adjust it to their preferences, playing styles, pace, and attitudes. This sense of freedom, self-control, personalisation, and autonomy seems to sustain the motivation, engagement, and active participation of the players. A variety of different types of available tasks and activities can trigger motivation, engagement and commitment of a wide range of players, support different cognitive processes, and promote collaboration. Players learn the game through their interactions with other players, through support, advice, guidance, peer-mentoring, confrontations, and commitment to the environment. A long-term group with strong social bonds, cohesion, shared goals, and the interconnection of the personal goals with the group goals seem to be more probable to foster collaboration, peer-mentoring, discussions and negotiation among group members, and the development of shared knowledge. Although the design defines most of these aspects of the environment, the community of the players - the social environment emerging - plays an equally important role. The players feel free to make their own rules of conduct, to set their own goals and select their mode and pace of playing. They develop their own communities, within and beyond the game environment, through websites and fora, with the main aim to learn the game and become experts.

Although our explorative approach does not allow for extrapolations to the design of all games and the general community of players, it nevertheless allowed us to identify patterns and issues for further research. Such issues are, for instance, the impact of sociability, cohesion, size, and structure of an MMOG group on collaborative learning processes, the relationship among player preferences and motivations, group type, and collaborative learning processes, the relationship among task types, cognitive processes, player interactions and the emergence of collaborative learning.

Learning the game, acquiring expertise and developing skills in MMOGs seems to be a long-term process, involving engagement, commitment, development of relationships with others, practice, and participation in the practices of the community. Further research on collaborative learning in MMOGs should consider this aspect. Longitudinal studies or case studies, for example, seem to be appropriate for capturing the process of the development of knowledge and expertise through sustained player interactions and collaboration.

References

1. Ang, C.S., Zaphiris, P.: Social learning in MMOG: an activity theoretical perspective. Interactive Technology and Smart Education 5(2), 84–102 (2008)
2. Bainbridge, W.S.: The Warcraft Civilization: Social Science in a Virtual World. The MIT Press (2010)
3. Barron, B.: When Smart Groups Fail. Journal of the Learning Sciences 12(3), 307–359 (2003)
4. Bartle, R.A.: Hearts, clubs, diamonds, spades: players who suit muds. Journal of MUD Research 1(1) (1996), http://www.mud.co.uk/richard/hcds.html
5. Boekaerts, M.: Context Sensitivity: Activated Motivational Beliefs, Current Concerns and Emotional Arousal. In: Volet, S., Järvelä, S. (eds.) Motivation in Learning Contexts: Theoretical and Methodological Implications, pp. 17–31. Pergamon, Elsevier (2001)
6. Charmaz, K.: Constructing Grounded Theory - A Practical Guide Through Qualitative Analysis. Sage (2006)
7. Creswell, J.W.: Research Design. Qualitative, Quantitative and Mixed Methods Approaches, 2nd edn. Sage Publications (2002)
8. Csikszentmihalyi, M.: Flow: the classic work on how to achieve happiness. Harper Perennial, New York (1992)
9. De Freitas, S., Griffiths, M.: Massively Multiplayer Online Role-Play Games for Learning. In: Ferdig, R.E. (ed.) Handbook of Research on Effective Electronic Gaming in Education, pp. 51–66. GI Global (2009)
10. Dickey, M.D.: Game design and learning: a conjectural analysis of how massively multiple online role-playing games (MMORPGs) foster intrinsic motivation. Educ. Tech. Res. 55(3), 253–273 (2007)
11. Dillenbourg, P., Järvelä, S., Fischer, F.: The Evolution of Research on Computer-Supported Collaborative Learning. In: Balacheff, N., Ludvigsen, S., Jong, T.D., Lazonder, A., Barnes, S. (eds.) Technology-Enhanced Learning Principles and Products, pp. 3–19. Springer, Dordrecht (2009)
12. Ducheneaut, N., Moore, R.J.: Gaining more than experience points: Learning social behavior in multiplayer computer games. In: Workshop on Social Learning Through Gaming (CHI 2004), Vienna, Austria, April 19 (2004)
13. Ducheneaut, N., Yee, N.: Collective Solitude and Social Networks in World of Warcraft. In: Romm-Livermore, C., Setzekorn, K. (eds.) Social Networking Communities and E-Dating Services: Concepts and Implications, pp. 78–100. Idea Group Inc., IGI (2008)
14. Ducheneaut, N., Yee, N., Nickell, E., Moore, R.J.: "Alone together?" Exploring the Social Dynamics of Massively Multiplayer Online Games. In: Proceedings of the SIGCHI Conference on Human Factors in computing systems (CHI 2006). ACM Press, New York (2006), doi:10.1145/1124772.1124834
15. Gagné, R.M.: Learnable aspects of problem solving. Educ. Psychol. 15(2), 84–92 (1980), doi:10.1080/00461528009529218
16. Garrison, D., Archer, T., Anderson, W.: Critical Inquiry in a Text-Based Environment. Computer Conferencing in Higher Education. The Internet and Higher Education 2(2-3), 87–105 (1999)

17. Gee, J.P.: Learning and Games. In: Salen, K. (ed.) he Ecology of Games: Connecting Youth, Games, and Learning. The John D. and Catherine T. MacArthur Foundation Series on Digital Media and Learning, pp. 21–40. The MIT Press, Cambridge (1999), doi:10.1162/dmal.9780262693646.021

18. Hjorth, L.: Games and gaming: an introduction to new media. Berg Pub. Ltd. (2011)

19. Hoepfl, M.C.: Choosing Qualitative Research: A Primer for Technology Education Researchers. Journal of Technology Education 9(1), 47–63 (1997)

20. Huffaker, D., Wang, J., Treem, J., Ahmad, M.A., Fullerton, L., Williams, D., et al.: The Social Behaviors of Experts in Massive Multiplayer Online Role-Playing Games. In: The 2009 International Conference on Computational Science and Engineering Proceedings, pp. 326–331. IEEE (2009)

21. Järvelä, S., Volet, S.: Motivation in Real-Life, Dynamic, and Interactive Learning Environments: Stretching Constructs and Methodologies. Eur. Psychol. 9(4), 193–197 (2004)

22. Jonassen, D.H.: Toward a design theory of problem solving. Educ. Tech. Res. 48(4), 63–85 (2000)

23. Jonassen, D.H.: Learning to solve problems: An instructional design guide. Pfeiffer (2004)

24. Jonassen, D.H., Kwon, H.: Communication patterns in computer mediated versus face-to-face group problem solving. Educ. Tech. Res. 49(1), 35–51 (2001)

25. Kolo, C., Baur, T.: Living a Virtual Life: Social Dynamics of Online Gaming. Game Studies 4(1) (2004), http://gamestudies.org/0401/kolo/

26. Malone, K.L.: Dragon Kill Points: The Economics of Power Gamers. Games and Culture 4(3), 296–316 (2009)

27. Manninen, T.: Rich Interaction in the Context of Networked Virtual Environments - Experiences Gained from the Multi-player Games Domain. In: Blanford, A., Vanderdonckt, J., Gray, P. (eds.) Joint Proceedings of HCI 2001 and IHM 2001 Conference, pp. 383–398 (2001)

28. Mayer, R.E.: Cognitive, metacognitive, and motivational aspects of problem solving. Instr. Sci. 26(1-2), 49–63 (1998)

29. McGrath, J.E.: Groups: Interaction and Performance. Prentice-Hall, Inc. (1984)

30. Myers, D.: Play and punishment: The sad and curious case of Twixt. In: Proceedings of The [Player] Conference, The Center for Computer Games Research, The IT University of Copenhagen, Copenhagen (2008)

31. Nardi, B., Harris, J.: Strangers and friends: collaborative play in World of Warcraft. In: Proceedings of the 2006 20th Anniversary Conference on Computer Supported Cooperative Work (CSCW 2006). ACM Press, New York (2006)

32. Reeves, S., Brown, B., Laurier, E.: Experts at Play: Understanding Skilled Expertise. Games and Culture 4(3), 205–227 (2009)

33. Schrader, P.G., McCreery, M.: The acquisition of skill and expertise in massively multiplayer online games. Educ. Tech. Res. 56(5-6), 557–574 (2008), doi:10.1007/s11423-007-9055-4

34. Sfard, A.: On Two Metaphors for Learning and the Dangers of Choosing Just One. Educational Researcher 27(2), 4–13 (1998), doi:10.3102/0013189X027002004

35. Silverman, M., Simon, B.: Discipline and Dragon Kill Points in the Online Power Game. Games and Culture 4(4), 353–378 (2009)

36. Slavin, R.: Research on Cooperative Learning and Achievement: What We Know, What We Need to Know. Contemp. Educ. Psychol. 21(1), 43–69 (1996)

37. Steinkuehler, C.A.: Learning in Massively Multiplayer Online Games. In: Kafai, Y.B., Sandoval, W.A., Enyedy, N., Nixon, A.S., Herrera, F. (eds.) Proceedings of the 6th International Conference on Learning Sciences, pp. 521–528. Erlbaum, Mahwah (2004)

38. Steinkuehler, C.A.: Why Game (Culture) Studies Now? Games and Culture 1(1), 97–102 (2006)

39. Van Den Bossche, P., Gijselaers, W.H., Segers, M., Kirschner, P.A.: Social and Cognitive Factors Driving Teamwork in Collaborative Learning Environments: Team Learning Beliefs and Behaviors. Small Gr. Res. 37(5), 490–521 (2006)

40. Voulgari, I., Komis, V.: Factors and Processes Involved in Collaborative Learning and Problem Solving in Massively Multiplayer Online Games: Aspects of the Designed and the Social Environment. In: IFIP Workshop, New Developments in ICT and Education, Amiens, 28-30 June (2010)

41. Voulgari, I., Komis, V.: Bending the Rules ... and Adding Some New Ones: Legal and Illegal Behaviours of Players in Massively Multiplayer Online Games. In: International Workshop DEG 2011 Involving End Users and Domain Experts in the Design of Educational Games (IS-EUD 2011). Torre Canne (Brindisi), Italy, June 7 (2011)

42. Wang, J., Huffaker, D., Treem, J., Fullerton, L., Ahmad, M.A., Williams, D., et al.: Focused on the Prize: Characteristics of Experts in Virtual Worlds. Presented to the Annual Meeting of the International Communication Association (ICA), Chicago, IL (2009)

43. Wenger, E.: Communities of practice: Learning, meaning, and identity, 1st edn. Cambridge Univ. Pr. (1998)

44. Wenger, E.: Communities of practice a brief introduction (2006), http://www.ewenger.com/theory/ (accessed April 2011)

45. Williams, D., Ducheneaut, N., Xiong, L., Zhang, Y., Yee, N., Nickell, E.: From Tree House to Barracks: The Social Life of Guilds in World of Warcraft. Games and Culture 1(4), 338–361 (2006)

46. WOWWiki. Dragon kill points. WoWWiki - Your guide to the World of Warcraft (2010), http://www.wowwiki.com/Dkp (accessed July 11, 2010)

47. Yee, N.: The Demographics, Motivations and Derived Experiences of Users of Massively-Multiuser Online Graphical Environments. PRESENCE: Teleoperators and Virtual Environments 15(3), 309–329 (2006)

48. Yee, N.: Social Architectures in MMOs (2008), http://www.nickyee.com/daedalus/archives/print/001625.php (accessed September 2011)

49. Zimmerman, B.J.: A social cognitive view of self-regulated academic learning. Journal of Educational Psychology 81(3), 329–339 (1989)

Validating Empirically a Rating Approach for Quantifying the Quality of Collaboration

Georgios Kahrimanis, Irene-Angelica Chounta, and Nikolaos Avouris

Human-Computer Interaction Group, University of Patras, Rio-Patras, Greece
{kahrimanis,houren}@ece.upatras.gr, avouris@upatras.gr

Abstract. Interdisciplinarity in the Computer Supported Collaborative Learning (CSCL) research field involves the application of several methodological approaches towards analysis that range from deep-level qualitative analyses of small interaction-rich episodes of collaboration, to quantitative measures of suitably categorized events of interaction used as indicators of the success of collaboration in some of its facets. This article adopts an alternative approach to CSCL analysis that aims at taking advantage of some desired properties of each of these diverse methodological trends, involving the use of a rating scheme for the assessment of collaboration quality. After defining a set of dimensions that cover the most important aspects of collaboration, it employs appropriately trained human raters basing their assessments on substantial aspects of collaboration that are not easily formalisable. The activities studied here regard 228 collaborating dyads, working synchronously on a problem-solving task. Based on this large dataset, relations between dimensions of collaboration quality are unraveled on empirical grounds, by elaborating ratings statistically using a multidimensional scaling technique.

1 Introduction

Computer Supported Collaborative Learning (CSCL) constitutes one of the most extensively developed paradigms of research and practice in intelligent networking and collaborative systems technology. Under specific conditions, collaborative interactions can trigger collaborative knowledge building (Scardamalia and Bereiter 1996) that is beneficial for learners participating in collaborative processes.

Apart from the study of the conditions that can lead to fruitful CSCL processes, and the "learning gains" that students may obtain, analysis of collaborative interactions per se constitutes one of the core aspects of the study of CSCL (Dillenbourg et al. 1995). Interdisciplinarity in the research field involves the application of several methodological approaches towards analysis of CSCL inspired, adopted, or developed based on diverse research disciplines. Most commonly used analysis studies range from deep-level qualitative analyses of small interaction-rich episodes of collaboration, to quantitative measures of suitably categorized events of interaction that are used as indicators of the success of collaboration in some of its facets (Stahl et al. 1996; Kahrimanis et al. 2011).

Whereas the latter approach to CSCL analysis offers possibilities for practical facilities such as quick or even automated assessments of collaboration as in (Avouris et al. 2004; Strijbos et al. 2006), it is claimed that it is often based on measuring "surface" aspectzs of collaboration (Stahl et al. 1996). On the other hand, cases that belong to the former approach, such as (Roschelle 1992; Stahl 2006), may be rightly considered the most suitable for in-depth CSCL analysis, they are, however, arduous and time-consuming, since they demand much effort for analysing relatively small episodes of collaboration. Moreover, it is difficult that they scale-up to analysis of extended datasets when dealing with large-scale studies.

This article adopts an alternative approach to CSCL analysis that aims at taking advantage of some desired properties of each of these diverse methodological trends. It involves the use of a rating scheme for the assessment of collaboration quality (Kelringer and Lee, 2000). After defining a set of dimensions that cover the most important aspects of collaboration, it employs appropriately trained human agents to assign ratings of collaboration quality to each dimension, basing their assessments on substantial aspects of collaboration which are not easily formalisable. Still, the outcome of the evaluation process is provided in quantitative form, suitable for statistical manipulation. This way, the results obtained can serve as a point of reference for intelligent techniques that are based on automatable formalisations, providing new information that takes account of deeper aspects of collaboration than most top-down approaches for evaluation, personalisation and adaptive feedback.

The activities studied here regard 228 collaborating dyads, working synchronously on a computer science problem-solving task with the use of the Synergo tool (Avouris et al. 2004). Based on this large dataset, relations between dimensions of collaboration quality are determined on empirical grounds, based on the ratings of collaboration quality applied for each dimension in each collaborative session. The technique selected, which allows the systematic view of these associations in statistical means, is multidimensional scaling (Kruskal and Wish 1978; De Leeuw and Heiser, 1982; Schiffman et al. 1981; Davison 1983; Young and Hamer 1994; Borg and Groenen 1997; Cox and Cox 2001), using dimensions of collaboration quality as the unit of analysis. That way, associations between core aspects of collaboration are represented in a two-dimensional space, with distances between dimensions denoting their dissimilarities. The results obtained are in accordance with the initial design of the rating scheme used, and further particularize the relations between its dimensions. General conclusions and further steps made possible by current findings are discussed in the last section of this article.

2 Collaborative Setting

Collaborative activities studied in this article involved about 350 computer science students at the department of Electrical and Computer Engineering of the University of Patras, Greece, engaged in jointly building the diagrammatic representation of an algorithm as an assignment of a two-hour laboratory session that was part of the first-year of studies course "Introduction to Computers". These activities took place in a single laboratory room, equipped with one computer per student. Students interacted through Synergo (Avouris et al. 2004), communicating via an integrated chat tool, and jointly designing a flow-chart representation of an algorithm in Synergo's shared workspace. A capture of a user's screen while collaborating with Synergo is shown in Fig. 1. Synergo provides libraries of objects supporting the notation of several diagrammatic models. Collaborative sessions lasted from 45 to 75 minutes and students worked in dyads, which were selected randomly. They were free to use their own resources such as textbooks or the web and were permitted to ask questions to a teacher, who restricted her feedback to technical or other minor aspects. In order to motivate students to work on the exercises collaboratively, they were informed that the grade they would get for the particular lab session would be determined by both the quality of their collaboration and the completeness and correctness of their joint solution. Dyads were arranged in space in a way that it was impossible for the students to use any other means of communication apart from these provided by Synergo.

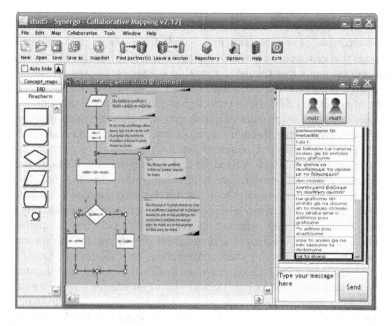

Fig. 1 Synergo in use: a capture of a user's screen collaborating on an algorithm problem

The problem domain of the task was basic algorithms in computer science. Students were asked to solve elementary algorithm exercises that are widely used for training basic algorithmic skills. For example, a basic piece of knowledge in algorithms is the concept of the variable (Samurcay 1989), which should be discriminated from the understanding of the variable that students may have from mathematics. Another major learning object was also the proper handling of algorithmic structures, such as the loop structure (Soloway et al. 1989). Participants were asked to solve specific algorithm problems requiring the use of these concepts by developing flowchart diagrams, a widely used modelling practice that provides a semiotic space for the design of algorithms (Bohl 1971). The task given to students can be considered an "intellective task" with a "demonstrably correct solution" (Laughlin 1980). The correctness of the solution is concretely defined, based on the notation of algorithms used. There are, of course, alternative ways to develop parts of the solution that are equally acceptable and correct, however, in each case, arguments on correctness and the pros and cons of each alternative can be based on solid criteria. All students were taught the knowledge demanded in order to handle the task sufficiently in university lectures before the lab sessions took place, although some of them may have been already familiar with the task domain from secondary education curricula.

3   Determining Relations between Dimensions of Collaboration Quality

The first step of analysis for the current study dealt with the definition of a conceptual framework of collaboration that is multidimensional, i.e. it defines several dimensions that cover core aspects of collaboration, and, on a second level, can be operationalised into a tool suitable for making assessments of specific instances of collaborative processes. An extended literature search led to the adoption of the work by Meier et al. (2007), who proposed a multidimensional conceptual framework suitable for the assessment of collaboration quality in synchronous interdisciplinary problem-solving through videocoenfencing systems, in a study that significantly influenced the work presented here.

The conceptual framework defined several dimensions of collaboration quality, further categorized into broader aspects of collaboration. The first broader aspect defined regards communication. On a first level, one dimension of collaboration deals with the need for the establishment of common ground of mutually shared concepts, assumptions and expectations (Clark 1996) between the participants. Common ground can be achieved and sustained if both partners (in the case of a dyad) work towards grounding their conversations on a moment-to-moment basis (Clark and Brennan 1991). Good practices in this dimension regard extra effort from the participants in order e.g. for a sender of an utterance to try to make her contribution understandable to their peer, or, for a receiver, to try to indicate understanding of what has been uttered. On a more elementary level, the framework covers, apart from the *content* of communication, the *process* of communicating as well. Practices of participants such as ensuring mutual attention (Clark 1996), and the proper management of the turn-taking mechanism are considered appropriate for the success of a collaborative process.

The second broader aspect of the framework is generally described as information processing. It covers collaborative activity that is tightly related to the task. On a first level, what is of major importance for successful collaboration in this aspect is that participants exchange and process information based on their complementary knowledge, so that they can build a shared knowledge base. In social psychology, the terms of information pooling and transactive memory (Wegner 1987) play a crucial role in describing such processes. From a collaborative learning research standpoint, similar processes can be studied under the term of knowledge acquisition. Knowledge acquisition can be achieved either by externalization of a participant's personal knowledge, or by its elicitation by their peer by asking for explanations (Fischer and Mandl 2003). On another level, after the pooling of information, collaborators have to reach a common decision on the best solution to the problem. In order to achieve that, collaborators have to evaluate the information exchanged, by stating arguments for and against the options at hand, and critically discussing different perspectives (Tindale et al. 2003).

Another important aspect of collaboration regards the coordination of participants on a broader level than the one mentioned before: on the task rather than the communicational level. It deals with practices of efficient structuring of the problem-solving process that involve issues such as the intelligent division into subtasks between participants (e.g. the optimal handling of interdependencies that may occur when subtasks build up on each other, or conflicts when group members need to access the same shared resources (Meier et al 2007)), the efficient management of the time resources available, and the proper handling of coordination demands imposed by the mediating tool's technicalities.

In addition to communication, information processing, and task coordination aspects, social aspects of collaboration are also given due attention by the conceptual framework. Under the frame of the managing interpersonal relationships aspect, the framework covers issues such as interpersonal support, helpfulness and friendliness that can be constructive for successful collaboration. Such desired practices can be reflected in the symmetry of the relationship, the extent of supportive communication and the way the conflicts are handled (Meier 2005).

Finally, the last category covered by Meier et al.'s conceptual framework (2007) deals with motivational aspects of collaboration. Orientation and dedication to the task on behalf of both participants rather than on task-irrelevant issues is considered to be a prerequisite of successful collaboration.

The conceptual framework described above was motivated and led to the definition of a rating scheme suitable for applying assessments of collaboration quality in all of its core dimensions. The rationale of this evaluation approach is described in more detail in the subsequent section.

Table 1 Meier et al.'s general aspects of collaboration

General aspect of collaboration
Communication
Joint information processing
Coordination
Interpersonal relationship
Motivation

4 The Rating Approach

4.1 Rating Scheme

The framework described in the previous section was operationalised through a rating scheme that was used as for assessing collaboration quality in its core dimensions. As an analysis tool, the rating scheme combines desirable properties of qualitative and quantitative techniques. Observed behavior can be compared to a predefined standard of exemplary collaboration that has been formed based on established CSCL theory and thorough empirical analyses of typical collaborative sessions. This can then lead to quantitative judgements of the quality of collaboration. The main advantage of the rating approach compared to common quantitative analysis is that it offers quantitative results that measure subtle aspects of collaboration (mainly being the object of study of in-depth qualitative analyses), rather than gross metrics that are usually based on quantities of events of users' interactions with the mediating tools. Simultaneously, in contrast to the approach adopted in this work, common qualitative approaches that demand in-depth analysis of collaboration usually can not be extended to more than a few rich episodes of collaboration, whereas issues of reliability and generalisability are more difficult to overcome in such cases.

Concerning the practical efficiency of analysis, rating processes are relatively time-effective, since assessment using the rating scheme is rather quick, provided that raters have been sufficiently trained. Therefore, rating schemes can be efficient when dealing with large datasets. Another important characteristic of the rating scheme analysis tool, is that it is multi-dimensional, i.e. it is used for assessing several dimensions of collaboration distinctly. In that way, the results of rating can be used for indicating which dimensions of collaboration are the most problematic and provide e.g. the opportunity for giving adaptive feedback and specialized instructions according to the demands of each separate group. This practice can be followed between different phases of an educational process that take place in sequence in e.g. an academic semester including several related lab sessions as in Meier et al. (2008). Moreover, the approach can still be useful as a first point of analysis for further, more detailed evaluation studies that necessitate more thorough research work.

Due to significant differences between the setting that lead to the definition of Meier, Spada, and Rummel's rating scheme and the current setting, a laborious process of generalising and adapting the initial conceptual framework to the current setting was followed (reported in detail in (Kahrimanis et al. 2009)). The adaptation of the rating scheme was done in two main phases: the first resulted in an adapted definition of the rating scheme's dimensions, and the second served to fine-tune the rating instructions. In the first phase of adaptation a bottom-up approach, which involved identification of "best practice" examples in the sample data, was combined with a top-down process, during which the definitions of all original dimensions were reformulated taking into account constraints and affordances characterizing the specific collaboration setting. In the second phase of adaptation, the dimensions' definitions were fine-tuned and illustrated with more detail, grounding each dimension's theoretical concepts in specific examples of collaborative practice from the data pool of the first round of adaptation (Kahrimanis et al. 2009). The resultant rating scheme specifies seven core dimensions of collaboration quality presented in Table 2.

The structure of the new scheme is in accordance with the rationale of the initial one, while definitions of dimensions do not only aim at fitting the current setting, but at being more generalisable as well.

Table 2 The adapted rating scheme

General aspect of collaboration	Dim. Num.	Dimension of collaboration
Communication	D1	Collaboration flow
	D2	Sustaining mutual understanding
Joint information processing	D3	Knowledge exchange
	D4	Argumentation
Coordination	D5	Structuring the Problem Solving Process
Interpersonal relationship	D6	Cooperative orientation
Motivation	D7	Individual task orientation

4.2 Rating Process

All dimensions were rated on the level of a collaborating dyad with the exception of individual task orientation (D7) which was rated for each participant separately. As in the initial approach by Meier et al. (2007), the ratings were applied in the scale from -2 to 2 with a step of 1 unit. One rating was assigned for each dimension of collaboration quality per collaborative session. A handbook was also developed in order to assist raters providing a rich source of detailed definitions of all dimensions, along with rating instructions and illustrative examples of episodes from the dataset. The rating process was based on video-like reproductions of the activities facilitated by the Synergo's playback tool (Kahrimanis et al. 2009).

The rating procedure was carried out in two main phases. The first one, reported in detail in (Kahrimanis et al. 2009), consisted of 101 dyads which were rated for each dimension by two raters with prior experience with the current setting, after an extended pilot phase of training. Inter-rater reliability scores were very good. The second phase, which was deemed necessary in order to extend the population of collaborative sessions, consisted of additional 149 dyads, for which the design and setting of the labs was identical with the one used in the first phase, varying only in minor aspects of task details (e.g initial values of variables were changed in order to avoid totally repeating the tasks that were given to students in the previous year's academic semester), and was thus appropriate for integrated analysis. This way, the dataset was significantly augmented and large-scale statistical elaborations from several points of view, such as the one presented later in this article, became possible. In the second phase, the ratings were applied by the same persons as in the first one and inter-rater reliability was examined for this phase as well.

4.3 Reliability of Ratings

Results of inter-rater reliability of the second rating procedure for each dimension of the scheme are illustrated in Table 3. For D7, the reliability scores for the average rating between the two students (D7a) and their absolute difference (D7b) are provided. The table contains also reliability scores for the average of the six first dimensions of the scheme (CQ).

Inter-rater reliability scores are good in reference to all empirical rules found in the literature (Fleiss 1981; Cicchetti and Sparrow 1981; Wirtz and Caspar 2002; George and Mallery 2003), although lying at somewhat lowers level than the excellent scores achieved in the first phase. Therefore, the ratings of any one of the raters could be reliably used for further elaborations.

Table 3 Inter-rater reliability scores for each dimension

General aspect of collabora- tion	Dimen- sion of collabo- ration		ICC	ICC (adj. = r)	Cro nb. α	Spe ar. ρ
Communication	Coll. flow	D1	.76	.77	.87	.74
	Sust. mut. underst.	D2	.79	.82	.89	.80
Joint information processing	Knowl. exchange	D3	.81	.81	.89	.74
	Argument ation	D4	.77	.77	.87	.75
Coordination	Struct. the Probl. Solv.Proc.	D5	.70	.69	.82	.71
Interpersonal relationship	Coop. orient.	D6	.82	.83	.90	.79
Motivation	Ind. task orient. (mean)	D7a	.71	.75	.84	.59
	Ind. task orient. (diff.)	D7b	.87	.87	.93	.61

5   Relations between Dimensions of Collaboration Quality

To some extent, relations between dimensions of collaboration quality can be roughly conceived from the definition of the conceptual framework and the rating scheme. Nevertheless, the extended dataset gathered offers the opportunity to validate such top-down assumptions empirically, by applying suitable statistical manipulations on the data. Moreover, the exact relations between dimensions can be detected based on the ratings applying in 228 cases.[1]

5.1   Multidimensional Scaling of Dimensions of Collaboration Quality

A systematic way to obtain an overall elaborate view on the associations between collaborative dimensions and empirically evaluate the use of the scheme regards the conduction of a MultiDimensional Scaling (MDS) analysis based on the bivariate correlations between them (Kruskal and Wish 1978; De Leeuw and Heiser, 1982; Schiffman et al. 1981; Davison 1983; Young and Hamer 1994; Borg and Groenen 1997; Cox and Cox 2001). MDS analysis is based on measures of similarities between variable pairs and no assumptions are presupposed on the distribution of the values which the variables take, the types of their similarity relations, or the way the similarity measures are obtained.

In the specific case of this study, the unit of analysis of the technique is the collaborative dimension as it is defined by the rating scheme. The algorithm takes as input the rating assigned to each dimension for 228 instances of collaborative sessions. The technique provides insightful two-dimensional diagrams representing the position of collaborative dimensions in such a way that dimensions correlated tightly are placed closer to each other in space than dimensions that do not relate that much. For the current application of the technique, disparities between correlations are represented with a spatial Euclidian distance.

The MDS algorithm used was SMACOF (Scaling by MAjorizing a COnvex Function) (De Leeuw) as it is implemented by XLSTAT (2009). This iterative algorithm aims at minimizing the normalized differences between a similarity matrix given as input (converted to a dissimilarity matrix) and the corresponding distance matrix that is represented as the outcome of the process.

5.2   Results and Internal Validation of the MDS Algorithm

The results of the application of the technique are depicted in Fig. 2 (using Kendall's τ scores for the calculation of correlations) and Fig. 3 (using Spearman's ρ for the calculation of correlations). The two diagrams are very similar and lead to the same interpretations.

---

[1] The cases rated in the two phases of training minus 32 cases that were left out of the final dataset due to technical problems in the logfile captures related to them (other kinds of analysis followed later than the work presented here used the logfiles and it was decided that these 32 dyads should be omitted in order to maintain a consistent dataset throughout related works).

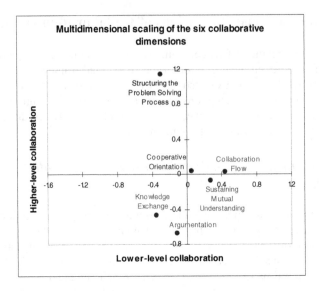

Fig. 2 Multidimensional scaling with the 6 collaborative dimensions using a similarity matrix based on Kendall's correlations

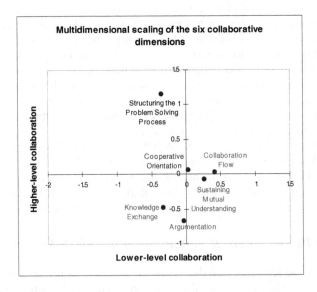

Fig. 3 Multidimensional scaling with the 6 collaborative dimensions using a similarity matrix based on Spearman's ρ correlations

The following Fig. 4 and Fig. 5 illustrate the Shepard diagrams of the application of the technique using Kendall's τ and Spearman's ρ values respectively.

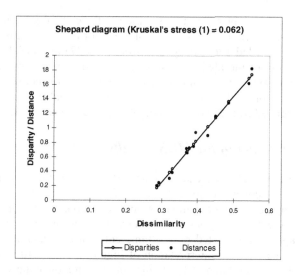

Fig. 4 Shepard diagrams of MDS algorithm using Kendall's τ correlations

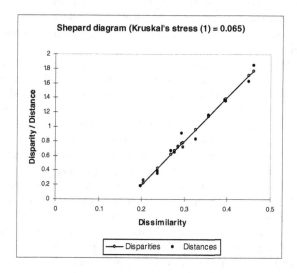

Fig. 5 Shepard diagrams of MDS algorithm using Spearman's ρ correlations

The Shepard diagram (Shepard 1962) is a scatter-plot that depicts the configured distances for the two-dimensional model in relation to the observed distances used as input (Steyvers 2002). The filled circles in the diagram represent
the Euclidean distances presented by the MDS algorithm, whereas the empty circles represent the distances calculated by the monotonic regression function of the
algorithm (De Leeuw 1977). The latter's slope is represented by the lines of Figures X.4 and X.5. The square root of the normalized sum of squared residuals
between the filled circles and the straight line is measured by Kruskal's stress

(Kruskal 1964), which provides an estimation of the goodness-of-fit of the results. Kruskal's stress for the first application of the algorithm was measured at the acceptable level (Kruskal 1964; Borg and Groenen 1997) of 0.062 for the concluding 28th iteration of the algorithm, whereas for the second application it took a similar value (0.065) at the 30th iteration. In both cases, the convergence criterion for the final version of the model stated that the stress value of the final iteration should be improved from the previous iteration by less than 0.00001.

5.3 Interpretation of the MDS Results

As is evident from Figures 2 and 3, dimensions covering different aspects of collaboration quality cover four different parts of the two-dimensional space (the dimension belonging to the motivational aspect is not contained in the diagram since it is rated differently than other dimensions). Dimensions covering the same aspect of collaboration (denoted by the same color in the diagram) stand close to each other. Regarding the interpretation of Figure 2, the coordinates of each dimension do not denote the quality of collaboration in a quantitative manner; they are used for the representation of its distance from other dimensions. Therefore, the range of each axis should be thought of as representing aspects of collaboration that differentiate dimensions on the way they reflect different facets of this specific axis. The rationale followed in order to reach meaningful interpretations is described below.

Higher-order dimensions of collaboration are reported with higher absolute values on the vertical axis, while lower-level ones have higher absolute values on the horizontal axis (cooperative orientation, D6, which is placed near the zero-point does not straightforwardly relate to any of these axes). Thereby, the vertical axis can be considered to stand for high-level collaboration aspects and the horizontal axis to stand for lower-level collaboration aspects.

Concerning the horizontal axis, the two communicational dimensions (D1 and D2) are placed on the right of the diagram, taking positive values, whereas the two information processing dimensions are placed on the left, taking negative values. Thus, from left to right, the horizontal axis can be considered to designate the range from task-related low-level facets of collaborative activity to task-unrelated facets of collaborative activity (task-related in this case refers to these aspects of collaboration that are significantly shaped by the specific task to be solved). In the case of lower level collaborative activity, task-unrelated facets mostly refer to communicational aspects. Collaboration flow (D1) takes the largest positive (and absolute) value on the axis, since it constitutes the lowest-level dimension of the scheme. Sustaining mutual understanding (D2), on the other hand, is placed closer to the zero point and has a more noticeable Y coordinate. Among the information processing dimensions, knowledge exchange (D3) has the biggest negative value because argumentation (D4) is related more to high-level collaborative activity. Structuring the problem solving process (D5) is also placed left from the Y axis. According to the interpretation of the axes developed above, this reflects the fact that structuring the problem solving process is shaped by task-related issues in the lower level of collaboration. For example, an algorithm development problem favors practices of task coordination such as the development of different small parts of the algorithm by participants in parallel (which are arranged according to the task's demands), the proper placement of flow-chart objects by each student so that the two parts can be then combined, or the development of a part of the algorithm by one student while their partner is checking for its correctness by assigning values to variables. Such practices are highly task-dependent and would not be reproduced in the case of a task of a different kind.

A similar rationale applies to the vertical axis concerning higher-level collaboration: structuring the problem solving process (D5) takes the highest absolute value on the upper part of the diagram, while the two information processing dimensions lie on the negative part of the axis. So, in a similar way with the vertical axis, the horizontal one can be considered to denote from up to down the range from task-unrelated high-level facets of collaborative activity to task-related high-level facets collaborative activity. Among the two joint information processing dimensions, argumentation (D4) takes a significantly higher value than information processing (D3), due to the fact that the former reflects higher-level facets of this aspect of collaborative activity. Structuring the problem solving process (D5), on this axis, takes a significantly positive value. Contrary to lower level aspects that the same dimension covers, higher level aspects of structuring the problem solving process are not tight to the specific task. They mainly refer to general strategies of collaborative problem solving, such as the division of labour between participants, the evaluation of one student of the other students work, and time management concerns so that a complete solution can be delivered on time. Communicational dimensions take approximately zero values on the vertical axis, since they are related to lower level aspects of collaboration. Among the two dimensions, sustaining mutual understanding (D2) appears to have a small load on the vertical axis, due to the fact that it is, even limitedly, related more to higher level issues than collaboration flow (D1) is.

Concerning both axes, cooperative orientation (D6) is located very close to the zero point. Social aspects of collaboration that relate to cooperative orientation do not have a straightforward mapping with higher or lower task-related or non task-related facets of collaboration, even though the dimension is correlated highly with all other dimensions of the scheme, something denoted in the diagram with its central position.

In general, results obtained from the MDS algorithm are in accordance with the definition of the dimensions of the rating scheme. Distances represented by the algorithm are reasonable: a diagram of a similar rationale, applied by the researchers in a top-down manner, would probably resemble the one found empirically. Moreover, the approach offers subtler information on the exact associations between dimensions.

Concluding, it should be noted that some properties of the definition of the axes are to some extent arbitrary and their descriptions related to higher and lower-level aspects of collaboration constitute an interpretation rather than an "objective" result of the technique. The examined instance of the MDS technique could lead to the same information with the axes rotated, or their sings inversed. What would remain the same is the relative position of the dimensions (not regarding minimal differences attributed to the goodness-of-fit of the algorithm). Therefore, for the output of the algorithm presented in the figures above, the algorithm was initialized in such a way that the axes would be more interpretable, something that constitutes a common practice when applying MDS or other techniques of similar purpose in several research domains. (Gutmann 1968; Borg and Lingoes 1987).

6 Conclusions and Further Research

The work reported in the current article implemented a rating scheme based approach for the evaluation of synchronous problem-solving collaborative activities in order to gain insight into the relations between distinct dimensions of collaboration quality. The multidimensional scaling approach, which was applied in a large dataset of collaborative sessions, largely confirmed on empirical grounds the rationale of the conceptual framework of the rating tool used, as regards the relations between core dimensions of collaboration quality. Furthermore, it provided additional insight on the exact placement of each dimension of collaboration in reference to two general axes of collaborative activity.

This kind of validation adds evidence that the rating approach can be a reliable tool for evaluating aspects of CSCL activities that are not easily grasped when using strict formalizations, and allow researchers to take full advantage of the practical opportunities that it can offer: the more feasible analysis of large datasets; the provision of a research aid for the conduction of further, more focused research; and the provision of adaptive feedback to students based on their collaborative performance. A pilot study that investigates the tool's application for the latter case is reported in (Meier et al. 2008).

Future research directions related to this work can follow several paths: statistical analysis reported here can be supplemented by in-depth qualitative investigations of collaborative activities, which can shed more light into the way different dimensions of collaboration are interlinked with each other in collaborative practice. Common trends that determine the placement of dimensions of collaboration close or far from each other in the MDS representation can serve as the initial point for further qualitative analysis based on interesting instances of collaboration that may reveal subtler associations between aspects of different dimensions, or recurring patterns of the simultaneous occurrence of good or bad practices in specific dimensions.

Furthermore, the current approach can be replicated using different versions of the rating scheme or applied in different settings of collaboration. Such efforts would help to indicate the extent to which current findings are indicatory of the way dimensions of collaboration are associated with each other in general, or if they mostly pertain to the specific CSCL setting under study.

References

1. Avouris, N., Margaritis, M., Komis, V.: Modelling interaction during small-group synchronous problem-solving activities: The Synergo approach. In: 2nd Int. Workshop on Designing Computational Models of Collaborative Learning Interaction, ITS 2004, Maceio, Brasil (September 2004)
2. Bohl, M.: Flowcharting Techniques. Science Research Associates, Chicago (1971)
3. Borg, I., Lingoes, J.: Multidimensional Similarity Structure Analysis. Springer, Beverley Hills (1987)
4. Borg, I., Groenen, P.: Modern Multidimensional Scaling. Springer, Berlin (1997)
5. Cicchetti, D.V., Sparrow, S.S.: Developing criteria for establishing the interrater reliability of specific items in a given inventory. American Journal of Mental Deficiency 86, 127–137 (1981)
6. Clark, H., Brennan, S.: Grounding in communication. In: Resnick, L.B., Levine, J., Teasley, S. (eds.) Perspectives on Socially Shared Cognition, pp. 127–149. APA Press, Washington, DC (1991)
7. Clark, H.: Using language. Cambrigde University Press, Cambridge (1996)
8. Cox, T.F., Cox, M.A.A.: Multidimensional Scaling. Chapman and Hall, London (2001)
9. Davison, M.L.: Multidimensional Scaling. John Wiley and Sons, New York (1983)
10. De Leeuw, J.: Applications of convex analysis to multidimensional scaling. In: Barra, J., Brodeau, F., Romier, G., van Cutsem, B. (eds.) Recent Developments in Statistics, pp. 133–145. North Holland Publishing Company, Amsterdam (1977)
11. De Leeuw, J., Heiser, W.J.: Theory of multidimensional scaling. In: Krishnaiah, P.R., Kanal, L.N. (eds.) Handbook of Statistics, vol. 2, pp. 285–316. North-Holland, Amsterdam (1982)
12. Dillenbourg, P., Baker, M., Blaye, A., O'Malley, C.: The evolution of research on collaborative learning. In: Reimann, P., Spada, H. (eds.) Learning in Humans and Machines, pp. 189–211. Springer, Berlin (1995)
13. Fischer, F., Mandl, H.: Being there or being where? Videoconferencing and cooperative learning. In: van Oostendorp, H. (ed.) Cognition in a Digital World, pp. 205–223. Lawrence Erlbaum Associates, Mahwah (2003)
14. Fleiss, J.L.: Statistical Methods for Rates and Proportions, 2nd edn. Wiley, New York (1981)
15. George, D., Mallory, P.: SPSS for Windows Step by Step: A Simple Guide and Reference. 11.0 Update. Allyn & Bacon, Boston (2003)
16. Guttman, L.A.: A general non-metric technique for finding the smallest coordinate space for a configuration of points. Psychometrika 33, 495–506 (1968)
17. Kahrimanis, G., Meier, A., Chounta, I.-A., Voyiatzaki, E., Spada, H., Rummel, N., Avouris, N.: Assessing Collaboration Quality in Synchronous CSCL Problem-Solving Activities: Adaptation and Empirical Evaluation of a Rating Scheme. In: Cress, U., Dimitrova, V., Specht, M. (eds.) EC-TEL 2009. LNCS, vol. 5794, pp. 267–272. Springer, Heidelberg (2009)
18. Kahrimanis G., Avouris, A., Komis, V.: Interaction analysis as a tool for supporting collaboration. An overview. In: Daradoumis, T., Caballe, S., Juan, A.A., Xhafa, F. (eds.) Technology-Enhanced Systems and Tools for Collaborative Learning Scaffolding (in press)
19. Kerlinger, F.N., Lee, H.B.: Foundations of behavioral research. Harcourt College Publishers, New York (2000)

20. Kruskal, J.B.: Multidimensional scaling by optimizing goodness-of-fit to a non-metric hypothesis. Psychometrica 29, 1–27 (1964)

21. Kruskal, J.B., Wish, M.: Multidimensional Scaling. Sage Publications, London (1978)

22. Laughlin, P.R.: Social combination processes of cooperative, problem-solving groups on verbal intellective tasks. In: Fishbein, M. (ed.) Progress in Social Psychology, vol. 1, pp. 127–155. Lawrence Erlbaum, Hillsdale (1980)

23. Meier, A., Spada, H., Rummel, N.: A rating scheme for assessing the quality of computer-supported collaboration processes. International Journal of Computer-Supported Collaborative Learning 2, 63–86 (2007)

24. Meier, A., Voyiatzaki, E., Kahrimanis, G., Rummel, N., Spada, H., Avouris, N.: Teaching students how to improve their collaboration: Assessing collaboration quality and providing adaptive feedback in a CSCL setting. In: Rummel, N., Weinberger, A. (eds.) New Challenges in CSCL: Towards Adaptive Script Support, Worshop in Proceedings of the Eighth International Conference of the Learning Sciences (ICLS 2008), Utrecht, vol. 3, pp. 338–345. International Society of the Learning Sciences (June 2008)

25. Roschelle, J.: Learning by collaboration: Convergent conceptual change. Journal of the Learning Sciences 2, 235–276 (1992)

26. Scardamalia, M., Bereiter, C.: Computer support for knowledge-building communities. In: Koschmann, T. (ed.) CSCL: Theory and Practice of an Emerging Paradigm, pp. 249–268. Lawrence Erlbaum Associates, Hillsdale (1996)

27. Samurcay, R.: The concept of variable in programming: Its meaning and use in problem solving by novice programmers. In: Soloway, E., Spohrer, J.C. (eds.) Studying the Novice Programmer, pp. 161–178. Lawrence Erlbaum, Hillsdale (1989)

28. Schiffman, S.S., Reynolds, M.L., Young, F.W.: Introduction to Multidimensional Scaling - Theory, Methods, and Applications. Academic Press, New York (1981)

29. Shepard, R.N.: Analysis of proximities: Multidimensional scaling with an unknown distance function I & II. Psychometrika 27, 125–140 & 219–246 (1962)

30. Soloway, E., Bonar, J., Ehrlich, K.: Cognitive strategies and looping constructs. In: Soloway, E., Spohrer, J.C. (eds.) Studying the Novice Programmer, pp. 191–207. Lawrence Erlbaum, Hillsdale (1989)

31. Stahl, G.: Sustaining group cognition in a math chat environment. Research and Practice in Technology Enhanced Learning (RPTEL) 1(2), 85–113 (2006)

32. Stahl, G., Koschmann, T., Suthers, D.: Computer-supported collaborative learning: An historical perspective. In: Sawyer, R.K. (ed.) Cambridge Handbook of the Learning Sciences, pp. 409–426. Cambridge University Press, Cambridge (2006)

33. Steyvers, M.: Multidimensional Scaling. In: Encyclopedia of Cognitive Science. Macmillan, London (2002)

34. Strijbos, J.W., Martens, R.L., Prins, F.J., Jochems, W.M.G.: Content analysis: What are they talking about? Computers and Education 46(1), 29–48 (2006)

35. Tindale, R.S., Kameda, T., Hinsz, V.B.: Group decision making. In: Hogg, M.A., Cooper, J. (eds.) Sage Handbook of Social Psychology, pp. 381–403. Sage, London (2003)

36. Wegner, D.M.: Transactive memory: A contemporary analysis of the group mind. In: Mullen, B., Goethals, G.R. (eds.) Theories of Group Behaviour, pp. 185–208. Springer, New York (1987)

37. Wirtz, M., Caspar, F.: Beurteilerübereinstimmung und Beurteilerreliabilität. Verlag für Psychologie, Göttingen (2002)

38. XLSTAT, Addinsoft (2009), http://www.xlstat.com

39. Young, F.W., Hamer, R.M.: Theory and applications of multidimensional scaling. Erlbaum, Hillsdale (1994)

Internet-Mediated Communities of Practice: Identifying a Typology of Critical Elements

Apostolos Kostas and Alivisos Sofos

Dept. of Primary Education, University of the Aegean
Dimokratias 1, Rhodes 85100, Greece
{apkostas,lsofos}@aegean.gr

Abstract. A community of practice is a group of people who share common concerns, problems or passions for a domain and who deepen their knowledge and expertise through interaction and collaboration on an ongoing basis. More and more people, groups and organizations are looking to develop Internet-mediated communities of practice in order to realize specific goals on informal learning and professional development. Harnessing the perceived values of communities across educational sector requires well-designed settings and procedures to achieve a sustainable level of functionality, communality, collaboration and knowledge sharing. Because current research supports the notion that there is not a systematic theory or a blueprint for design of online communities, this work aims to define a basic typology of various critical elements for successful and sustainable Internet-mediated communities of practice, via a meta-analysis and critical synthesis of related literature.

1 Introduction

The evolution of the World Wide Web and the advent of more collaborative ICTs technologies has created a new paradigm of Media Knowledge: classic one-way production of information on the Web is substituted by a dynamic process of information co-{*production, organization, discovery, sharing*}. Technologies are driving changes in human behavior, interaction, information creation and consumption and knowledge acquisition and sharing. The paradigms for learning have already evolved from traditional face-to-face and in-classroom patterns to open and distance learning activities with high level of collaboration, enhanced by Web 2.0 social practices, which promote interaction in distributed online learning environments and formation of virtual or internet-mediated communities of practice (IMCoP).

However, critical elements leading to successful knowledge sharing and learning within communities are not well understood, especially those for the design and establishment phase of an IMCoP [1]. While the literature tends to treat the concept of communities of practice as one-dimensional construct, a closer look at real settings reveals that their various characteristics make each community unique, even if they share some common features. Within a particular context, different characteristics or configurations or combinations of characteristics (e.g. target-group, size, technology,

T. Daradoumis et al. (Eds.): Intelligent Adaptation & Personalization Techniques, SCI 408, pp. 311–334.
springerlink.com   © Springer-Verlag Berlin Heidelberg 2012

age, etc.) lead to many variations of types [2] and may be more or less conductive to success, thus highlighting the need for a more contingent approach to design and management.

Moreover, IMCoPs as a sub-category of virtual communities should not be considered as a transparent catalyst capable of enabling an organization to seamlessly disseminate or capitalize socially generated knowledge, but rather be understood in terms of theoretical and functional limitations [3].

While it is accepted in the literature that there is no formula, nor recipe for successful community development, i.e. there is not a "one-fits-all" design strategy, there are a number of well-founded practical and theoretical studies investigating various aspects of IMCoPs. For example:

- Dube et al. [4] investigate the impact of structuring characteristics on intentionally created IMCoPs inside organizational contexts.
- Fontaine and Milllen [5] try to measure benefits as results from community activities, members' use of content and technology resources and overall participation.
- Gannon-Leary and Fontainha [6] discuss various benefits, barriers and success factors for communities and virtual communities.
- Hara et al. [7] examine a typology for IMCoP that are formed and operate outside of organizational contexts.
- Ardichvili [1] explore motivational factors, barriers and enablers of participation in online communities.
- Schwen and Hara [8] examine theoretical limitations in Wenger's theory of social learning.

Moreover, by examining the cumulative body of knowledge on IMCoPs, we find that most of the studies tend to be descriptive in nature, focus either on specific aspects of the communities or in specific professional settings (i.e. teacher education, identify formation, etc.) and typically includes in-depth case studies of specific groups in specific knowledge domains [7].

Based on this ascertainment, our work aims at a literature review on communities of practice, virtual communities and learning communities for:

- *Practical and theoretical issues on design, development and sustainability.*
- *Synthesis of a basic typology of structural characteristics, motivators, success factors and barriers*, as the dominant critical elements for successful learning and knowledge sharing within communities.

We believe that the provision of a basic typology of critical elements will facilitate a better understanding of this form of collaboration and will assist various stakeholders in better designing and operating IMCoPs within the educational domain.

This chapter is organized as follows. The next section provides the conceptual foundations of CoPs and IMCoPs. Then the research methodology of the meta-analysis is described and results are presented and discussed. The chapter ends with a short discussion and conclusions.

2   Conceptual Foundations

Community is the fundamental social environment where learning and knowledge construction takes place on the basis of meaning negotiation via peer's interactions. Typically it refers to a group of people who live and act in the same geographical area sharing common aims and values and is differentiated from any other group formation based on: (a) Core characteristics such as *people, common ties* and *purpose, social interactions* and *activity in time/space* [9, 10, 11], (b) Hallmarks such as *agency, belonging, cohesion* and *diversity* [12] and (c) Processes within the community such as *acting together, dialoguing, collaborating* and *bridging* [13].

By extending the metaphor of community in the early days of Internet, Rheingold [14] first coined the term *Virtual community* as social aggregations in the network, where people join public conversations thus forming a web of interpersonal relationships within the Cyberspace.

Because Preece [15] considered the term *Internet Community* as terminological weak due to the fact that any form of communication between two or more individuals on the Internet may be considered as a community formation, she then provided a non-ambiguous definition by stating that Internet community is a group of people interacting in a virtual environment, having common goals, specific rules and behavioral norms.

A special case of communities where learning and community are interrelated under the concept of social constructivism is the *Community of Learning and Community of Practice* (CoP). *Community of learning* is defined by Reinmann-Rothmeier et al. [16] as a community in which members are tied together by a common interest to inquire a certain case in depth and learn together, share knowledge and solve problems collaboratively by this process and by Kilpatrick et al. [17] as group of people who shares a common goal, works together, respect different opinions, promote opportunities for active learning and develop a collaborative environment for empowerment of membership and new knowledge formation.

Moreover, *learning communities* as a learning paradigm is a superset of the *communities of learners* according to Watkins [12], where the process of enquiry and knowledge-generation characterizes community of learners, while the process of collective reflection and meta-learning characterizes learning communities.

Palloff and Pratt [18] define 4 fundamental elements of learning communities:

- *People*, i.e. social presence
- *Purpose* and *policies*
- *Interactivity*, i.e. interaction, communication, team work, collaboration
- *Reflective/transformative learning*

Moreover, Riel and Polin [19] analyzed 3 types of learning communities:

- *Task-based* communities, where the shared goal is completing tasks that are beyond the capabilities of any single person.
- *Knowledge-based* communities, where the shared goal is to design external representations of the community thought and created knowledge.

- *Practice-based* communities, where a shared practice yields knowledge and learning. In this classification, CoPs are considered as a type of learning communities (Fig. 1). Indeed, in real settings communities of learning and practice are two interrelated and overlapping learning paradigms even if learning within a CoP can be thought as a by-product of activities during the practice and not the main objective, according to its theoretical foundation [13].

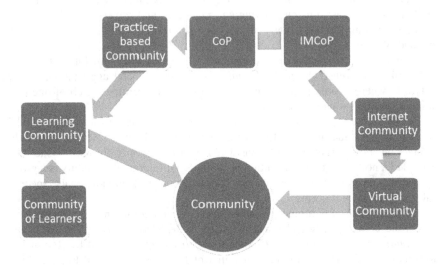

Fig. 1 IMCoP and Typology of Communities

2.1 Historical Overview and Theory of Communities of Practice

CoPs trace their roots to constructivism a learning theory, which shifts control from instructor to learner, is more relevant to adult experience and is characterized by concepts such as ill-structured problems, learning in social and physical context of real-world problems, shared goals, cognitive tools, an instructor's role as facilitator or coach, etc. [20, 21, 22, 23, 24].

Many of the actual notions about CoPs have their base on studies in the Work Practice and Technology group at the Institute for Research on Learning (IRL) at the Xerox Palo Alto Research Center (PARC) in the late 1980s [25]. During this period an increasing interest on Social Constructivist models of learning had already began to acquire an increasing interest, shifting learning from the simple knowledge transfer from educator to student (Behaviorist model) to the process of knowledge co-construction, a theoretical model based on Vigotsky's Socio-Cultural theory. During the early period, there were 2 dominant approaches on CoPs theory evolution:

- *Situated learning*, based on the Legitimate Peripheral Participation (LLP) to describe the underlying process where learning in this context involves becoming an "insider" (i.e. a member of the community) [26].
- *Organizational learning*, aiming in bringing together theories of working, learning and innovation, where Brown and Duguid [27] identified 3 basic concepts describing features of work practice: *narration, collaboration* and *social constructions*. CoP as a term was first coined by Lave and Wenger [26] who defined it as "*...an activity system about which participants share understandings concerning what they are doing and what that means in their lives and for their community...*" (p.98).

Despite the differences of these approaches, both agree that CoPs are autonomous groups primarily concerned with learning as an ongoing process within the community and knowledge is situated, mutable and socially constituted.

During the latest period of CoPs evolution the concept is extended to other contexts related to more organizational settings and is directly linked to process of knowledge management.

Moreover, there is an explicit view that CoPs can be geographically distributed and benefit from having a technological infrastructure to support their activities. Wenger et al. [28] coin CoPs as an evolutionary process for learning in groups, completion of shared tasks and production of artifacts. He distinguishes CoPs from other organizational structures (such as Formal Work Group, Project Team, Informal Network) describing 3 interrelated dimensions supporting CoPs: *Joint enterprise*, what CoP is about and understood, continually renegotiated by its members, *Mutual engagement* between members that binds them in a single entity, *Shared repertoire*, related with the artifacts, routines, sensibilities, vocabulary, etc. developed by the members over time.

Moreover, the term Community of Practice is re-defined by Wenger et al. [28] as a group of people who share a concern, a set of problems, or a passion about a topic and who deepen their knowledge and expertise in this area by interacting on an ongoing basis and by Hildreth et al. [29] as group of professionals tied together in an informal manner via their engagement in a common class of problems and efforts to find solutions, thus comprising a potential inventory of shared knowledge and expertise.

According to Wenger et al. [28] "*...a CoP is not just a Web site, a database, or a collection of best practices. It is a group of people who interact, learn together, build relationships, and in the process develop a sense of belonging and mutual commitment...*"(p.34). CoPs are now regarded mainly as:

- platforms for sharing and internalizing tacit knowledge, which is hard to communicate because it is mostly intuitive and embedded in a specific context and learning is promoted via communication among the community members.
- as peer-to-peer collaborative networks driven by willing participation of their members, who are focused on learning and building capacity and engaged in knowledge sharing, while developing expertise and solve problems [30].

Key concepts of CoPs are:

(a) *Community knowledge*, which is greater from the sum of individual participant knowledge [31] and advances while simultaneously advancing the individual's knowledge [32].

(b) *Facilitation*, the process that fine-tunes and nudges discussion and learning in the right direction.

Moreover, CoPs allow 2 key aspects of collaboration to be developed [33, 34]: peer interaction for negotiation and co-construction of knowledge and (b) expert-to-apprentice interaction.

According to Wenger et al. [28], even if CoPs may have various forms, they all share a basic structure, which is a unique combination of three fundamental elements that "...*make a CoP an ideal knowledge structure – a social structure that can assume responsibility for developing and sharing knowledge...*"(p.29).:

- **Domain:** A domain defines a set of issues and legitimizes the community by affirming its purpose and value to the community's members. It can range from common know-how to specialized expertise. In general, it is easier to define the domain based on existing discourse or professional discipline, but it is possible that members will come together without a specified topic. Without commitment to a domain, a community is just a group of friends, because it guide members how to organize their knowledge, present ideas, contribute and participate, help to sort out what to share and how to distinguish innovative ideas from trivial ones.

- **Community:** Individuals form a community when they interact, learn together and build relationships in the context of a domain. Thus a social learning system is created through sharing of peers' view of the domain and reveals various perspectives on a common problem. Membership and participation have various forms within the community and is a subject of personal, design and facilitation issues.

- **Practice:** Practice denotes a set of socially defined ways of doing things in a specific domain. It is a set of common approaches and standards that create a basis for action, communication, problem solving, performance and accountability. Practice is the specific knowledge that community develops shares and maintains, whereas the domain denotes the topic the community focus on.

2.2 Internet-Mediated Communities of Practice

Although, early research focus on face-to-face communities, the proliferation of new online collaboration tools coupled with a rapidly increasing interest of business and learning organizations, resulted in the emergence of a new form of collective learning and knowledge sharing through IMCoPs [35]. Because *virtual communities of practice* are often confused with other online formations such as *portals*, *networks* or *information communities*, *interest groups* or *blogs* in terms of the level of collaboration and engagement, in this work IMCoPs are defined as

"*...online social networks in which people with common interests, goals, or practices interact to share information and knowledge, and engage in social interactions...*" [36: p.1880]. It is recognized that IMCoPs by nature involve more design than face-to-face CoPs [37]. While both types share similar characteristics (learning communities with members who are mutually engaged in shared practice aiming to develop a repertoire of communal resources), they are operationally distinguishable (Table 1).

Table 1 Cop vs. IMCoP

	CoP	IMCoP
Design	Emerge from existing groups	Top-down
Membership	Closed, follows certain norms	Open, without certain norms. Identification is based on an idea or a task, rather than the place, with more fluid formal boundaries, less dominated norms and greater flexibility [18, 21]
Leadership	Emerge from the community	Pre-Assigned
Communication	Face-to-face	ICT-mediated
Development	Less Time	More Time: virtual community is the designed community, whereas the CoP is what emerges from the designed community
ICT	Plays no role	"Place" in virtual communities is substituted by the Web

Designing and sustaining an IMCoP is still a "balancing" act capable of blending the social capital issues with ICT. Internet technologies can be used to complement CoPs and learning communities and not replace them, because the question of whether CoPs can be partially or fully online is a debatable issue [40] arising two important questions:

- *Can relationship and trust developed and sustained in IMCoPs?*

The trust between members of a CoP depends on the existing levels of trust in the organization itself, which takes time to develop [41]. Since virtual connections are brief and intermittent and serendipity is limited online, the opportunities for members to develop relationships and build trust are also limited. Brown and Duguid [42] argued that it is difficult to form an IMCoP, as online or electronic communication can only strengthen existing face-to-face CoPs. Further challenges facing online CoPs are the replication of elements of face-to-face meetings and group dynamics through computer-mediated communication and the development of norms for interacting as a group [43]. Also, online activities have to be designed so that members will keep coming back, and so that they can derive value from these community activities. According to Riel and Polin [19] the challenge is that "*...virtual networks must also be able to support subtle cultural mechanisms that shape interaction, identity, and access, such as rituals and traditions that distinguish newcomers from old-timers in communities that rely on face-to-face* encounters..." (p.32). It is important to negotiate norms (goals, ethics, liabilities, and communication styles) in an online community, as according to Palloff and Pratt [21] *"...in the online environment, those collaboratively negotiated norms are probably even more critical as they form the foundation on which the community is built. Agreement about how a group will interact and what the goals are can help move that group forward..."* (p. 23).

On the other hand, space-time flexibility of the Internet helps create more space for social interaction: relationships and trust can be developed and sustained in online environments [44] and relationships can be developed in a virtual environment as *"...The archiving of online interactions makes possible forms of interaction that can be both more flexible and more durable than face-to-face interactions..."* [45: p. 2]. From a technical point of view, if the online environment is enhanced with more collaborative technology (Web 2.0 social networking communities) then the degree of transparency and trust within this environment is more likely to increase over time (Fig. 2).

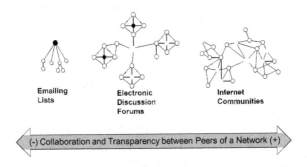

Fig. 2 From emailing lists to Web 2.0: Degrees of Collaboration, Trust and Transparency

Finally, as suggested by Palloff and Pratt [21] *"...although face-to-face contact at some point in the community-building process can be useful and facilitate community development, the contact is not likely to change the group dynamic created online. It is possible, however, to build community without it..."* (p. 23).

– ***Can tacit knowledge and practice be shared within IMCoPs?***

ICT can provide the essential authentic context required by situated learning, as they have designed a multimedia learning environment to support authentic learning situations and their study found that situated learning and knowledge sharing could be successfully supported by this environment [22]. But whether doing and learning can take place in the virtual environment and whether tacit and situated learning can take place online, because learning in a CoP is tacit (learning 'to be') rather than explicit (learning 'to know'), it has to be situated [8, 41]. Moreover, while IMCoPs are difficult to be established and successfully operated, if carefully designed, online communities can be formed and they will grow [19, 37, 43, 46].

3 Research Framework

The above open issues must guide the organizations in taking pro-active actions to maximize the possibility of built relationship and trust in online settings and manage to share tacit knowledge and practice within IMCoP, in order to achieve an effective and sustainable design. For this reason, it is necessary to have a clear understanding of which are the critical elements of online communities of practice and learning.

All of the above studies and more found in the literature are research cases presenting a wide range of results, from practical hints and tips to meta-issues expressed at a conceptual level. These cases reveal an attempt of the researchers to identify and categorize all the potential issues relative to community's life-cycle. The term community life-cycle, i.e. the evolution patterns of communities' development, has been theoretically examined, but it is still under research in terms of empirical longitudinal studies:

- Palloff and Pratt [21] talked about the "forming, norming, storming, performing, and adjourning" cycles of online learning environments.
- Levin and Cervantes [47] described communities as having gone through life cycles of 'proposal, refinement, organizational, pursuit, wrap-up, and publication stages'.
- Wenger et al. [28] have identified a three-phase 'formation, growing, transformation' and five-stage 'potential, coalescing, maturing, stewardship, and transformation' life cycle of CoPs.
- Preece [15] described a four-stage 'pre-birth, early life, maturity, and death' model.

According to the literature, community development and operation can be divided into three general phases: *formation, sustaining/maturing, transformation or disengaging* and is therefore a necessary condition, for a successful intervention through those phases and the identification of the critical elements. The purpose of this study is to identify a basic typology about IMCoPs design and development issues. So, the initial research question driving this study is:

"Which are the critical elements of IMCoPs?"

3.1 Methodology

In order to identify critical elements we adopted a three-step methodology (Table 2):

STEP 1: We conducted a literature review on a large heterogeneous sample of journal and proceedings articles (presenting miscellaneous research cases in order to achieve "maximum variation sampling" [48: p.28] on two areas of research highly relevant to IMCoPs: communities of practice and learning and virtual communities.

STEP 2: We analyzed the articles against our initial research question. Through a careful analysis of the literature found, we systematically extracted all information that could be considered structural characteristic, motivator, success factor or barrier, thus resulting in four distinctive sets of raw data.

STEP 3: We employed an iterative process on each set of raw data from STEP 2 in order to develop a group of high level categories and subcategories for each type of critical element. With this process, we managed to identify redundancies between raw data and to clarify elements that were too broad or not well-defined.

Table 2 Research Overview

Research Question	What are the critical elements of an IMCoP?
Field of Investigation	• Communities of Practice
	• Learning and Virtual Communities
Data Collection	• Journal and proceedings papers
	• "maximum variation sampling" [48] (p. 28)
Data Analysis	• Step 1: Row Data Extraction
	• Step 2: Analysis and Basic Categorization
	• Step 3: Clustering – Categories Refinement
Data References	Elements were traced in a total of 35 papers.
	References: [1 - 8], [11, 13, 15], [28], [49], [50-59], [60-69], [70, 72, 73]

Moreover, the final lists of critical elements are an aggregation of all possible elements found in the literature and for each list entry we have cited every possible reference in an adequate manner, in order to achieve soundness and validity of the analysis process.

3.2 Results

The first level of critical analysis and synthesis of the results revealed the following 4 distinctive groups (or categories) of **critical elements** (Fig 3) for the community's life-cycle:

- **Structural characteristics** of IMCoPs.
- **Success Factors** that a-posteriori characterize, or a-priori predicts a successfully IMCoP design and operation.
- **Motivators** for membership and active participation.
- **Barriers** that potentially lead an IMCoP to failure or earlier disengaging.

Fig. 3 Critical elements of IMCoPs life-cycle

3.2.1 Structural Characteristics

The term "structural characteristic" refers to the rather stable element that defines an IMCoPs unique identity [2]. Most of these characteristics are subject of design decisions taken during the formation phase and will either affect the life cycle positively, or will create challenges that will need to be acted upon. Some of these characteristics may evolve over time while others are settled during the launching stage and remain stable during the whole life cycle. Research focusing on the dynamics taking place during the launching phase of an IMCoP reveals that some structuring characteristics are more conductive to success at this stage than others [49].

Clustering revealed four categories of structural characteristics:

– **Demographics**
– **Organizational**
– **Functional**
– **Learning/Knowledge.**

3.2.1.1 Demographics Structural Characteristics

Structural characteristics are defined by seven categories of elements describing the identity of an IMCoP (Table 3).

Table 3 Demographics Structural Characteristics

Title	Description
Cultural Diversity	An IMCoP according to its members may face a *homogeneous* or *heterogeneous* level of cultural diversity and there are 3 levels of potential diversity within an IMCoP: national, organizational and developmental cultures.
	References: *[2], [3], [4], [13], [28], [50], [52], [55], [57], [58], [59], [60]*
Age	Refers to the period of time that IMCoP has been through and the level of maturity. This characteristic is related with the maturing and transformation phase of the life cycle. It is expected that an IMCoP may evolve to a higher level of maturity slower or faster, or stall at one phase of the life cycle.
	References: *[2], [3], [4], [13], [28], [50], [51], [52], [53], [54]*
Life span	Refers to the working time period: an IMCoP could operate on a *temporal* or *permanent* basis.
	References: *[2], [3], [4], [13], [28], [50], [51], [52], [53]*
Size	Refers to the number of active members of the community. According to ([28]) this characteristic may vary from only a few people to more than thousands of people.
	References: *[2], [13], [28], [51], [52], [55], [56]*
Community experience	Prior community experience may vary from *extensive* when community is based on an existing network, *medium* when members have already work in groups, to *low* when members are newcomers.
	References: *[2], [4], [13], [55]*
Orientation	Refers to the overall objective and mission of the community: an *operational* orientation of the community focuses on sort term goals, while the *strategic* one supports the overall mission of the community in long-term.
	References: *[2], [4], [28]*.
Geographic dispersion	Refers to the geographical span of the members: *low dispersion* refers to very small geographical span (for example an inter-building community) while high dispersion refers to global span (for example an international community).
	References: *[2], [4]*

3.2.1.2 Organizational Structural Characteristics

This category contains a list of 4 distinctive elements characterizing the organizational context of an IMCoP (Table 4).

Table 4 Organizational Structural Characteristics

Title	Description
Formation Type	An IMCoP can be established on the basis of a *"process"* or a *"product"* and also can be established from the members (*bottom-up design*) or *intentionally* from the organization (*top-down design*). References: *[2], [3], [4], [8], [28], [50], [51], [55], [57], [60], [61], [64]*
Purpose	Community has to serve a general purpose, such as *informal learning, professional development, knowledge management, professional identity formation, special purpose community, etc.* References: *[8], [13], [28], [50], [51], [54], [55], [60], [61], [62], [63]*
Organization type	The dimensions that force the organizational context of the IMCoP may shape the environment of the community. So, community may be: • *Self-organized* or *managed* • *Low-boundary crossing* (members belong to the same organization) to *high-boundary crossing* (members belong to various organizations) and • Faces a *low level of formalisation* (the community has not been integrated into the formal structure of the organization) to *high level of formalisation* (the community is part of the organization) References: *[2], [4], [8], [50], [54], [55], [57], [60], [61], [64]*
Management	If the IMCOP is managed, then the *facilitation, monitoring, scaffolding* and *coaching* level has to be considered as a structuring characteristic, because it will affect the sustaining/maturity phase of the community's life cycle. References: *[2], [4], [50], [51], [55]*

3.2.1.3 Functional Structural Characteristics

This category contains a list of 3 distinctive elements characterizing the functional and technological infrastructure of an IMCoP (Table 5).

Table 5 Functional Structural Characteristics

Title	Description
Interrelation	Interrelations between members of the community may be based on common: vocabulary of the members, world perception by the members, interests between the members, history of the members References: *[3], [13], [28], [49], [50], [52], [55], [57], [61]*
ICT	*Availability*: Small vs. big *Reliance*: Small vs. Big *Alignment*: Level of ICT infrastructure alignment with the IMCoP context and information/knowledge flow References: *[2], [49], [51], [55]*
Membership	Membership patterns in IMCoPs may be *open* vs. *close, voluntary* vs. *compulsory, stable* vs. *fluid.* Moreover, mutuality between members in the community may vary from harmonious to conflicting, working or activity relationships and collective vs. tentative. References: *[49], [51]*

3.2.1.4 Learning & Knowledge Sharing Structural Characteristics

This category contains a list of 6 distinctive elements characterizing learning and knowledge sharing within an IMCoP (Table 6).

Table 6 Learning & Knowledge Sharing Structural Characteristics

Title	Description
Collaboration	Level and types of collaboration activities
	References: *[13], [52], [64]*
Domain	Domain of practice and community's knowledge
	References: *[28], [60]*
Practice	Practice field of community members
	References: *[28], [65]*
Social capital	Relations, participation, history, identity, belonging, cohesion, etc.
	References: *[52], [54]*
Context	Level of contextualization within IMCoP
	References: *[64]*
Knowledge type	Implicit (tacit) vs. explicit, codified vs. un-codified, diffused vs. un-diffused, personal vs. public, structured vs. unstructured
	References: *[55]*

3.2.2 Motivators

The potential benefits of the participation in an IMCoP define in an interdependent manner the motivators for active participation and collaboration and the overall sustainability of the community.

Literature review and synthesis revealed that potential motivators (Table 7) can be:

- Capitalized in *short* or *long-term*.
- On a *personal level* according to individual needs, or may be *associated with the whole community*.
- Identified either *explicit* or *implicit*.

Clustering revealed five categories of motivators:

- Professional Development
- Personal Development
- Resource and Information Sharing
- Ethics

Table 7 Motivators

Title	Short-term	Long-term	Personal	Community	Explicit	Implicit
Professional Development						
Professional challenges encounter	•		•		•	
Improve quality of processes			•		•	
Reach learning environments			•			•
Development of synergies			•	•		•
Learning curve reduction			•			•
Personal Development						
Emotional improvement			•	•		•
Informal learning activities			•			•
Reduce project completion time			•		•	
Creativity & innovation			•	•		•
Peer relations & social networking	•	•	•	•	•	
Resource and Information Sharing						
Get answers from experts	•		•		•	
Experts network formation			•	•		•
Promotion & dissemination of practices	•	•		•	•	
Resources re-use	•	•	•	•	•	
Information and knowledge sharing	•	•	•	•	•	•
Ethics						
Recognizability	•	•	•	•	•	
Altruism	•	•	•	•	•	
Common values/vision/objectives		•	•	•		•

References: *[1], [5], [6], [28], [49], [57], [60], [61], [66], [67], [68], [69], [70]*

3.2.3 Success Factors

"Successful CoPs are well-balanced systems that oscillate between exploring new practices and exploiting existing ones" ([54]:pp.345). Answering the question *"which are the main success factors for an IMCoP"* our literature review and synthesis revealed six categories of success factors (Table 8):

- **Personal**
- **Administration**
- **Technology**
- **Content**
- **Operational**
- **Learning/Knowledge**

Table 8 Success Factors

Title	Description
Administration	Facilitation, moderation or organizational formal supervision is factors influencing the operation of a community.
	Elements
	✓ Efficient and capable facilitation by experienced and skilled facilitators with the support of domain experts.
	✓ Clear operational rules, behavioral norms and active moderation in order to sustain communication and content generation.
	✓ Effective roles assignment and monitoring good and bad practices in order to promote active participation, flexibility and members reward.
	References: *[11], [28], [54], [57], [58], [60], [61], [64], [65], [69], [72]*
Technology	ICT provides the infrastructure or moreover, the environment in which the "eco-system" of the community will evolve.
	Elements
	✓ Registration system with active profiling and social networking characteristics.
	✓ Security & privacy infrastructure to guarantee secure communication and exchange of information.
	✓ Role management system and security levels according to specific roles and tasks assigned to members.
	✓ Push-pull mechanisms, in order to sustain a minimum level of behavioral norms.
	✓ Content Management System (CMS), in order to manage information and processes in a dynamic and flexible manner, with efficient search engine and filters.
	✓ Usability of user-interface.
	✓ Simulation of face-to-face communication with Video chatting, virtual rooms, video-conference and virtual presentations infrastructure.
	✓ 24/7 reliability and availability of the technological infrastructure.
	References: *[1], [8], [15], [49], [57], [58], [61], [62], [64], [69], [72]*

Table 8 (*continued*)

Operational	This category of elements is related mostly with the maturing and sustaining phase of the community life-cycle and is primarily based on the quality of communication and collaboration among members.
	Elements
	✓ Security and trust environment among members and cultivation of "feeling of belonging".
	✓ Empowerment of common values/vision/objectives/understanding.
	✓ Netiquette acceptance by the members.
	✓ Development of mutuality/sociability/trust.
	✓ Efficient members support
	✓ Clear definition of common goals focusing on the strategic objectives and sub-objectives of the community.
	✓ Balance on typical and non-typical procedures with pluralism and rhythm on communities' activities.
	✓ Community assessment and evaluation mechanisms with dissemination of common activities and public reward of accomplishments of collaborating activities
	✓ Regularity on communication activities with fun on communication
	✓ Face-to-face meetings and events.
	References: *[1], [8], [28], [49], [52], [54], [58], [60], [69], [72]*
Learning & Knowledge Sharing	In virtual learning environments, learning activities are structured to promote knowledge negotiation among learners towards the creation of new body of knowledge.
	Elements
	✓ Adequate time for social capital development, as it inculcates value to communities and high returns to individuals and the community as a whole
	✓ Well-defined knowledge domain.
	✓ Active participation with low level of lurking.
	✓ Member cognitive impulse, alignment and awareness.
	✓ Competitive activities for production of common artifacts, based on the objectives of the community with meaningful dialogue within the community.
	✓ Design balance between "learning to be" and "learning about".
	✓ Common feeling of response against communities tasks.
	✓ Members understanding of knowledge as a common value.
	References: *[1], [8], [11], [52], [53], [54], [62], [64], [68], [69]*

Table 8 (*continued*)

Content	Not only content on the Internet is important ("content is the king" as it is coined in various web sites…), but the way that content is produced, consumed and presented as well.

Elements

✓ Integrity and authenticity of information via work-flow management and regular feed with outer expertise.

✓ Value and relevancy with the domain ad practice of the community.

✓ Resources sufficiency: members of the community need to have multiple resources of information for a more objective view.

✓ Multimedia knowledge objects and multimodal presentation of information.

✓ Audience-oriented web design, following specific guidelines.

References: *[11], [28], [54], [55], [61], [64], [72]*

3.2.4 Barriers

The last case of our research question is the list of potential barriers influencing the operation of an IMCoP. Clustering revealed four categories of barriers (Table 9):

– **Theoretical**
– **Design/Organizational/Functional,**
– **Communication**
– **Personal**

Table 9 Barriers

Category: **Design/Organizational/Function**

Elements

- Insufficient design and development of ICT infrastructure and services, lack of support.
- Social capital growth usually takes time, while the web environment favors short-terms connections.
- There isn't any evaluation metric for the design phase of an IMCoP due to the lack of a specified design methodology.
- Low interest due to ill design, lack of objectives and common values.
- Indistinct practice.
- Lack of a core group to serve as motivator for the other members and produce a critical mass of content and collaboration activities.
- Information hoarding/lack of trust and insufficient flow of information.
- Cohesiveness reduction through time.

References: *[1] , [15] [49], [54], [62], [63], [65], [68]*

Table 9 (*continued*)

Category: **Personal**

Elements

- Lack of time for a member to engage in various activities.
- Cultural diversities.
- No motives or/and low professional confidence and modesty or/and personal particularities leads to lack of self-trust.
- High level of competitiveness.
- Reluctance to innovation and use of new practices and methods.
- Low digital skills.
- Common understandings diversities.
- Members are not familiar with knowledge sharing values and have diverse common understandings.

References: *[1], [15], [49], [53], [57], [58], [63], [68]*

Category: **Communication**

Elements

- Lack of communication norms.
- Insufficient communication tools.
- Low face-to-face communication level.
- High larking/ read-only member's ratio.
- Asynchronous communication without efficient moderation.
- Trust formation in virtual environments i shard to achieve.
- Lack of non-verbal communication.
- Lack of interpersonal connections lead to a typical communication model, insufficient to produce reflection, conflict and social learning.

References: *[1], [15], [57], [58], [62] [63]*

Category: **Theoretical Open Issues**

Elements

- Initial Wenger's theory of social learning is a descriptive, social, middle-level theory and not a prescriptive one. Thus, it cannot serve as a design guide for the development of an IMCoP.
- The use of Wenger's theory for organizational reform may lead to a paradox with various side-effects for the community's consistency, because it is based on a non-typical, non-hierarchical, social construction.
- Instructional design for learning in communities must be balanced between middle-level social theory and micro-level normative models for personal learning.

References: *[8], [49], [73]*

4 Discussion

An increasing number of studies have debated on *"what makes an effective IM-CoP?"* To answer this question one has to define, against which design goals should evaluate the success of a CoP, within which design and evaluation framework and at which point of the community life cycle.

The main purpose of this work has been to further our understanding of Internet-mediated Communities of Practice. We have identified a typology of design issues by "framing" IMCoP in its historical and theoretical context, against which we presented a detailed list of critical elements. Those elements may identify the nature of the community and affects in various ways the formation, sustaining/maturing and transformation or disengaging of the community. Analysis revealed that IMCoPs may benefit organizations in educational sector by encouraging social learning through participation and collaboration with peers. Because forming a CoP in virtual setting intentionally is not a straight forward process, the implication of this work is that it provides to practitioners a "check list" of critical elements, a clear understanding of which, may serve as guidelines and principles to follow in order to establish an IMCoP. Through literature meta-analysis and synthesis, critical elements were clustered in four basic categories and for each category the relating elements were presented and analyzed:

- Structural characteristics
- Motivators
- Success factors
- Barriers

Structural characteristics of IMCoPs identify and differentiate communities and may serve as guidelines to group and compare them. In conjunction with potential success and failure factors, designers of IMCoPs may predict a-priori or evaluate the success of the community. Moreover, understanding potential motivators and enablers that drive each individual's engagement in community's "eco-system", one can adapt the design and development of various structural elements of the community, thus increasing the chances to be successful. Also, analysis showed that the design and operation of an IMCoP is a very complicated, non-deterministic process, which has to take into account a lot of parameters and keep the balance between state of the art ICT and personal/communal cultures and motives. Building this typology of design issues, namely critical elements, was a step in the development of a finer understanding of theoretical and practical issues of IMCoPs. In order to increase its external validity, future research should:

- Investigate which of the critical elements are the most influential and at which phase of the IMCoP life cycle, through cross case analysis of a large number of existing communities on the Web, and
- Explain individual elements, tensions and dualities and their role in IMCoPs evolution and function, through activity and cognition theory.

References

[1] Ardichvili, A.: Learning and Knowledge Sharing in Virtual Communities of Practice: Motivators, Barriers, and Enablers. Advances in Developing Human Resources 10(4), 541–554 (2008)
[2] Bouhris, A., Dube, L., Jacob, R.: The Success of Virtual Communities of Practice: The Leadership Factor. Electronic Journal of Knowledge Management 3(1), 23–34 (2005)

[3] Kerno, S.: Limitations of communities of practice: a consideration of unresolved issues and difficulties in the approach. Journal of Leadership and Organizational Studies 5(1), 69–78 (2008)

[4] Dube, L., Bouhris, A., Jacob, R.: The impact of structuring characteristics on the launching of virtual communities of practice. Journal of Organizational Change Management 18(2), 145–166 (2005)

[5] Fontaine, M., Millen, R.: Understanding the Benefits and Impact of Communities of practice. In: Hildreth, P., Kimble, C. (eds.) Knowledge Networks Innovation through Communities of Practice, pp. 1–13. Idea Group Publishing (2004)

[6] Gannon-Leavy, P., Fontainha, E.: Communities of Practice and virtual learning communities: benefits, barriers and success factors, eLearning Papers, vol. 5 (2007), http://tinyurl.com/3by7e78 (last visited September 2010)

[7] Hara, N., Shachaf, P., Stoerger, S.: Online communities of practice typology revisited. Journal of Information Science XX(X), 1–18 (2009)

[8] Schwen, T., Hara, N.: Communities of practice: A metaphor for online design? In: Barab, S., Kling, R., Gray, J. (eds.) Designing for Virtual Communities in the Service of Learning, pp. 154–178. Cambridge University Press (2004)

[9] Hillery, G.: Definitions of community: areas of agreement. Rural Sociology 20, 111–123 (1955)

[10] Poplin, D.E.: Communities: a survey of theories and methods of research, 2nd edn. MacMillan, New York (1979)

[11] Stuckey, B.E.: Making the Most of the Good Advice: Meta-Analysis of Guidelines for Establishing an Internet-Mediated Community of Practice. In: The IADIS International Conference on Web-based Communities (WBC 2004), Lisbon, Portugal, pp. 91–98 (March 2004)

[12] Watkins, C.: Classrooms as Learning Communities. Routledge, New York (2005)

[13] Mendes, C., da Silva, A., Tribolet, J.: Learning Communities and Communities of practice Organizational learning Systems. In: CAPSI 2008, 8th Conferenceia Associacao Portuguest de Sistemas de Informacao, Setubal, Portugal (2008)

[14] Rheingold, R.: The Virtual Community: Homesteading on the electronic frontier. Addison-Wesley, New York (1993)

[15] Preece, J.: Online Communities: Usability, Sociability, Theory and Methods. In: Earnshaw, R., Guedj, A., Vince, T. (eds.) Frontiers of Human-Centered Computing, Online Communities and Virtual Environments, pp. 263–277. Springer (2001)

[16] Reinmann-Rothmeier, G., Mandl, H., Prenzel, M.: Computer supported learning environments. In: Planning, Formation and Assessment. Wiley-VCH, München (2000)

[17] Kilpatrick, S.I., Barrett, M.S., Jones, T.A.: Defining Learning Communities. In: Australian Association for Research in Education AARE 2003 Conference Papers: International Education Research Conference, Auckland, New Zealand EJ (2003) ISBN 1176-4902

[18] Palloff, R., Pratt, K.: Online Learning Communities in Perspective. In: Luppicini, R. (ed.) Online Learning Communities, pp. 3–16. Information Age Publishing (2007)

[19] Riel, M., Polin, L.: Online learning communities. In: Barab, S., Kling, R., Gray, J. (eds.) Designing for Virtual Communities in the Service of Learning, pp. 16–52. Cambridge University Press, Cambridge (2004)

[20] Knowles, M., Holton, E., Swanson, R.: The adult learner: the definitive classic in adult education and human resource development. Gulf Publishing, Houston (1998)

[21] Palloff, R., Pratt, K.: Building learning communities in cyberspace: effective strategies for the online classroom. Jossey-Bass, San Francisco (1999)

[22]   Oliver, R., Herrington, J.: Using situated learning as a design strategy for web-based learning. In: Abbey, B. (ed.) Instructional and Cognitive Impacts of Web-Based Education. Idea Publishing (2000)

[23]   Persichitte, K.: A case study of lessons learned for the Web-based educator. In: Abbey, B. (ed.) Instructional and Cognitive Impacts of Web-Based Education. Idea Publishing, Hersey (2000)

[24]   Squire, K., Johnson, C.: Supporting distributed communities of practice with interactive television. Educational Technology Research and Development 48(1), 23–43 (2000)

[25]   Kimble, C.: Communities of Practice: Never Knowingly Undersold. In: Tomadaki, E., Scott, R. (eds.) Innovative Approaches for Learning and Knowledge Sharing, EC-TEL 2006 Workshops Proceedings, pp. 218–234 (2006)

[26]   Lave, J., Wenger, E.: Situated learning: legitimate peripheral participation. Cambridge University Press, New York (1991)

[27]   Brown, J., Duguid, P.: Organizational learning and communities of practice. Organization Science 2(1), 40–57 (1991)

[28]   Wenger, E., McDermott, R., Snyder, W.: Cultivating Communities of Practice: A guide to Managing Knowledge. Harvard Business Review, 139–145 (2002)

[29]   Hildreth, P., Kimble, C., Wright, P.: Communities of Practice in the Distributed International Environment. Journal of Knowledge Management 4(1), 27–38 (2000)

[30]   Wick, C.: Knowledge management and leadership opportunities for technical communicators. Technical Communication 47(4), 515–529 (2000)

[31]   Cherardi, S., Nicolini, D.: The organizational learning of safety in communities of practice. Journal of Management Inquiry 9(1), 7–18 (2000)

[32]   Winsor, D.: Learning to do knowledge work in systems of distributed cognition. Journal of Business and Technical Communication 15(1), 5–28 (2001)

[33]   Bielaczyc, K., Collins, A.: Learning in classrooms: a reconceptualization of educational practice. In: Reigeluth, C., Mahwah, E. (eds.) Instructional-Design Theories and Models A New Paradigm of Instructional Theory, vol. 2. Lawrence Erlbaum Associates (1999)

[34]   Soden, R., Halliday, J.: Rethinking vocational education: a case study in care. International Journal of Lifelong Education 19(2), 172–182 (2000)

[35]   Wartburg, I., Rost, K., Teichert, T.: The creation of social and intellectual capital in virtual communities of practice: shaping social structure in virtual communities of practice. International Journal of Learning and Change 1(3), 299–316 (2006)

[36]   Chiu, C., Hsu, M., Wang, E.: Understanding knowledge sharing in virtual communities: An integration of social capital and social cognitive theories. Decision Support Systems 42(3), 1872–1888 (2006)

[37]   Barab, S., MaKinster, J., Scheckler, R.: Designing system dualities: Characterizing an online professional development community. In: Barab, S., Kling, R., Gray, J. (eds.) Designing for Virtual Communities in the Service of Learning, pp. 53–90. Cambridge University Press (2004)

[38]   Nachmias, R., Mioduser, D., Oren, A., Ram, J.: Web-supported emergent-collaboration in higher education courses. Educational Technology and Society 3(3), 94–104 (2000)

[39]   Robey, D., Kho, H., Powers, C.: Situated learning in cross-functional virtual teams. Technical Communication 47(1), 51–66 (2000)

[40]   Ellis, D., Oldridge, R., Vasconcelos, A.: Community and virtual community. Annual Review of Information Science and Technology 38, 146–186 (2004)

[41] Nichani, M., Hung, D.: Can a CoP exist online? Educational Technology 42(4), 49–54 (2002)

[42] Brown, J.S., Duguid, P.: Universities in the digital age. Change: The Journal of the Academy of Higher Education 28(4), 10–19 (1995)

[43] Schlager, M., Fusco, J., Schank, P.: Evolution of an online education community of practice. In: Renninger, K., Shumar, W. (eds.) Building Virtual Communities: Learning and Change in Cyberspace, pp. 129–158. Cambridge University Press (2002)

[44] Haythornthwaite, C., Kazmer, M., Robins, J., Shoemaker, S.: Community development among distance learners: Temporal and technological dimensions. Journal of Computer-Mediated Communication 6(1) (2000)

[45] Shumar, W., Renninger, K.: Introduction: On conceptualizing community. In: Renninger, K., Shumar, W. (eds.) Building Virtual Communities: Learning and Change, pp. 1–17. Cambridge University Press (2002)

[46] Cuthbert, A., Clark, D., Linn, M.: WISE learning communities: Design considerations. In: Renninger, K., Shumar, W. (eds.) Building Virtual Communities: Learning and Change, pp. 215–246. Cambridge University Press (2002)

[47] Levin, J., Cervantes, R.: Understanding the life cycles of network-based learning communities. In: Renninger, K., Shumar, W. (eds.) Building Virtual Communities: Learning and Change, pp. 269–292. Cambridge University Press (2002)

[48] Miles, M., Huberman, A.: Qualitative Data Analysis. Sage Publications (1994)

[49] Dube, L., Bouhris, A., Jacob, R.: Towards a Typology of Virtual Communities of Practice. Interdisciplinary Journal of Information, Knowledge and Management 1, 69–93 (2006)

[50] Koch, M., Fusco, J.: Designing for Growth: Enabling Communities of Practice to Develop and Extend Their Work Online. In: Kimble, C., Hildreth, P. (eds.) Communities of Practice: Creating Learning Environments for Educators, vol. 2, Information Age Publishing (2008)

[51] Backroad Connections Pty Ltd., "What are the conditions for and characteristics of effective online learning communities?", Australian Flexible Learning Framework Quick Guides series, Australian National Training Authority (2010), http://tinyurl.com/25kv8vb (last visited September 2010)

[52] Schwier, R.A.: A Typology of Catalysts, Emphases and Elements of Virtual Learning Communities. In: Luppicini, R. (ed.) Trends in Distance Education: A Focus on Communities of Learning. Information Age Publishing (2007)

[53] Garber, D.: Growing Virtual Communities. International Review of Research in Open and Distance Learning 5(2) (2004)

[54] Probst, G., Borzillo, S.: Why CoPs succeed and why they fail? European Management Journal 26, 335–347 (2008)

[55] Scarso, E., Bolisani, E., Salvador, L.: A systematic framework for analyzing the critical success factors of communities of practice. Journal of Knowledge Management 13(67), 431–447 (2009)

[56] Toral, S., Martanez-Torres, M., Barrero, F., Cortas, F.: An empirical study of the driving forces behind online communities. Internet Research 9(4), 378–392 (2009)

[57] MacLeod, P.: A Discussion of the Problems of Identity and of Establishing Communities of Practice in Online Learning Contexts. In: Kommers, P., Richards, G. (eds.) Proceedings of World Conference on Educational Multimedia, Hypermedia and Telecommunications, pp. 2894–2899 (2005)

[58]   Hasanali, F.: Critical Success Factors of Knowledge Management, APQC 2002 (2002), `http://tinyurl.com/6flgofr` (last visited September 2010)

[59]   Stanoevska-Slabeva, K., Schmid, B.: A Typology of Online Communities and Community Supporting Platforms. In: Proceedings of the 34th Annual Hawaii International Conference on System Sciences (HICSS-34), vol. 7. IEEE Computer Society (2001)

[60]   Stein, W.: A Qualitative Study of the Characteristics of a Community of Practice for Knowledge Management and Its Success Factors. International Journal of Knowledge Management 1(3), 1–24 (2005)

[61]   Hibbert, K., Rich, S.: Virtual Communities of Practice. In: Weiss, J., Nolan, J., Hunsinger, J., Trifonas, P. (eds.) The International Handbook of Virtual Learning Environments. Springer International Handbooks of Educations, vol. 14, pp. 563–579 (2006)

[62]   Hung, D., Nichani, M.: Differentiating between CoPs and Quasi-Communities: Can CoPs Exist Online? International Journal on E-learning 1(3), 23–29 (2008)

[63]   Johnson, C.M.: A survey of current research on online communities of practice. The Internet and Higher Education 4, 45–60 (2001)

[64]   Restler, S., Woolis, D.: Actors and Factors: Virtual Communities for Social Innovation. The Electronic Journal of Knowledge Management 5(1), 89–96 (2007)

[65]   Bruck, B.: Creating and Maintaining an Online Community a primer for community organizers and moderators (2004), `http://tinyurl.com/6bhqbg5` (last visited September 2010)

[66]   Wasko, M., Faraj, S.: It is what one does: why people participate and help others in electronic communities of practice. Journal of Strategic Information Systems 9, 155–173 (2000)

[67]   Millen, D.R., Fontaine, M.A., Muller, M.J.: Understanding the Benefit and Costs of Communities of Practice. Communications of the ACM 45(4), 69–73 (2002)

[68]   Correia, M., Paulos, A., Mesquita, A.: Virtual Communities of Practice: Investigating Motivations and Constraints in the Processes of Knowledge Creation and Transfer. Electronic Journal of Knowledge Management 8(1), 11–20 (2010)

[69]   Vrasidas, C., Zembylas, M., Chamberlain, R.: The Design of Online Learning Communities: Critical Issues. Educational Media International 41(2), 135–143 (2004)

[70]   UNDP, Building Online Communities of practice: The International Open Source Network Model, e-Note, vol. 5 (2005), `http://tinyurl.com/256mpgw` (last visited September 2010)

[71]   Dabbagh, N.: The online learner: Characteristics and pedagogical implications. Contemporary Issues in Technology and Teacher 7(3), 217–226 (2007)

[72]   Hernades, C.A., Fresneda, P.S.: Main Critical Success Factors for the Establishment and operation of Virtual Communities of Practice (2003), `http://tinyurl.com/67yq758` (last visited September 2010)

[73]   Storberg-Walker, J.: Wenger's Communities of Practice Revisited: A (Failed?) Exercise in Applied Communities of Practice Theory-Building Research. Advances in Developing Human Resources 10(4), 555–577 (2008)

[74]   Sofos, A.: Digital Literacy as a Category of Media competence and Literacy – an Analytical Approach of Concepts and Presuppositions for Supporting Media Competence at School. In: Bauer, P., Hoffmann, H., Mayrberger, K. (eds.) Fokus Medienpädagogik – Aktuelle Forschung-und Handlungsfelder, München, pp. 62–82 (2010)

Author Index